CLASSICAL
ELECTROMAGNETISM
VIA
RELATIVITY

CLASSICAL ELECTROMAGNETISM
VIA
RELATIVITY

An Alternative Approach
to
Maxwell's Equations

W. G. V. ROSSER, M.Sc., Ph.D.,

Senior Lecturer in Physics, University of Exeter

NEW YORK
PLENUM PRESS
LONDON
BUTTERWORTHS

Published in the U.S.A. by
PLENUM PRESS
a division of
PLENUM PUBLISHING CORPORATION
227 West 17th Street, New York, N.Y. 10011

First published by
Butterworth & Co. (Publishers) Ltd.

Suggested U.D.C. number: 538.3 : 530.12
Library of Congress Catalog Card Number 68–29072

PHYSICS

Printed in Great Britain by Bell and Bain Limited, Glasgow

CONTENTS

v

CONTENTS

CONTENTS

vii

PREFACE

One of the arts of teaching is to attack a problem from a different point of view, if the conventional approach has not been fully appreciated. This often helps the class to appreciate the conventional approach more fully. Maxwell's equations are generally introduced during a long course on electromagnetism, in which the correct interpretations and experimental bases of Maxwell's equations are sometimes obscured by the large amount of time of necessity devoted to the important applications of individual laws. The displacement 'current' is often introduced in a way that gives the wrong impression of its role in contemporary electromagnetic theory. There have been refinements of interpretation since Maxwell's time, particularly following the abandonment of the ether theories and the introduction of the theory of relativity. This is particularly true of the displacement 'current'. In this monograph, Maxwell's equations are approached from a different point of view from the conventional approach, namely using Coulomb's law and the transformations of the theory of special relativity. This approach illustrates the essential unity of Maxwell's equations, and the modern interpretation of the various terms in Maxwell's equations. It is the reverse of the historical approach, since the theory of special relativity arose out of optics and electromagnetism.

In order to keep the discussion of Maxwell's equations in Chapter 4 as simple as possible, the discussion is restricted to systems of 'point' charges moving with uniform velocities. The case of accelerating charges is discussed in Appendix 3. A brief review is given in Appendix 2 of the experimental evidence on which Maxwell's equations are normally developed, and some of the extra assumptions implicit in the way they are sometimes applied. The magnetic and electric fields, associated with electric circuits, are discussed from an atomistic viewpoint in Appendices 4 and 5 respectively. Appendices 2, 3, 4 and 5 are self-contained. They form an important part of the text. This material is included in the form of Appendices, so as to keep the main body of the text in Chapter 4 as simple as possible. In Chapter 5, a review is given of the scalar and vector potential from the atomistic point of view, leading up to the development of the Liénard–Wiechert potentials. In Chapter 6 a review is given of the conventional approach to relativistic electromagnetism. After illustrating the relativistic invariance of Maxwell's equations, an account is given of the electrodynamics of moving media, and unipolar induction to illustrate the importance of relativistic effects.

PREFACE

The subject matter of this monograph can be used to give supplementary lectures *following* an undergraduate course on Electricity in which Maxwell's equations are developed. The monograph is meant to complement conventional text books, not to replace them. It is assumed that, at this stage, the students are already familiar with Maxwell's equations and with vector analysis, which are used freely throughout the text. In M.K.S. units Maxwell's equations for the macroscopic fields take the form

$$\mathbf{\nabla} . \mathbf{D} = \rho$$
$$\mathbf{\nabla} . \mathbf{B} = 0$$
$$\mathbf{\nabla} . \mathbf{E} = -\partial \mathbf{B}/\partial t$$
$$\mathbf{\nabla} . \mathbf{H} = \mathbf{J} + \partial \mathbf{D}/\partial t$$

where $\mathbf{D}$ is the electric displacement, $\mathbf{B}$ the magnetic induction, $\mathbf{E}$ the electric intensity, $\mathbf{H}$ the magnetizing force, ρ the charge density and $\mathbf{J}$ the current density at the field point. A summary of the relevant formulae of vector analysis is given in Appendix 1. Derivations of the necessary relativistic transformations are given in Chapter 1. As time may not always be available in an already crowded curriculum for the necessary supplementary lectures, the discussions and in particular the mathematical steps have been given in full, so that the students themselves can read the monograph *following* a conventional course on electromagnetism. Problems are set in most chapters. It is hoped that the monograph will prove useful to electrical engineers and applied mathematicians, as well as physicists.

It is hoped that some of the material in this monograph will find its way into conventional courses on electromagnetism. For example, the forces between convection currents, discussed in Section 2.2, could be introduced into first year undergraduate courses. The electric and magnetic fields of a charge moving with uniform velocity developed in Chapter 3, and the Biot–Savart law developed in Appendix 4, could be introduced as soon as the students have done a course on special relativity. Electromagnetic induction can then be illustrated in terms of *Figure 4.7*. The development of the equation $\varepsilon_0 c^2 \mathbf{\nabla} . \mathbf{B} = \mathbf{J} + \varepsilon_0 \partial \mathbf{E}/\partial t$ using Sections 4.6(a) and 4.9 is probably the most satisfactory way of introducing the displacement 'current', which can then be illustrated using the examples given in Sections 4.7, 4.8, 4.11 and 4.12 and in Appendix 2(h).

This monograph was developed from a series of articles in *Contemporary Physics* **1** (1959) 134; **1** (1960) 453 and **3** (1961) 28. The author would like to thank Professor G. K. T. Conn for his encouragement throughout.

W. G. V. Rosser

Exeter

1

A SURVEY OF THE THEORY OF SPECIAL RELATIVITY

1.1. HISTORICAL INTRODUCTION

In this chapter, a brief review is given of the transformations of the theory of special relativity. A reader interested in a more comprehensive discussion is referred to the author's books *An Introduction to the Theory of Relativity* (Butterworths, London, 1964), and *Introductory Relativity* (Butterworths, London, 1967). In this chapter, these books will be referred to as I and II respectively, followed by the appropriate section numbers. In this monograph, we shall be concerned only with the theory of special relativity, or the restricted theory of relativity as it is sometimes called. This theory relates phenomena to inertial reference frames only. Effects associated with accelerating reference frames will not be considered.

The definition of an inertial reference frame is the same in the theory of special relativity as in Newtonian mechanics. If in a reference frame, a particle under the influence of no forces travels in a straight line with constant speed, then that reference frame is called an inertial reference frame. The laboratory system is generally a very satisfactory approximation to an inertial reference frame for the description of electromagnetic phenomena.

Consider an inertial reference frame Σ' moving with uniform velocity v relative to an inertial reference frame Σ along their common x axis. Let the co-ordinates of an event relative to Σ be at the point x, y, z, and let the co-ordinates of the same event relative to Σ' be at x', y', z'. Let the origins of Σ and Σ' coincide at $t = t' = 0$. These are the same conditions as illustrated in *Figure 1.1(a)*. If it is assumed that time is absolute, it can be shown that

$$x' = x - vt; \quad y' = y; \quad z' = z; \quad t' = t \qquad (1.1)$$

These are the Galilean transformations (Reference: I, Section 1.4 or II, Section 1.4). The appropriate velocity transformations are

$$u'_x = u_x - v; \quad u'_y = u_y; \quad u'_z = u_z \qquad (1.2)$$

1

where **u** is the velocity of a particle relative to Σ and **u**′ its velocity relative to Σ′.

According to the principle of relativity, the laws of physics are the same in all inertial reference frames. Newton's laws of motion satisfy the principle of relativity in inertial reference frames, provided the co-ordinates and time are transformed using the Galilean transformations (Reference: I, Section 1.4 or II, Section 1.4). For example, if one were in a ship steaming with *uniform* velocity out to sea on a calm day, provided all velocities were very much less than the velocity of light, one would find from experiments on mechanics

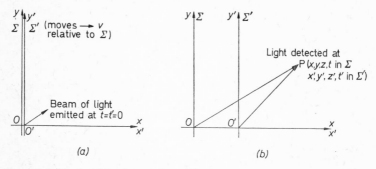

(a)

(b)

Figure 1.1. The inertial reference frame Σ′ moves with uniform velocity v relative to Σ along the common x axis. The origins O and O′ coincide at $t=t'=0$. The y and y′ and the z and z′ axes coincide at $t=t'=0$. Light emitted from the origins O and O′ at $t=t'=0$ is detected at P

that Newton's laws were valid, at least to an extremely good approximation, and one would not be able to determine the speed of the ship by means of mechanical experiments, without looking at something external to the ship. It is now known that for velocities comparable with the velocity of light, Newtonian mechanics is inadequate, and that the co-ordinate transformations appropriate to the new theory are the Lorentz transformations, which are developed in Section 1.2.

The laws of classical electromagnetism, such as Coulomb's law, the Biot–Savart law, Faraday's law etc., are summarized by Maxwell's equations. In addition to describing the behaviour of electric and magnetic phenomena, Maxwell's equations can be used to interpret classical physical optics in terms of electromagnetic waves. Now, if the co-ordinates and time are transformed using the Galilean transformations, Maxwell's equations do not obey the principle of relativity. In the last quarter of the nineteenth century

it was assumed that Newtonian mechanics and the Galilean transformations were correct, since at that time, no experiments on very
high speed atomic particles had been carried out. Thus no significant
deviations from Newtonian mechanics had been observed. If the
Galilean transformations were assumed to be correct, Maxwell's
equations could only hold in one reference frame, and it should have
been possible to identify this absolute system by electrical experiments
such as the Trouton–Noble experiment (Reference: I, Section 2.4 or
II, Appendix 1) and optical experiments such as the Michelson–
Morley experiments (I, Section 2.3 or II, Appendix 1). However,
in practice it proved impossible to identify by means of experiments
any absolute reference frame for the laws of optics and electromagnetism. What Einstein did in the theory of special relativity was
to postulate that the laws of optics and electricity and magnetism
obeyed the principle of relativity. This meant abandoning the
Galilean transformations in favour of the Lorentz transformations.
In his 1905 paper entitled 'On the Electrodynamics of Moving
Bodies', in which the theory of special relativity was first developed,
Einstein[1] wrote

> Examples of this sort, together with the unsuccessful attempts to
> discover any motion of the earth relatively to the 'light medium',
> suggest that the phenomena of electrodynamics as well as of mechanics
> possess no properties corresponding to the idea of absolute rest. They
> suggest rather that, as has already been shown to the first order of small
> quantities, the same laws of electrodynamics and optics will be valid
> for all frames of reference for which the equations of mechanics hold
> good. We will raise this conjecture (the purport of which will hereafter
> be called the 'Principle of Relativity') to the status of a postulate, and
> also introduce another postulate, which is only apparently irreconcilable
> with the former, namely, that light is always propagated in empty space
> with a definite velocity c which is independent of the state of motion
> of the emitting body. These two postulates suffice for the attainment of
> a simple and consistent theory of the electrodynamics of moving bodies
> based on Maxwell's theory for stationary bodies.

Einstein's two main postulates were the principle of relativity and
the principle of the constancy of the speed of light.
According to the principle of relativity, all physical laws are the
same in all inertial reference frames. As an example of the scope of
the principle of relativity in the theory of special relativity, consider
a ship going out to sea with *uniform* velocity on a calm day. If
experiments in mechanics, optics, electromagnetism or nuclear
physics were performed in a room inside the ship, according to the
principle of relativity, the laws determined on the basis of these

3

experiments would be precisely the same as if they were carried out in a laboratory at rest relative to the earth. Without looking at anything external to the ship, one could not determine the speed of the ship by means of any of these experiments.

According to the principle of the constancy of the speed of light, the speed of light in empty space has the same numerical value in all directions in all inertial reference frames. The statement that the speed of light has the same numerical value in all inertial reference frames implies that the fundamental units of length and time are defined in the same way in all inertial reference frames.

At the time Einstein[1] introduced the theory of special relativity (1905), the generally accepted theories of optics and electromagnetism were based on Maxwell's equations. Having postulated the principle of relativity for all physical laws, it would have been reasonable for Einstein to assume as his second postulate that the appropriate laws of electromagnetism and optics were Maxwell's equations. Instead, Einstein chose the principle of the constancy of the speed of light as his second postulate. It is illustrated in I, Section 3.3 or II, Appendix 2 that, if one did assume that Maxwell's equations were correct and obeyed the principle of relativity, then the principle of the constancy of the speed of light follows. The choice of the principle of the constancy of the velocity of light as a postulate by Einstein[1] simplified the derivation of the Lorentz transformations. It also enabled Einstein to discuss the measurement of the times of spatially separated events and to criticize the concept of absolute time before developing the Lorentz transformations. After developing the Lorentz transformations, Einstein[1] went on to show that the Lorentz transformations did in fact leave Maxwell's equations invariant in mathematical form.

At the time the theory was introduced, there was no direct experimental verification of the principle of the constancy of the speed of light. Recently, several direct experimental checks of the postulate have been performed. For example in 1964 Alväger, Farley, Kjellman and Wallin[2], determined the velocities of photons arising from the decay of π°-mesons into two photons each. The velocity of the π°-mesons, estimated using the equations of the theory of special relativity, was $0.99975c$. The measured velocity of the photons arising from π°-meson decays was $(2.9977 \pm 0.0004) \times 10^8$ metres/sec. This value is consistent with the accepted value of 2.9979×10^8 m/sec for the velocity of light emitted by a stationary source. This result shows that the velocity of light does not add on to the velocity of the source according to the Galilean transformations (eqn 1.2). Thus there is now direct experimental evidence for the principle of the constancy of the speed of light. Our development of the theory of special relativity in this chapter will be based on the principle of relativity and the principle of the constancy of the speed of light.

It has been illustrated how the theory of special relativity evolved out of classical electromagnetism, and we shall be concerned exclusively with the application of the theory to electromagnetism. It should be remembered that many of the applications of the theory of special relativity have been in the fields of nuclear physics and cosmic rays, where its use is indispensable, since the deviations from Newtonian mechanics are very large at the high energies involved. The theory of special relativity and the Lorentz transformations can now be developed from experiments on high speed mechanics. (Reference: II, Appendices 3(b) and 4.)

1.2. THE LORENTZ TRANSFORMATIONS

In our discussions of relativistic transformations we shall be concerned, initially, with the transformation of the co-ordinates and times of events from one inertial reference frame Σ to another inertial reference frame Σ', which is moving with uniform velocity v relative to Σ along their common x axis as shown in *Figure 1.1(a)*. Let the co-ordinates of an event relative to Σ be represented by x, y and z, and the co-ordinates of the same event relative to Σ' by x', y', and z'. The times of the event relative to Σ and Σ' will be represented by t and t' respectively. It will be assumed that the origins of Σ and Σ' coincide at a time $t = t' = 0$, and that the directions of the y' and z' axes of Σ' coincide with the y and z axes of Σ at $t = 0$ as shown in *Figure 1.1(a)*.

The Lorentz transformations will be developed for a simplified case. Consider a beam of light emitted from the origins of Σ and Σ' at the instant $t = t' = 0$ when they coincide, as shown in *Figure 1.1(a)*. Let the light reach a light detector at the point P. Let this event be measured, to be at the position x, y, z at a time t relative to Σ, using rulers and synchronized clocks at rest relative to Σ. An observer at rest in Σ would say that the light travelled along the path OP such that

$$\frac{OP}{t} = \frac{\sqrt{x^2 + y^2 + z^2}}{t} = c \qquad (1\ 3)$$

or

$$x^2 + y^2 + z^2 - c^2 t^2 = 0 \qquad (1.4)$$

where c is the velocity of light.

5

An observer at rest in Σ' would agree that the light did reach the light detector at P (which may be moving relative to both Σ and Σ'). Using rulers and synchronized clocks stationary in Σ', let an observer at rest in Σ' record the same event at a position x', y', z' at a time t'. Relative to Σ' the light would appear to have travelled the path $O'P$, such that

$$\frac{O'P}{t'} = \frac{\sqrt{x'^2 + y'^2 + z'^2}}{t'} = c \qquad (1.5)$$

or,

$$x'^2 + y'^2 + z'^2 - c^2 t'^2 = 0 \qquad (1.6)$$

where c is the velocity of light.

Notice the same value is used for the speed of light in both Σ and Σ', so as to be in accord with the principle of the constancy of the speed of light. The co-ordinates x, y, z and t in Σ and x', y', z' and t' in Σ' refer to the *same* event, namely the detection of the light at P in *Figure 1.1(b)*. The required transformations must transform eqn (1.6) into eqn (1.4). Direct substitution in eqn (1.6) verifies that the transformations which do transform eqn (1.6) into eqn (1.4) are

$$x' = \gamma(x - vt) \qquad (1.7)$$

$$y' = y \qquad (1.8)$$

$$z' = z \qquad (1.9)$$

$$t' = \gamma(t - vx/c^2) \qquad (1.10)$$

where,

$$\gamma = 1/\sqrt{1 - v^2/c^2} \qquad (1.11)$$

These are the Lorentz transformations. (Derivations are given in I, Section 3.4 and II, Appendix 3.)

The inverse transformations which change eqn (1.4) into eqn (1.6) are

$$x = \gamma(x' + vt') \qquad (1.12)$$

$$y = y' \qquad (1.13)$$

$$z = z' \qquad (1.14)$$

$$t = \gamma(t' + vx'/c^2) \qquad (1.15)$$

Notice the inverse transformations can be obtained from eqns (1.7),

6

(1.8), (1.9) and (1.10) by interchanging primed and unprimed quantities and replacing v by $-v$. The inverses of all relativistic transformations can be obtained by this procedure.

1.3. RELATIVITY OF SIMULTANEITY OF EVENTS

Let two events occur at two separated points x_1 and x_2 in the inertial frame Σ. Let them be measured to occur at the same time t in Σ. According to the Lorentz transformations these events would be recorded at times t_1' and t_2' by clocks at rest in Σ' where t_1' and t_2' are given by

$$t_1' = \gamma\left[t-\frac{v}{c^2}x_1\right] \quad \text{and} \quad t_2' = \gamma\left[t-\frac{v}{c^2}x_2\right] \qquad (1.16)$$

Since x_1 is not equal to x_2, t_1' cannot be equal to t_2', so that, if the Lorentz transformations are correct, then two spatially separated events which are simultaneous in Σ, would not be measured to be simultaneous in Σ'. Similarly, if two events occur simultaneously at two spatially separated points x_1' and x_2' in Σ', according to the Lorentz transformations, they would not be measured to be simultaneous in Σ. Thus, according to the theory of special relativity the simultaneity of spatially separated events is not an absolute property, as it was assumed to be in Newtonian mechanics.

1.4. TIME DILATION

Consider a clock at rest at the point $(x', y' = 0, z' = 0)$ in Σ'. Let it emit ticks at times t_1' and t_2' in Σ'. In Σ, the events, associated with the successive ticks emitted by the clock at rest in Σ', are recorded at

$$x_1 = \gamma(x'+vt_1'); \quad t_1 = \gamma(t_1'+vx'/c^2) \qquad (1.17)$$

and

$$x_2 = \gamma(x'+vt_2'); \quad t_2 = \gamma(t_2'+vx'/c^2) \qquad (1.18)$$

Hence,

$$(t_2-t_1) = \frac{(t_2'-t_1')}{\sqrt{1-v^2/c^2}} \qquad (1.19)$$

The time interval between two events, taking place at the same point in an inertial reference frame and measured by one clock stationary at that point, is called the proper time interval between the events. In eqn (1.19), $t_2'-t_1'$ is the proper time interval between the events. The time interval t_2-t_1 in any other inertial frame is longer than the

B 7

proper time interval. In eqns (1.17) and (1.18), x_1 is not equal to x_2 so that the two events are not at the same point in Σ and so t_1 and t_2 must be measured by spatially separated synchronized clocks. Einstein's[1] prescription for synchronizing two stationary separated clocks is to send a light signal from clock 1 at a time t_1 measured by clock 1. Let the light be reflected from clock 2 at a time t_2 measured by clock 2. If the reflected light returns to clock 1 at a time t_3 measured by clock 1 then clocks 1 and 2 are synchronous if

$$t_2 - t_1 = t_3 - t_2$$

that is if the time measured for light to go one way is equal to the time measured for light to go in the opposite direction.

Time dilation has been confirmed by experiments on the decay of π and μ-mesons.

1.5. LORENTZ CONTRACTION

Consider a body moving parallel to the x axis with velocity v relative to Σ. In Σ', which moves with velocity v relative to Σ, the body is at rest. Let the length of the body be measured relative to Σ, by recording the positions of its extremities at x_1 and x_2 at the same time t in Σ. In Σ' the corresponding events are

$$x_1' = \gamma(x_1 - vt); \quad t_1' = \gamma(t - vx_1/c^2)$$
$$x_2' = \gamma(x_2 - vt); \quad t_2' = \gamma(t - vx_2/c^2)$$

Subtracting

$$x_1' - x_2' = \gamma(x_1 - x_2)$$

Though t_1' is not equal to t_2', since the body is at rest in Σ', $x_1' - x_2'$ is equal to l_0, the proper length of the body measured when it is at rest in Σ'. If l is the length of the moving body measured in Σ

$$l = l_0\sqrt{1 - v^2/c^2} \tag{1.20}$$

Thus a body moving with velocity v relative to an observer is measured to be shorter in the ratio $\sqrt{1 - v^2/c^2}$ in its direction of motion relative to the observer.

A body of proper volume V_0, moving with velocity $\mathbf{v}$ relative to an observer, can be divided into thin rods parallel to $\mathbf{v}$. Each one of these rods is reduced in length by a factor $\sqrt{1 - v^2/c^2}$ so that the measured volume of the moving body is

$$V = V_0\sqrt{1 - v^2/c^2} \tag{1.21}$$

The results developed in Sections 1.3, 1.4 and 1.5 illustrate how the theory of special relativity necessitated a revision of the classical

8

ideas of space and time. For a discussion of the physical implications and the experimental confirmations of these predictions the reader is referred to I, Chapter 3, or II, Chapter 3.

1.6. THE VELOCITY TRANSFORMATIONS

Let a particle be measured to be at the point x, y, z at a time t in Σ and at the point $x+\delta x$, $y+\delta y$, $z+\delta z$ at a time $t+\delta t$ in Σ. The velocity of the particle, relative to Σ, is defined as a vector $\mathbf{u}$ having components

$$u_x = \frac{\delta x}{\delta t}; \quad u_y = \frac{\delta y}{\delta t}; \quad u_z = \frac{\delta z}{\delta t} \qquad (1.22)$$

The magnitude of the velocity is given by

$$u^2 = u_x^2 + u_y^2 + u_z^2 \qquad (1.23)$$

In the inertial frame Σ' which is moving with uniform velocity v relative to Σ along their common x axis, corresponding to the measurement of the position of the particle at x, y, z at a time t in Σ, one has in Σ'

$$x' = \gamma(x-vt); \quad y' = y; \quad z' = z; \quad t' = \gamma(t-vx/c^2) \qquad (1.24)$$

Corresponding to $x+\delta x$, $y+\delta y$, $z+\delta z$, $t+\delta t$ in Σ one has in Σ'

$$x'+\delta x' = \gamma[x+\delta x - v(t+\delta t)]; \quad y'+\delta y' = y+\delta y$$

$$z'+\delta z' = z+\delta z; \quad t'+\delta t' = \gamma[t+\delta t - v(x+\delta x)/c^2] \qquad (1.25)$$

Subtracting eqns (1.24) from eqns (1.25),

$$\delta x' = \gamma(\delta x - v\delta t); \quad \delta y' = \delta y; \quad \delta z' = \delta z; \quad \delta t' = \gamma(\delta t - v\delta x/c^2)$$

The velocity of the particle relative to Σ' has components

$$u_x' = \frac{\delta x'}{\delta t'} = \frac{(\delta x - v\delta t)}{(\delta t - v\delta x/c^2)} = \frac{(\delta x/\delta t - v)}{[1 - v(\delta x/\delta t)/c^2]}$$

$$u_x' = \frac{(u_x - v)}{(1 - vu_x/c^2)} \qquad (1.26)$$

$$u_y' = \frac{\delta y'}{\delta t'} = \frac{\delta y}{\gamma(\delta t - v\delta x/c^2)} = \frac{(\delta y/\delta t)\sqrt{1-v^2/c^2}}{[1 - v(\delta x/\delta t)/c^2]}$$

or

$$u_y' = \frac{u_y\sqrt{1-v^2/c^2}}{(1 - vu_x/c^2)} \qquad (1.27)$$

9

Similarly,

$$u'_z = \frac{u_z\sqrt{1-v^2/c^2}}{(1-vu_x/c^2)} \tag{1.28}$$

The inverse transformations are

$$u_x = \frac{(u'_x+v)}{(1+vu'_x/c^2)} \tag{1.29}$$

$$u_y = \frac{u'_y\sqrt{1-v^2/c^2}}{(1+vu'_x/c^2)} \tag{1.30}$$

$$u_z = \frac{u'_z\sqrt{1-v^2/c^2}}{(1+vu'_x/c^2)} \tag{1.31}$$

Now

$$u'^2 = u'^2_x+u'^2_y+u'^2_z = \frac{(u_x-v)^2+(u^2_y+u^2_z)(1-v^2/c^2)}{(1-vu_x/c^2)^2}$$

Since,

$$u^2_y+u^2_z = u^2-u^2_x,$$

$$1-u'^2/c^2 = 1-\frac{\left(\dfrac{u_x}{c}-\dfrac{v}{c}\right)^2+\left(\dfrac{u^2}{c^2}-\dfrac{u^2_x}{c^2}\right)\left(1-\dfrac{v^2}{c^2}\right)}{(1-vu_x/c^2)^2}$$

$$(1-u'^2/c^2) = \frac{(1-v^2/c^2)(1-u^2/c^2)}{(1-vu_x/c^2)^2}$$

or,

$$\sqrt{1-u'^2/c^2} = \frac{\sqrt{(1-v^2/c^2)(1-u^2/c^2)}}{(1-vu_x/c^2)} \tag{1.32}$$

Similarly, for the inverse transformation

$$\sqrt{1-u^2/c^2} = \frac{\sqrt{(1-v^2/c^2)(1-u'^2/c^2)}}{(1+vu'_x/c^2)} \tag{1.33}$$

1.7. TRANSFORMATION OF THE MOMENTUM AND ENERGY OF A PARTICLE

It is shown in textbooks on relativity (e.g. I, Section 5.2 and II, Section 5.2) that if the laws of mechanics are to obey the principle of relativity, when the co-ordinates and time are transformed

10

according to the Lorentz transformations, then the laws of mechanics must be modified, such that the inertial mass of a moving particle varies with its velocity. If a particle of rest mass m_0 is moving with a velocity $\mathbf{u}$ relative to Σ, its inertial mass relative to Σ is given by

$$m = \frac{m_0}{\sqrt{1-u^2/c^2}} \qquad (1.34)$$

The quantity m is called the relativistic mass. Eqn (1.34) has been confirmed by experiments such as those of Bucherer (Reference: I, Section 5.4.3 or II, Section 5.4.4) who determined the ratio of charge to mass for β-rays.

Similarly, if the particle has velocity $\mathbf{u}'$ relative to Σ', its relativistic inertial mass relative to Σ' is

$$m' = \frac{m_0}{\sqrt{1-u'^2/c^2}} \qquad (1.35)$$

The momentum of the particle moving with velocity $\mathbf{u}$ relative to Σ is defined as

$$\mathbf{p} = m\mathbf{u} = \frac{m_0\mathbf{u}}{\sqrt{1-u^2/c^2}} \qquad (1.36)$$

having components

$$p_x = \frac{m_0 u_x}{\sqrt{1-u^2/c^2}} \quad \text{etc.} \qquad (1.37)$$

The momentum of the particle relative to Σ' is defined as

$$\mathbf{p}' = m'\mathbf{u}' = \frac{m_0\mathbf{u}'}{\sqrt{1-u'^2/c^2}} \qquad (1.38)$$

The force acting on a particle is defined as the rate of change of the momentum of the particle, so that in Σ, we have

$$\mathbf{f} = \frac{d\mathbf{p}}{dt} = \frac{d}{dt}(m\mathbf{u}) = \frac{d}{dt}\left(\frac{m_0\mathbf{u}}{\sqrt{1-u^2/c^2}}\right) \qquad (1.39)$$

and in Σ'

$$\mathbf{f}' = \frac{d\mathbf{p}'}{dt'} = \frac{d}{dt'}(m'\mathbf{u}') = \frac{d}{dt'}\left(\frac{m_0\mathbf{u}'}{\sqrt{1-u'^2/c^2}}\right) \qquad (1.40)$$

Let a force $\mathbf{f}$ acting on a particle give rise to a displacement $d\mathbf{l}$ in a time dt relative to Σ. The work done is defined as $\mathbf{f} \cdot d\mathbf{l}$. If all the

11

work done goes into increasing the kinetic energy T of the particle, then,

$$dT = \mathbf{f} \cdot d\mathbf{l}$$

and,

$$\frac{dT}{dt} = \mathbf{f} \cdot \frac{d\mathbf{l}}{dt} = \mathbf{f} \cdot \mathbf{u} \tag{1.41}$$

Using eqn (1.39), this can be written as

$$\frac{dT}{dt} = \frac{d}{dt}(m\mathbf{u}) \cdot \mathbf{u} = m\mathbf{u} \cdot \frac{d\mathbf{u}}{dt} + \mathbf{u} \cdot \mathbf{u}\frac{dm}{dt} \tag{1.42}$$

Now

$$u^2 = u_x^2 + u_y^2 + u_z^2$$

Differentiating with respect to time, and using eqn (A1.1) of Appendix 1,

$$2u\frac{du}{dt} = 2u_x\frac{du_x}{dt} + 2u_y\frac{du_y}{dt} + 2u_z\frac{du_z}{dt} = 2\mathbf{u} \cdot \frac{d\mathbf{u}}{dt}$$

Also,

$$\mathbf{u} \cdot \mathbf{u} = u^2$$

and,

$$\frac{dm}{dt} = \frac{du}{dt}\frac{dm}{du} = \frac{du}{dt}\frac{d}{du}\left(\frac{m_0}{\sqrt{1-u^2/c^2}}\right) = \frac{du}{dt}\frac{m_0 u/c^2}{(1-u^2/c^2)^{3/2}}$$

Substituting in eqn (1.42)

$$\frac{dT}{dt} = \frac{m_0}{\sqrt{1-u^2/c^2}} u\frac{du}{dt} + m_0\frac{du}{dt}\frac{u^3/c^2}{(1-u^2/c^2)^{3/2}}$$

or

$$\frac{dT}{dt} = \frac{d}{dt}\left[\frac{m_0 c^2}{\sqrt{1-u^2/c^2}}\right]$$

Integrating, assuming $T = 0$ when $u = 0$, we have

$$T = \frac{m_0 c^2}{\sqrt{1-u^2/c^2}} - m_0 c^2$$

or,

$$T = mc^2 - m_0 c^2 \tag{1.43}$$

or,

$$mc^2 = T + m_0 c^2 \tag{1.44}$$

The quantity mc^2, which is equal to the sum of the kinetic energy of the particle and the quantity m_0c^2, is called the total energy of the particle and will be denoted by W, so that

$$W = mc^2 \qquad (1.45)$$

From eqn (1.38)

$$p'_x = \frac{m_0 u'_x}{\sqrt{1-u'^2/c^2}}$$

Substituting for u'_x from eqn (1.26) and for $(1-u'^2/c^2)^{-1/2}$ from eqn (1.32), we have

$$p'_x = m_0 \frac{(u_x-v)}{(1-vu_x/c^2)} \frac{(1-vu_x/c^2)}{\sqrt{1-v^2/c^2}\sqrt{1-u^2/c^2}}$$

$$= \frac{m_0}{\sqrt{1-u^2/c^2}} \frac{(u_x-v)}{\sqrt{1-v^2/c^2}}$$

$$= \gamma\,(mu_x - mv)$$

But,

$$mu_x = p_x \quad \text{and} \quad m = W/c^2$$

Hence,

$$p'_x = \gamma(p_x - vW/c^2) \qquad (1.46)$$

Now,

$$p'_y = \frac{m_0 u'_y}{\sqrt{1-u'^2/c^2}}$$

Substituting for u'_y from eqn (1.27) and for $(1-u'^2/c^2)^{-1/2}$ from eqn (1.32), we obtain

$$p'_y = m_0 \frac{u_y\sqrt{1-v^2/c^2}}{(1-vu_x/c^2)} \frac{(1-vu_x/c^2)}{\sqrt{1-v^2/c^2}\sqrt{1-u^2/c^2}}$$

$$= m_0 u_y/\sqrt{1-u^2/c^2} = p_y$$

that is,

$$p'_y = p_y \qquad (1.47)$$

Similarly,

$$p'_z = p_z \qquad (1.48)$$

Now

$$W' = m'c^2 = \frac{m_0 c^2}{\sqrt{1 - u'^2/c^2}}$$

$$= \frac{m_0 c^2(1 - vu_x/c^2)}{\sqrt{1 - v^2/c^2}\sqrt{1 - u^2/c^2}} = \gamma mc^2(1 - vu_x/c^2)$$

or

$$W' = \gamma(W - vp_x) \tag{1.49}$$

Similarly, one has the inverse transformations

$$p_x = \gamma(p_x' + vW'/c^2); \quad p_y = p_y'; \quad p_z = p_z'; \quad W = \gamma(W' + vp_x') \tag{1.50}$$

1.8. THE FORCE TRANSFORMATIONS

Consider a particle of rest mass m_0 having velocity $\mathbf{u}'$ and momentum $\mathbf{p}'$ relative to Σ', and having velocity $\mathbf{u}$ and momentum $\mathbf{p}$ relative to Σ. In Section 1.7 the force acting on a particle measured relative to Σ' was defined as

$$f_x' = dp_x'/dt'. \tag{1.40}$$

Now from eqn (1.46),

$$p_x' = \gamma(p_x - vW/c^2)$$

Hence,

$$f_x' = \gamma \frac{d}{dt'}(p_x - vW/c^2)$$

Now,

$$\frac{d}{dt'} = \frac{dt}{dt'}\frac{d}{dt}$$

and

$$\frac{dt'}{dt} = \frac{d}{dt}\gamma\left(t - \frac{vx}{c^2}\right) = \gamma(1 - vu_x/c^2) \tag{1.51}$$

Hence,

$$f_x' = \frac{\gamma}{\gamma(1 - vu_x/c^2)}\frac{d}{dt}(p_x - vW/c^2)$$

$$= \frac{1}{(1 - vu_x/c^2)}\left(f_x - \frac{v}{c^2}\frac{dW}{dt}\right)$$

Using eqns (1.44) and (1.41)

$$\frac{\mathrm{d}W}{\mathrm{d}t} = \frac{\mathrm{d}}{\mathrm{d}t}(T+m_0c^2) = \frac{\mathrm{d}T}{\mathrm{d}t} = \mathbf{f} \cdot \mathbf{u}$$

Hence,

$$f'_x = \frac{1}{(1-vu_x/c^2)}\left(f_x - \frac{v}{c^2}\mathbf{f} \cdot \mathbf{u}\right) \tag{1.52}$$

Since

$$\mathbf{f} \cdot \mathbf{u} = f_xu_x + f_yu_y + f_zu_z,$$

$$f'_x = f_x - \frac{vu_y}{c^2(1-vu_x/c^2)}f_y - \frac{vu_z}{c^2(1-vu_x/c^2)}f_z \tag{1.53}$$

Since

$$p'_y = p_y$$

$$\frac{\mathrm{d}p'_y}{\mathrm{d}t'} = \frac{\mathrm{d}p_y}{\mathrm{d}t'} = \frac{\mathrm{d}t}{\mathrm{d}t'}\frac{\mathrm{d}p_y}{\mathrm{d}t}$$

But from eqn (1.51),

$$\frac{\mathrm{d}t}{\mathrm{d}t'} = 1/\gamma(1-vu_x/c^2)$$

Hence,

$$f'_y = \frac{f_y}{\gamma(1-vu_x/c^2)} = \frac{\sqrt{1-v^2/c^2}}{(1-vu_x/c^2)}f_y \tag{1.54}$$

Similarly, since

$$p'_z = p_z,$$

$$f'_z = \frac{f_z}{\gamma(1-vu_x/c^2)} = \frac{\sqrt{1-v^2/c^2}}{(1-vu_x/c^2)}f_z \tag{1.55}$$

The inverse transformations are

$$f_x = f'_x + \frac{vu'_y}{c^2(1+vu'_x/c^2)}f'_y + \frac{vu'_z}{c^2(1+vu'_x/c^2)}f'_z \tag{1.56}$$

$$f_y = \frac{\sqrt{1-v^2/c^2}}{(1+vu'_x/c^2)}f'_y \tag{1.57}$$

$$f_z = \frac{\sqrt{1-v^2/c^2}}{(1+vu'_x/c^2)}f'_z \tag{1.58}$$

B*

1.9. THE LORENTZ FORCE

The force transformations will be used extensively in the text, so that it is worth while reviewing how the concept of force is normally used in electromagnetic theory and the theory of special relativity. If a charge of magnitude q coulombs is moving with velocity $\mathbf{u}$ metres per second in a region of empty space where there are electric and magnetic fields of strength $\mathbf{E}$ volts/metre and $\mathbf{B}$ webers/metre^2 respectively, then, according to classical electromagnetic theory, the force on q measured in newtons is given by the expression for the Lorentz force, namely,

$$\mathbf{f} = q\mathbf{E} + q\mathbf{u} \times \mathbf{B} \tag{1.59}$$

The magnitude of the magnetic induction $\mathbf{B}$ can be calculated, if the positions of the magnetizing coils, the currents in the coils, the positions of magnetic materials, etc. are given. The electric intensity $\mathbf{E}$ can be calculated if the positions and potentials of all the conductors and dielectrics are given. These calculated values of $\mathbf{E}$ and $\mathbf{B}$ are used to calculate the force on the charge q. (Sometimes, it is more convenient to measure the magnetic field with a search coil and fluxmeter than to calculate it.) The motion of the charge is then predicted by equating the Lorentz force calculated in this way to the rate of change of the momentum of the charge. That is,

$$\mathbf{f} = q\mathbf{E} + q\mathbf{u} \times \mathbf{B}$$

and,

$$\mathbf{f} = \frac{d\mathbf{p}}{dt} = \frac{d}{dt}\left(\frac{m_0\mathbf{u}}{\sqrt{1-u^2/c^2}}\right)$$

It is not strictly necessary to introduce the term force, since one can write

$$\frac{d}{dt}\left(\frac{m_0\mathbf{u}}{\sqrt{1-u^2/c^2}}\right) = q\mathbf{E} + q\mathbf{u} \times \mathbf{B} \tag{1.60}$$

We shall, however, find it convenient to talk in terms of the Lorentz force acting on a moving charge, since it is simpler than calling it the rate of change of momentum.

It should be noted that in eqn (1.60), in addition to the variation of mass with velocity, it is assumed that the magnitude of the charge q is independent of its velocity. This is *the principle of constant charge*, according to which the *total* charge on a particle is independent of its velocity. It is shown in I, Section 3.3 that the principle

16

of constant charge is implicit in Maxwell's equations, if it is assumed that Maxwell's equations obey the principle of relativity. If the magnitude of the total charge on a particle did depend on the velocity of the particle, for example, in a way similar to the variation of mass with velocity, then hydrogen atoms and molecules would not be electrically neutral, since, on the average, the atomic electrons move faster than the nuclei of the atoms and molecules relative to the laboratory. If the total charge on a particle varied with velocity, 'neutral' atoms and molecules would be deflected in electric fields. In 1960, King[3] showed that the charges on the electrons and protons in hydrogen molecules were numerically equal, to an accuracy of 1 part in 10^{20}. King also showed that the charges of the positive protons in helium nuclei cancelled the charges of the orbital electrons, with nearly the same experimental accuracy. The principle of constant charge will be taken as axiomatic in Chapters 2 and 3.

Eqn (1.60) has been verified in many ways, such as Bucherer's determination of e/m for β-rays (Reference: I, Section 5.4.3 or II, Section 5.4.4). Eqn (1.60) has been used in the design of proton synchrotrons capable of accelerating protons up to energies of >25 GeV. However, if the charge is emitting electromagnetic radiation, eqn (1.60) must be modified so as to allow for the effect of radiation reaction. This is important in the design of very high energy electron synchrotrons. All the experimental results available are consistent with a speed dependent inertial mass and a constant total electric charge.

1.10. THE DEFINITIONS OF **E** AND **B**

The electric intensity **E** and the magnetic induction **B** at a point in empty space, measured relative to an inertial reference frame Σ, will be defined in terms of the Lorentz force acting on a moving test charge q at the point where the fields are required.

We have,

$$\mathbf{f} = q\mathbf{E} + q\mathbf{u} \times \mathbf{B} \qquad (1.59)$$

where **f** is in newtons, if q is in coulombs, **u** in metres/sec, **E** in volts/metre and **B** in webers/metre^2 (or teslas). The first term on the right hand side of eqn (1.59) is the electric force (denoted $\mathbf{f}_{elec}$) and the second term is the magnetic force (denoted $\mathbf{f}_{mag}$). If the test charge q were at rest in Σ, then the only force acting on q, measured relative to Σ, would be $\mathbf{f}_{elec}$ so that **E** can be defined by the relation

$$\mathbf{E} = \underset{q \to 0}{\text{Limit}} (\mathbf{f}_{elec})/q \qquad (1.61)$$

Notice **E** is parallel to $\mathbf{f}_{elec}$ so that the electric intensity at a point is in the direction in which a stationary test charge would start to move if placed at that point.

If the test charge q moves with velocity **u** relative to Σ, then according to eqn (1.59), there is an extra contribution to the total force on q, over and above the electric force $\mathbf{f}_{elec}$, namely the magnetic force which is given by

$$\mathbf{f}_{mag} = q\mathbf{u} \times \mathbf{B} \qquad (1.62)$$

This equation can be used to define **B**. The magnitude of the magnetic force depends on the magnitude and direction of **u**, the velocity of the test charge. It is a maximum when **u** is perpendicular to **B** and a minimum when **u** is parallel to **B**. The direction of **B** is defined as the direction in which the test charge q would be moving, when it experienced no magnetic force. The strength of the magnetic induction can be defined in terms of the maximum magnetic force, obtained when **u** is perpendicular to **B**, that is

$$B = \operatorname*{Limit}_{q \to 0} (f_{mag})_{max}/qu \qquad (1.63)$$

The sense of **B** is defined such that $\mathbf{f}_{mag}$, **B** and **u** obey the left hand rule with the thumb of the left hand pointing in the direction of $\mathbf{f}_{mag}$, the first finger in the direction of **B** and the second finger in the direction of **u** (for a positive charge). This sense is in agreement with the direction in which the north 'pole' of a compass needle would point in empty space.

It is customary to represent electric and magnetic fields on diagrams using *imaginary* field lines, drawn such that the direction of the field line at a point is in the same direction as the field at that point. The number of field lines is generally limited by spacing the lines, such that on the diagrams the number of lines per unit area crossing a surface at right angles to the field line is proportional to the strength of the field at that point. Thus the field lines on the diagrams are closest together where the field intensity is highest. This concept of using field lines to represent field intensities quantitatively on diagrams will be used extensively in the diagrams throughout the text. (Cf., however, the quotation from Hertz given on page 119.)

The electric intensity **E**$'$ and the magnetic induction **B**$'$ relative to the inertial reference frame Σ', which moves with uniform velocity v relative to Σ, can be defined in terms of the Lorentz force acting on the test charge q. Let it move with velocity **u**$'$ relative to Σ', corresponding to the velocity **u** relative to Σ. In Σ', we have

$$\mathbf{f}' = q\mathbf{E}' + q\mathbf{u}' \times \mathbf{B}' = \mathbf{f}'_{elec} + \mathbf{f}'_{mag} \qquad (1.64)$$

18

where the forces are measured relative to Σ'. The quantities $\mathbf{E}'$ and $\mathbf{B}'$ are defined in the same way relative to Σ' as $\mathbf{E}$ and $\mathbf{B}$ were defined relative to Σ using the relations $\mathbf{E}' = \mathbf{f}'_{elec}/q$ and $\mathbf{f}'_{mag} = q\mathbf{u}' \times \mathbf{B}'$. The force $\mathbf{f}'_{elec}$ would be the force on a charge q at rest relative to Σ'. Such a charge would be moving relative to Σ and would be acted upon by a magnetic force in Σ. Notice the same value was used for q in eqns (1.59) and (1.64), so as to be in accord with the principle of constant charge.

REFERENCES

[1] EINSTEIN, A. *Ann. Phys., Lpz.* **17** (1905) 891

[2] ALVÄGER, T., FARLEY, F. J. M., KJELLMAN, J. and WALLIN, I. *Phys. Letters* **12** (1964) 260

[3] KING, J. G., *Phys. Rev. Letters* **5** (1960) 562

2

ELECTROMAGNETISM AS A SECOND ORDER EFFECT

2.1. INTRODUCTION

It is important for the reader to appreciate from the outset that the magnetic forces between moving charges are generally very much smaller than the electric forces between the same moving charges, and that magnetic forces can be interpreted as second order relativistic effects. Before proceeding in Chapter 3 to consider the general case of two moving charges, in this chapter a simple example will be considered from the viewpoint of Maxwell's equations. The same example will then be interpreted in terms of the force transformations of the theory of special relativity.

2.2. FORCES BETWEEN TWO PARALLEL CONVECTION CURRENTS

Consider two infinitely long, straight, thin, uniformly charged, non-conducting wires a distance r apart *in vacuo* lying in the xy plane of an inertial reference frame Σ. Let both wires move in the positive x direction with uniform velocity v relative to Σ, as shown in *Figure 2.1*. Let the electric charge on the wires be λ coulomb/metre measured relative to Σ. Since the wires are moving relative to Σ, the charge distributions give rise to convection currents relative to Σ. The charge passing any point per second is λv, so that the magnitude of the convection currents is λv amperes.

The electric and magnetic fields due to wire 1, will be calculated at the position of wire 2 using Maxwell's equations. We have

$$\nabla \times \mathbf{E} = -\frac{\partial \mathbf{B}}{\partial t} \tag{2.1}$$

$$\nabla \cdot \mathbf{D} = \rho \tag{2.2}$$

When the infinitely long wires are moving with uniform velocity, $\partial \mathbf{B}/\partial t$ is zero everywhere and eqn (2.1) reduces to

$$\nabla \times \mathbf{E} = 0.$$

20

For a point in empty space $\mathbf{D} = \varepsilon_0\mathbf{E}$, where $\varepsilon_0 = 8\cdot85 \times 10^{-12}$ farads/metre is the electric space constant. Hence, eqn (2.2) can be written

$$\mathbf{\nabla} \cdot \mathbf{E} = \rho/\varepsilon_0$$

Hence Gauss' law can be used to calculate the electric field. By symmetry the electric field due to wire 1 must diverge radially from

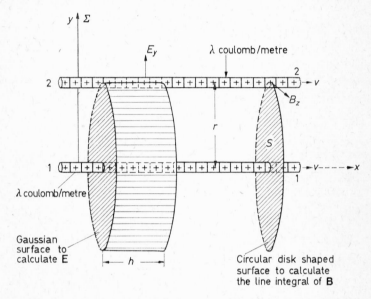

Figure 2.1. The calculation of the electric and magnetic forces between two convection currents using Maxwell's equations

wire 1. Considering the cylindrical Gaussian surface of radius r and height h shown in *Figure 2.1*, applying Gauss' law, we have

$$2\pi rhE_r = \lambda h/\varepsilon_0$$

$$E_r = E_y = \frac{\lambda}{2\pi\varepsilon_0 r} \qquad (2.3)$$

This electric field gives rise to a repulsion on wire 2 in the y direction of magnitude

$$f_{\text{elec}} = \frac{\lambda^2}{2\pi\varepsilon_0 r} \text{ newtons/metre length} \qquad (2.4)$$

21

Now consider the equations

$$\nabla \times \mathbf{H} = \mathbf{J} + \frac{\partial \mathbf{D}}{\partial t} \tag{2.5}$$

and

$$\nabla \cdot \mathbf{B} = 0 \tag{2.6}$$

When the wires are moving with uniform velocity $\partial \mathbf{D}/\partial t$ is zero, and for empty space $\mathbf{B} = \mu_0 \mathbf{H}$, where $\mu_0 = 4\pi \times 10^{-7}$ henrys/metre is the magnetic space constant.

Eqn (2.5) becomes

$$\nabla \times \mathbf{B} = \mu_0 \mathbf{J}$$

Integrating over the circular disk-like surface S of radius r, having wire 1 as centre, as shown in *Figure 2.1*,

$$\int (\nabla \times \mathbf{B}) \cdot \mathbf{n} dS = \mu_0 \int \mathbf{J} \cdot \mathbf{n} dS$$

where $\mathbf{n}$ is a unit vector normal to the element of area dS.

Using Stokes' theorem, eqn (A1.25) of Appendix 1, one obtains Ampère's circuital theorem:

$$\int (\nabla \times \mathbf{B}) \cdot \mathbf{n} dS = \oint \mathbf{B} \cdot d\mathbf{l} = \mu_0 \int \mathbf{J} \cdot \mathbf{n} dS = \mu_0 \lambda v$$

By symmetry, $\mathbf{B}$ has the same value at all points on the circumference of the surface S. Hence

$$2\pi r B = \mu_0 \lambda v$$

giving

$$B_z = \frac{\mu_0 \lambda v}{2\pi r} \tag{2.7}$$

The magnetic force on wire 2 is $\int I_2 \, d\mathbf{l}_2 \times \mathbf{B}_1$, so that the magnetic force on wire 2 per metre length is an attractive force given by

$$f_{\text{mag}} = -\frac{\mu_0 \lambda^2 v^2}{2\pi r} \text{ newtons/metre length}$$

The resultant force per unit length of wire 2 is given by

$$f_y = f_{\text{elec}} + f_{\text{mag}}$$

$$= \frac{\lambda^2}{2\pi \varepsilon_0 r} - \frac{\mu_0 \lambda^2 v^2}{2\pi r}$$

$$= \frac{\lambda^2}{2\pi \varepsilon_0 r} (1 - \mu_0 \varepsilon_0 v^2)$$

22

FORCES BETWEEN TWO PARALLEL CONVECTION CURRENTS

Numerical substitution of $8 \cdot 85 \times 10^{-12}$ for ε_0 and $4\pi \times 10^{-7}$ for μ_0 shows that $\mu_0 \varepsilon_0$ is equal to $1/c^2$, where $c = 3 \times 10^8$ metres/sec is the velocity of light in empty space.

Hence,

$$f_y = \frac{\lambda^2}{2\pi\varepsilon_0 r} (1 - v^2/c^2) \qquad (2.8)$$

In this simple case, the ratio of the magnetic force to the electric force is v^2/c^2, where v is the velocity of the charges relative to Σ, the inertial reference frame chosen. This illustrates that magnetic forces are of second order.

Consider a typical case, say a current of 1 ampere flowing in a copper wire of cross-sectional area, $a = 1$ mm^2(10^{-6} metre^2). Let N be the number of charge carriers per metre^3 ($\sim 8 \cdot 5 \times 10^{28}$ for copper), and let $q = 1 \cdot 6 \times 10^{-19}$ coulomb be the charge on each electron. The current is given by

$$I = qNav$$

where v is the mean drift velocity of the electrons. Substituting we find $v \sim 7 \times 10^{-5}$ metres/sec. If the charges shown in *Figure 2.1* moved with this speed, the ratio of the magnetic to the electric forces given by eqn (2.8) would be about 5×10^{-26}. Why then are the magnetic forces between electric circuits so important? In practice, in a copper conductor there is no resultant electric charge inside the conductor, since the charges on the moving electrons are compensated by the positively charged ions which are virtually at rest in a stationary conductor, as illustrated in *Figure 2.2*.

The force on the moving electrons in conductor 2 due to the moving electrons in conductor 1 is a repulsive force given by eqn (2.8), namely

$$f_{--} = \frac{\lambda^2}{2\pi\varepsilon_0 r} (1 - v^2/c^2) \text{ newtons/metre length} \qquad (2.9)$$

Since the positive ions in conductor 2 are at rest, there is no magnetic force on them, so that the total force on the positive ions in conductor 2 due to the moving electrons in conductor 1 is an attractive force given by

$$f_{-+} = -\frac{\lambda^2}{2\pi\varepsilon_0 r} \text{ newtons/metre length} \qquad (2.10)$$

23

The force on the moving electrons in conductor 2 due to the stationary positive charges in conductor 1 is an attractive force given by

$$f_{+-} = -\frac{\lambda^2}{2\pi\varepsilon_0 r} \text{ newtons/metre length} \qquad (2.11)$$

The force on the positive ions in conductor 2 due to the stationary positive ions in conductor 1 is a repulsive force given by

$$f_{++} = +\frac{\lambda^2}{2\pi\varepsilon_0 r} \text{ newtons/metre length} \qquad (2.12)$$

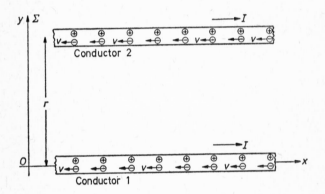

Figure 2.2. The forces between two conduction currents I. In the simplified model used, the positive ions are at rest and the negative electrons all move with the same uniform velocity v. The positive charge per unit length is $+\lambda$ and the negative charge per unit length $-\lambda$. The electric forces cancel leaving only the second order magnetic forces between the charges

Adding eqns (2.9), (2.10), (2.11) and (2.12), we find that the resultant force per unit length of conductor 2 due to all the charges in conductor 1 is an attractive force given by

$$f = -\frac{\lambda^2 v^2}{2\pi\varepsilon_0 c^2 r} = -\frac{I^2}{2\pi\varepsilon_0 c^2 r} \text{newtons/metre length} \qquad (2.13)$$

Thus, the electric forces between the charges in the two electrically neutral conductors in *Figure 2.2* cancel out to zero, leaving only the second order magnetic forces that are observed experimentally.

24

2.3. FORCES BETWEEN TWO PARALLEL CONVECTION CURRENTS USING RELATIVITY THEORY

The example illustrated in *Figure 2.1* and discussed in Section 2.2 using Maxwell's equations will now be considered from the viewpoint of the theory of special relativity. It will be assumed that λ, the charge

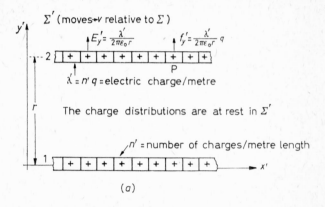

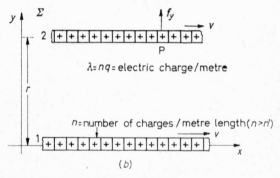

Figure 2.3. The calculation of the electric and the magnetic forces between two convection currents using the theory of special relativity. (a) The charge distributions are at rest in Σ'; there is only an electric force between the charges in Σ'. (b) In Σ, the charge distributions move with uniform velocity v, and there are both electric and magnetic forces between the charges

per metre length on the moving wires, measured relative to Σ, is made up of n discrete charges per unit length of magnitude q coulombs each as shown in *Figure 2.3(b)*. In Σ,

$$\lambda = nq \qquad (2.14)$$

25

According to the theory of relativity the laws of electromagnetism are the same in all inertial reference systems. Hence, the problem can be considered from the reference frame Σ', moving with uniform velocity v relative to Σ along the common x axis. In Σ' the wires are at rest, as shown in *Figure 2.3(a)*. Let λ', the charge per metre measured in Σ', be made up of n' charges per metre length of magnitude q Coulombs each so that

$$\lambda' = n'q \tag{2.15}$$

The same value q is used for the total charge on a particle in Σ and Σ', so as to be in accord with the principle of constant electric charge. The charge distributions are at rest in Σ'. According to the Lorentz contraction, eqn (1.20), a length l_0 of the wire at rest in Σ' is measured to be $l_0\sqrt{1-v^2/c^2}$ in Σ, since the wire is moving with velocity v relative to Σ. The number of charges in a length l_0 of wire in Σ', which is equal to $n'l_0$, is measured to be in a length $l_0\sqrt{1-v^2/c^2}$ relative to Σ, as illustrated in *Figure 2.3*. Hence the number of charges per unit length in Σ is $n'l_0/l_0\sqrt{1-v^2/c^2}$, so that

$$n = \frac{n'}{\sqrt{1-v^2/c^2}} \tag{2.16}$$

showing that the charge per unit length is greater in Σ than Σ', as illustrated in *Figure 2.3*. Hence, using eqn (2.14),

$$\lambda = nq = \frac{n'}{\sqrt{1-v^2/c^2}}q = \frac{\lambda'}{\sqrt{1-v^2/c^2}} \tag{2.17}$$

In Σ', since the charge distributions are at rest, Maxwell's equations reduce to Coulomb's law of force between electrostatic charges. Using Gauss' law as in Section 2.2, the electric field at the position of wire 2 due to wire 1 is

$$E'_y = \frac{\lambda'}{2\pi\varepsilon_0 r}$$

According to eqn (1.8), the separation of the wires r is the same in Σ and Σ', as it is measured in the y direction. The force on one of the charges (labelled P and of magnitude q) of wire 2 is equal to

$$f'_x = 0; \quad f'_y = \frac{\lambda'q}{2\pi\varepsilon_0 r}; \quad f'_z = 0 \tag{2.18}$$

It will now be assumed that the force transformations derived in

26

Section 1.9 are correct. For the single charge P of wire 2, since $\mathbf{u}' = 0$, we have from eqns (1.56), (1.57), and (1.58)

$$f_x = f'_x \tag{2.19}$$

$$f_y = \sqrt{1 - v^2/c^2}\, f'_y \tag{2.20}$$

$$f_z = \sqrt{1 - v^2/c^2}\, f'_z \tag{2.21}$$

Since

$$f'_x = f'_z = 0,$$

we have

$$f_x = f_z = 0.$$

Substituting from eqn (2.18) into eqn (2.21), we have for the force on the charge P measured in Σ

$$f_y = \sqrt{1 - v^2/c^2}\, \frac{\lambda' q}{2\pi\varepsilon_0 r} \tag{2.22}$$

The force per metre on wire 2, measured in Σ, is equal to the number of charges per metre length, measured in Σ, times the force on each charge given by eqn (2.22), that is,

$$\text{force/metre length} = n f_y = n \sqrt{1 - v^2/c^2}\, \frac{\lambda' q}{2\pi\varepsilon_0 r}$$

But from eqn (2.17), $\lambda' = \lambda\sqrt{1 - v^2/c^2}$, and from eqn (2.14) $nq = \lambda$. Hence in Σ,

$$\text{force/metre length} = \frac{\lambda^2}{2\pi\varepsilon_0 r}(1 - v^2/c^2) \tag{2.23}$$

This is in agreement with eqn (2.8), which was derived in Section 2.2 by applying Maxwell's equations in the inertial frame Σ. This example illustrates how the magnetic forces produced by electric currents can be calculated, in some cases, from Coulomb's law for the forces between electrostatic charges, if the principle of constant electric charge and the force transformations of the theory of special relativity are taken as axiomatic. One has to include all second order effects, since the magnetic forces between moving charges are themselves only of second order. For example, in the present case, the effects associated with the Lorentz contraction had to be included.

According to Newtonian mechanics, force should be absolute, that is have the same numerical value in all inertial reference frames.

27

ELECTROMAGNETISM AS A SECOND ORDER EFFECT

(Reference: I, Section 1.4 or II, Section 1.4). If this were correct, then the forces between two moving charges should be the same in all inertial reference frames, and equal to the force between them in the reference frame in which one of the charges is at rest. Hence, if force were absolute, electromagnetism would reduce to Coulomb's law, the forces between two moving charges being the same in all inertial reference frames. According to the theory of special relativity, the magnetic forces between moving charges can be interpreted as relativistic effects, and represent those parts of the transformed relativistic forces which depend on the velocity of the test charge relative to the observer.

PROBLEMS

Problem 2.1—Assume that the charge of magnitude q situated at P in Figure 2.3 (*b*) has a velocity **u** in an arbitrary direction relative to Σ. In Σ', wire 1 is at rest and even though the charge at P is moving in Σ', it is acted upon by an electric force only. By transforming this force to Σ, show that the force on the charge P, measured in Σ, can be expressed in the form $q\mathbf{E}+q\mathbf{u}\times\mathbf{B}$, where **E** and **B** are given by eqns (2.3) and (2.7) respectively.

Problem 2.2—Two 'point' charges of magnitude q coulombs are moving side by side with the same velocity **v** parallel to the x axis and in the xy plane of Σ. They have the same x co-ordinates. Do the charges attract or do they repel, that is, is the electric force of repulsion greater than the magnetic force of attraction?

(Hint: Consider the charges from the inertial reference frame Σ' in which the charges are at rest. In this reference frame they repel each other with a force given by Coulomb's law. Transform this force back to Σ, assuming, of course, that $v < c$.)

THE ELECTRIC AND MAGNETIC FIELDS OF A 'POINT' CHARGE MOVING WITH UNIFORM VELOCITY

3.1. FORCES BETWEEN MOVING CHARGES

The general case of the forces between two 'point' charges moving with *uniform* velocities will now be considered from the viewpoint of the theory of special relativity. The following will be taken as axiomatic:

(1) The force, velocity and co-ordinate transformations of the theory of special relativity.

(2) The principle of constant charge, according to which the numerical value of the total electric charge on a particle is the same in all inertial reference frames (cf. Section 1.9 of Chapter 1).

(3) Coulomb's law for the electric field of a stationary electric 'point' charge. By a 'point' charge, is meant a charge, such as an electron or proton, whose dimensions are small enough for the charge to be considered as being approximately at one point of space only, when applying the Lorentz transformations.

(4) The electromagnetic fields $\mathbf{E}$ and $\mathbf{B}$ are propagated from moving charges with a velocity c in empty space.

Consider two point charges q_1 and q_2 constrained to move with *uniform* velocities $\mathbf{u}$ and $\mathbf{v}$ respectively, relative to an inertial reference frame Σ. Choose the directions of the axes of Σ such that q_2 moves with uniform velocity v relative to Σ along the x axis as shown in *Figure 3.1(a)*. Choose the zero of time such that q_2 is at the origin of Σ at the time $t = 0$. Let the charge q_1 be at the point P at a distance r from the origin and have co-ordinates x, y, z at a time $t = 0$, and let its velocity at that instant be $\mathbf{u}$, having components u_x, u_y and u_z, as shown in *Figure 3.1(a)*. The force on the charge q_1 due to the charge q_2 will be calculated in the inertial frame Σ at the time $t = 0$, when q_2 is at the origin.

In the inertial frame Σ' moving with uniform velocity v relative to Σ along the common x axis, the charge q_2 is at rest at the origin of Σ' for all time as shown in *Figure 3.1(b)*. Let q_1 be at P' at a distance r'

from the origin O' and have co-ordinates x', y', z', t' in Σ' corresponding to the point P, which has co-ordinates x, y, z, $t = 0$ relative to Σ. From the Lorentz transformations, since $t = 0$, one has

$$x' = \gamma x \qquad (3.1)$$

$$y' = y \qquad (3.2)$$

$$z' = z \qquad (3.3)$$

$$t' = -\gamma vx/c^2 \qquad (3.4)$$

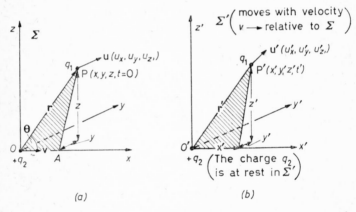

Figure 3.1. (a) In the inertial reference frame Σ, the charge q_2 moves with uniform velocity v. The force on q_1 is calculated at the time $t=0$ when q_2 is at the origin. (b) In the inertial reference frame Σ', the charge q_2 remains at rest at the origin

where
$$\gamma = 1/\sqrt{1-v^2/c^2} \qquad (3.5)$$

Hence, in *Figure 3.1(b)*
$$(O'P')^2 = r'^2 = x'^2 + y'^2 + z'^2$$
$$= \gamma^2 x^2 + y^2 + z^2$$
$$= \gamma^2[x^2 + (1-v^2/c^2)(y^2 + z^2)]$$

Hence,
$$r' = \gamma s \qquad (3.6)$$

where
$$s = \sqrt{x^2 + (1-v^2/c^2)(y^2 + z^2)} \qquad (3.7)$$

30

Now,

$$s^2 = x^2 + (1 - v^2/c^2)(y^2 + z^2)$$
$$= x^2 + y^2 + z^2 - (v^2/c^2)(y^2 + z^2)$$

But,

$$r^2 = x^2 + y^2 + z^2$$

Hence,

$$\frac{s^2}{r^2} = 1 - \frac{v^2}{c^2}\frac{(y^2 + z^2)}{r^2}$$

But in *Figure 3.1(a)*

$$(y^2 + z^2)/r^2 = (AP)^2/r^2 = \sin^2\theta$$

where θ is the angle between **v** and **r** in Σ.

Hence eqn (3.7) can be rewritten

$$s = r\left(1 - \frac{v^2}{c^2}\sin^2\theta\right)^{1/2} \tag{3.8}$$

Since the charge q_2 is at rest in Σ', its magnetic induction $\mathbf{B}'_2$ in Σ' is zero. It will be assumed that the electric field at the point P' in Σ' at $t' = -\gamma v x/c^2$ in Σ' (corresponding to $t = 0$ in Σ) is given by *Coulomb's law*, namely

$$\mathbf{E}'_2 = q_2\mathbf{r}'/4\pi\varepsilon_0 r'^3$$

The charge q_1 is moving with velocity $\mathbf{u}'$ in Σ'. Experiments in the laboratory system have confirmed that the force on a non-radiating moving charge is given by the Lorentz force (cf. Section 1.9). Since $\mathbf{B}'_2$ is zero in Σ', we expect the force on q_1 in Σ' to be

$$\mathbf{f}' = q_1\mathbf{E}'_2 = q_1q_2\mathbf{r}'/4\pi\varepsilon_0 r'^3$$

Writing this force into its components, one has

$$f'_x = \frac{q_1q_2x'}{4\pi\varepsilon_0 r'^3} \text{ etc.}$$

Substituting for r' from eqn (3.6) and for x' from eqn (3.1) one obtains

$$f'_x = \frac{q_1q_2x}{4\pi\varepsilon_0\gamma^2 s^3} = q_1gx \tag{3.9}$$

31

where,

$$g = \frac{q_2}{4\pi\varepsilon_0\gamma^2 s^3} = \frac{q_2(1-v^2/c^2)}{4\pi\varepsilon_0 s^3} \tag{3.10}$$

Similarly, using eqns (3.2), (3.3) and (3.6)

$$f_y' = \frac{q_1 q_2 y}{4\pi\varepsilon_0\gamma^3 s^3} = \frac{q_1 g y}{\gamma} \tag{3.11}$$

$$f_z' = \frac{q_1 q_2 z}{4\pi\varepsilon_0\gamma^3 s^3} = \frac{q_1 g z}{\gamma} \tag{3.12}$$

From eqn (1.54) for the force transformations,

$$f_y' = \frac{f_y}{\gamma(1-vu_x/c^2)} \tag{1.54}$$

Rearranging,

$$f_y = \gamma(1-vu_x/c^2)f_y'$$

Substituting for f_y' from eqn (3.11)

$$f_y = q_1 g y(1-vu_x/c^2) \tag{3.13}$$

Similarly,

$$f_z = q_1 g z(1-vu_x/c^2) \tag{3.14}$$

From eqn (1.53)

$$f_x' = f_x - \frac{vu_y}{c^2(1-vu_x/c^2)}f_y - \frac{vu_z}{c^2(1-vu_x/c^2)}f_z$$

Substituting for f_x', f_y and f_z from eqns (3.9), (3.13) and (3.14) respectively, we have

$$q_1 g x = f_x - \frac{vu_y}{c^2}q_1 g y - \frac{vu_z}{c^2}q_1 g z$$

Rearranging

$$f_x = q_1 g x + \frac{q_1 g v}{c^2}(yu_y + zu_z)$$

$$f_x = q_1 g x \left(1 - \frac{vu_x}{c^2}\right) + \frac{q_1 g v}{c^2}(xu_x + yu_y + zu_z)$$

or, using eqn (A1.1) of Appendix 1,

$$f_x = q_1 g \left(1 - \frac{vu_x}{c^2}\right)x + q_1 g \frac{v}{c^2}(\mathbf{u}\cdot\mathbf{r}) \tag{3.15}$$

From eqns (3.13) and (3.14)

$$f_y = q_1 g \left(1 - \frac{vu_x}{c^2} \right) y \qquad (3.13)$$

$$f_z = q_1 g \left(1 - \frac{vu_x}{c^2} \right) z \qquad (3.14)$$

Since $\mathbf{v}$ has an x component only, from eqn (A1.1),

$$vu_x = \mathbf{u} \cdot \mathbf{v}$$

Eqns (3.13), (3.14) and (3.15) can be combined into the single vector equation

$$\mathbf{f} = q_1 g \left(1 - \frac{\mathbf{u} \cdot \mathbf{v}}{c^2} \right) \mathbf{r} + q_1 g \frac{\mathbf{v}}{c^2} (\mathbf{u} \cdot \mathbf{r})$$

or,

$$\mathbf{f} = q_1 (g\mathbf{r}) + \frac{q_1 g}{c^2} [\mathbf{v}(\mathbf{u} \cdot \mathbf{r}) - \mathbf{r}(\mathbf{u} \cdot \mathbf{v})] \qquad (3.16)$$

Now for any three vectors $\mathbf{A}$, $\mathbf{B}$ and $\mathbf{C}$

$$\mathbf{A} \times (\mathbf{B} \times \mathbf{C}) = \mathbf{B}(\mathbf{A} \cdot \mathbf{C}) - \mathbf{C}(\mathbf{A} \cdot \mathbf{B}) \qquad (A1.6)$$

so that

$$\mathbf{v}(\mathbf{u} \cdot \mathbf{r}) - \mathbf{r}(\mathbf{u} \cdot \mathbf{v}) = \mathbf{u} \times (\mathbf{v} \times \mathbf{r})$$

Hence eqn (3.16) can be rewritten

$$\mathbf{f} = q_1 (g\mathbf{r}) + q_1 \mathbf{u} \times \left(\frac{\mathbf{v}}{c^2} \times g\mathbf{r} \right) \qquad (3.17)$$

The charge q_1 which is moving with velocity $\mathbf{u}$ relative to Σ can be treated as a test charge to measure the electric and magnetic fields produced by the charge q_2 at the point P in *Figure 3.1(a)*, at the instant $t = 0$ when q_2 is at the origin. Eqn (3.17) can be written in the form

$$\mathbf{f} = q_1 \mathbf{E}_2 + q_1 \mathbf{u} \times \mathbf{B}_2 \qquad (3.18)$$

where,

$$\mathbf{E}_2 = g\mathbf{r} \qquad (3.19)$$

and,

$$\mathbf{B}_2 = \frac{\mathbf{v}}{c^2} \times g\mathbf{r} = \frac{\mathbf{v}}{c^2} \times \mathbf{E}_2 \qquad (3.20)$$

Eqn (3.18) confirms that the Lorentz force is adequate to describe the forces between 'point' charges moving with uniform velocities.

From eqn (3.10)

$$g = \frac{q_2(1-v^2/c^2)}{4\pi\varepsilon_0 s^3}.$$

where, from eqn (3.8)

$$s^3 = r^3\left(1-\frac{v^2}{c^2}\sin^2\theta\right)^{3/2}$$

Substituting in eqn (3.19),

$$\mathbf{E}_2 = \frac{q_2\mathbf{r}(1-v^2/c^2)}{4\pi\varepsilon_0 r^3[1-(v^2/c^2)\sin^2\theta]^{3/2}} \tag{3.21}$$

Substituting for $\mathbf{E}_2$ in eqn (3.20),

$$\mathbf{B}_2 = \frac{\mathbf{v}}{c^2}\times\mathbf{E}_2 = \frac{\mathbf{v}\times\mathbf{r}q_2(1-v^2/c^2)}{c^2 4\pi\varepsilon_0 r^3[1-(v^2/c^2)\sin^2\theta]^{3/2}} \tag{3.22}$$

Eqns (3.21) and (3.22) give the electric and magnetic fields produced by the charge q_2 at a distance $(\mathbf{r}, \theta)$ away from the charge q_2 at the instant the charge is at the origin in *Figure 3.1(a)*. The quantities $\mathbf{r}$ and θ relate to the position of the charge at the time the field is determined at the field point P. Eqn (3.18) for the force on q_1 at $t = 0$ can be written

$$\mathbf{f} = q_1\mathbf{E}_2 + q_1\frac{\mathbf{u}\times(\mathbf{v}\times\mathbf{E}_2)}{c^2} \tag{3.23}$$

The ratio of the second term to the first term on the right hand side of eqn (3.23) is less than uv/c^2. According to the theory of special relativity both u and v must be less than c, the velocity of light in empty space, so that the magnetic force between the two moving charges is always less than the electric force between the charges.

3.2. ELECTRIC AND MAGNETIC FIELDS OF A MOVING 'POINT' CHARGE

Eqns (3.21) and (3.22) will now be rewritten for the general case of any charge of magnitude q moving with *uniform* velocity $\mathbf{u}$ relative to an inertial reference frame Σ. The symbol $\mathbf{u}$ will be used, henceforth, for the velocity of a particle relative to Σ, and the ratio u/c will be denoted by β. The symbol v will be reserved for the relative velocity of Σ and Σ' and v will only be equal to the velocity of a particle, if the particle is at rest relative to one of the reference

34

frames. The electric intensity due to a charge q moving with uniform velocity $\mathbf{u}$ is

$$E = \frac{q\mathbf{r}(1-\beta^2)}{4\pi\varepsilon_0 r^3(1-\beta^2\sin^2\theta)^{3/2}} \qquad (3.24)$$

where $\mathbf{r}$ is a vector from the charge to the field point where the field is required, $\beta = u/c$ and θ is the angle between $\mathbf{r}$ and $\mathbf{u}$ as shown in *Figure 3.2* The position of the charge is its position at the time the field is required. This is called the 'present' position of the charge. If spherical polars are used, with the polar angle measured from the direction of the velocity of the charge, and $\mathbf{a}_r$, $\mathbf{a}_\theta$ and $\mathbf{a}_\phi$ are unit

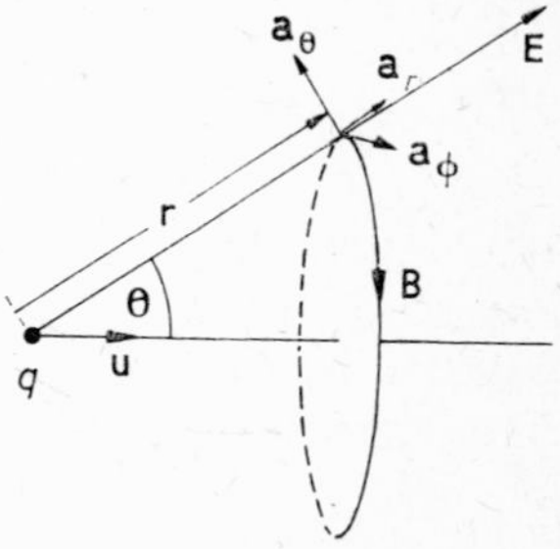

Figure 3.2. Spherical polar co-ordinates are used to represent the electric and magnetic fields of a charge q moving with uniform velocity $\mathbf{u}$. The electric intensity $\mathbf{E}$ is in the direction of $\mathbf{a}_r$ and the magnetic induction $\mathbf{B}$ is in the direction of $\mathbf{a}_\phi$

vectors in the directions of increasing r, θ and ϕ respectively as shown in *Figure 3.2*, then $\mathbf{E}$ can be rewritten

$$\mathbf{E} = \mathbf{a}_r E_r + \mathbf{a}_\theta 0 + \mathbf{a}_\phi 0 = \mathbf{a}_r E_r \qquad (3.25)$$

where,

$$E_r = \frac{q(1-\beta^2)}{4\pi\varepsilon_0 r^2(1-\beta^2\sin^2\theta)^{3/2}} \qquad (3.26)$$

The magnetic induction is given by

$$\mathbf{B} = \frac{\mathbf{u}}{c^2}\times\mathbf{E} = \frac{q\mathbf{u}\times\mathbf{r}(1-\beta^2)}{4\pi\varepsilon_0 c^2 r^3(1-\beta^2\sin^2\theta)^{3/2}} \qquad (3.27)$$

In spherical polars

$$\mathbf{B} = \mathbf{a}_\phi B_\phi \qquad (3.28)$$

where

$$B_\phi = \frac{qu\sin\theta(1-\beta^2)}{4\pi\varepsilon_0 c^2 r^2(1-\beta^2\sin^2\theta)^{3/2}} \qquad (3.29)$$

Eqns (3.24) and (3.27) are the same as the equations derived via

35

the Liénard–Wiechert potentials using classical electromagnetic theory (reference: Panofsky and Phillips[1]).

When $\mathbf{u}$, the velocity of the charge is zero, that is $\beta = 0$, eqn (3.24) reduces to

$$E = q/4\pi\varepsilon_0 r^2$$

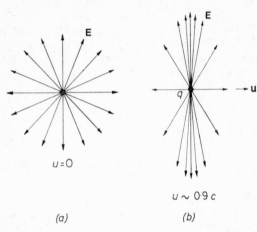

Figure 3.3. (a) *The electric field of a stationary positive charge is spherically symmetric. (b) If the charge is moving with uniform velocity, the electric field diverges radially from the 'present' position of the charge. The electric intensity is increased in the direction perpendicular to* $\mathbf{u}$, *but decreased in the directions parallel to and antiparallel to* $\mathbf{u}$

in agreement with Coulomb's law. When $\mathbf{u} = 0$, the electric intensity is the same in all directions as illustrated in *Figure 3.3(a)*. The number of lines of $\mathbf{E}$ is limited in *Figure 3.3* such that the number of lines of $\mathbf{E}$ per unit area perpendicular to $\mathbf{E}$ is proportional to the magnitude of $\mathbf{E}$. This gives a visual picture of the strength of $\mathbf{E}$, the lines of $\mathbf{E}$ being closest together in *Figure 3.3*, where the field strength is greatest.

When $\mathbf{u}$ is finite, the electric field is still radial, that is $\mathbf{E}$ diverges radially from the 'present' position of the charge, as shown in *Figure 3.3(b)*. The electric field strength is still proportional to $1/r^2$. However, according to eqn (3.24), the intensity is not the same in all directions, though it is still symmetric about $\theta = \pi/2$. When $\theta = 0$ or $\theta = \pi$, eqn (3.24) reduces to

$$E = q(1-\beta^2)/4\pi\varepsilon_0 r^2$$

36

Thus the electric intensity is reduced in the direction of **u**, the direction of motion of the charge, and in the direction opposite to **u** in the ratio of $(1-\beta^2)$ compared with the electrostatic case. When $\theta = \pi/2$

$$E = q/4\pi\varepsilon_0 r^2\sqrt{1-\beta^2}$$

Hence, the electric intensity is increased in the direction perpendicular to **u** in the ratio $1/\sqrt{1-\beta^2}$ compared with the electrostatic case. It will be shown in Section 4.2 that the total flux of **E** remains equal to q/ε_0, when the charge is moving. A typical case of the electric

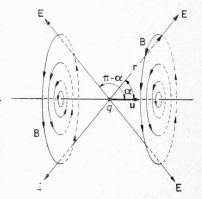

Figure 3.4. The magnetic field lines of a charge moving with uniform velocity are circles concentric with the direction of **u**. *The direction of* **B** *is given by the right-handed corkscrew rule. The field is in the same direction for values of* $\theta = \alpha$ *and* $\theta = \pi - \alpha$, *where* θ *is the angle between* **r** *and* **u**

field due to a charge moving with uniform velocity $\sim 0\cdot 9c$ is shown in *Figure 3.3(b)*. This illustrates how, though the total number of lines of **E** is the same as for the electrostatic case, the lines of **E** are bunched towards the direction perpendicular to **u**. For $\beta = 0\cdot 9$ the electric field is $0\cdot 19$ times the electrostatic value for $\theta = 0$ and $2\cdot 3$ times the electrostatic value for $\theta = \pi/2$. For $\beta = 0\cdot 99$ the ratios are $0\cdot 02$ and $7\cdot 1$ respectively. Thus in the extreme relativistic case the lines of **E** are all nearly perpendicular to **u**.

The lines of magnetic induction **B** are perpendicular to both **u**, the velocity of the charge q, and the lines of **E** which diverge radially from the 'present' position of the charge. The lines of **B** therefore form closed circles in the plane perpendicular to the direction of motion of q, which is the direction of **u**. The circles are concentric with the direction of **u**. The lines of **B** in two planes are sketched in *Figure 3.4*. The intensity of **B** decreases as $1/r^2$. For a given r, the intensity is the same for the positions $\theta = \alpha$ and $\theta = \pi - \alpha$. The sense of rotation of the lines of **B** about the direction of **u** is the same in both cases.

The direction of **B** is consistent with the right-handed corkscrew rule, if $q\mathbf{u}$ is treated as a current element $I\,d\mathbf{l}$. According to the right-handed corkscrew rule the direction of **B** is the direction in which a right-handed corkscrew would have to be rotated in order that it should advance in the direction of current flow.

As the velocity of the charge tends to the velocity of light, the lines of **E** bunch closer and closer to the direction perpendicular to the direction of motion of the charge, with **B** remaining perpendicular to **E**. Hence, at extremely high velocities ($u \rightarrow c$) the electromagnetic field of a charge moving with uniform velocity resembles the field of a plane wave. In the limit ($u = c$), **E** and **B** would be perpendicular to each other and to **u**, with $E = cB$ as for a plane electromagnetic wave in empty space.

If $u \ll c$, $\beta \ll 1$, eqns (3.24) and (3.27) reduce to

$$\mathbf{E} \simeq \frac{q\mathbf{r}}{4\pi\varepsilon_0 r^3} \tag{3.30}$$

$$\mathbf{B} \simeq \frac{q\mathbf{u} \times \mathbf{r}}{4\pi\varepsilon_0 c^2 r^3} \tag{3.31}$$

Eqn (3.30) is the same as Coulomb's law. Eqn (3.31) is sometimes called the Biot–Savart law, particularly when it is applied to a current element. Eqn (3.31) can be used to develop the expression for the magnetic field due to a steady conduction current flowing in an electric circuit and the expression for the forces between two current elements. Derivations are given in Appendix 4. For eqn (3.31) to be consistent with the normal expression for the Biot–Savart law, which includes the magnetic space constant μ_0 [cf. eqn (A2.22) of Appendix 2], we must have $\mu_0 = 1/\varepsilon_0 c^2$ [cf. eqn (A4.15) of Appendix 4].

3.3. THE ELECTRIC AND MAGNETIC FIELDS RELATIVE TO THE RETARDED POSITION

According to classical electromagnetic theory, the electric and magnetic fields of a moving charge are propagated from the charge with a speed c in empty space. This is illustrated by the retarded potentials which can be developed from classical electromagnetic theory in the way outlined in Section 5.7 of Chapter 5. Consider two stationary charges at a distance l apart. The force between them is given by Coulomb's law. If one of the stationary charges starts to move, according to classical electromagnetic theory, the other charge should not experience a change in the force acting on it until a time l/c later. This finite time for propagating signals is an integral part of the theory of special relativity, where it is assumed that energy and momentum and hence signals and information cannot be transmitted with a velocity exceeding the velocity of light. It will be assumed that in empty space the electric field $\mathbf{E}$ and the magnetic field $\mathbf{B}$ are propagated away from a moving charge with the velocity of light.

If a charge q is at the point O at a time t in *Figure 3.5(a)* then according to eqns (3.24) and (3.27) the electric and magnetic field strengths at the field point P at a distance $\mathbf{r_0}$ from O at the *same* time t as q is at O are, for a charge moving with uniform velocity $\mathbf{u}$,

$$\mathbf{E} = \frac{q\mathbf{r_0}(1-u^2/c^2)}{4\pi\varepsilon_0 s^3} \tag{3.32}$$

$$\mathbf{B} = \frac{\mathbf{u}\times\mathbf{E}}{c^2} \tag{3.33}$$

where according to eqn (3.8)

$$s = r_0[1-(u^2/c^2)\sin^2\theta]^{1/2} \tag{3.34}$$

The position O in *Figure 3.5(a)* is called the 'present' position of the charge, that is the position of the charge when the fields at P are determined. If the fields are propagated with a speed c away from the charge, the fields determined at P must have 'left' the charge at some earlier time, when the charge was at its retarded position R in *Figure 3.5(a)*. After the fields have left the charge at its retarded position the charge may be accelerated as illustrated in *Figure 3.5(b)*. It is only when the charge continues to move with uniform velocity, that the fields can be related to the 'present' position of the charge. The electric and magnetic field strengths will now be related to the retarded position of the charge in *Figure 3.5(a)*. Let the distance from the retarded position R to the field point be $[\mathbf{r}]$. The electric and magnetic fields take a time $[r]/c$ to go from the retarded position R to the field point P. The charge is at its retarded position at the retarded time $t-[r]/c$. Square brackets will be placed around quantities, if the quantities are measured at the retarded time $t-[r]/c$.

In the time $[r]/c$ the fields take to go from the retarded position R to the field point P in *Figure 3.5(a)*, the charge q moves a distance $[\mathbf{u}]\,[r]/c$ with uniform velocity $[\mathbf{u}]$ to its 'present' position O. Hence in *Figure 3.5(a)*, we have from the law of vector addition,

$$\mathbf{r_0} = [\mathbf{r}]-\frac{[\mathbf{u}][r]}{c} = \left[\mathbf{r}-\frac{r\mathbf{u}}{c}\right] \tag{3.35}$$

Now for the ΔROP in *Figure 3.5(a)*,

$$\frac{[r]}{\sin(\pi-\theta)} = \frac{[ur/c]}{\sin\alpha}$$

$$\sin\alpha = \frac{[u]}{c}\sin\theta$$

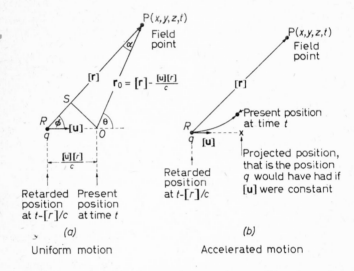

Figure 3.5. *Geometric relations between the retarded position, the 'present' position and the 'projected' position of a moving charge q and the field point P, (a) for a charge moving with uniform velocity, (b) for an accelerating charge*

Now from eqn (3.34),

$$s = r_0\left(1-\frac{[u^2]}{c^2}\sin^2\theta\right)^{1/2}$$

so that

$$s = r_0(1-\sin^2\alpha)^{1/2} = r_0\cos\alpha$$

In ΔOSP, in *Figure 3.5(a)*

$$SP = r_0\cos\alpha = s \qquad (3.36)$$

But,

$$RS = OR\cos\phi = [ur/c]\cos\phi = [\mathbf{r}\cdot\mathbf{u}/c]$$

Hence,

$$S\mathcal{V} = RP - RS = [r] - [\mathbf{r} . \mathbf{u}/c] \qquad (3.37)$$

From eqns (3.36) and (3.37),

$$s = \left[r - \frac{\mathbf{r} . \mathbf{u}}{c} \right] \qquad (3.38)$$

Using eqn (3.35), eqns (3.32) and (3.33) can be rewritten

$$\mathbf{E} = \frac{q}{4\pi\varepsilon_0 s^3} \left[\mathbf{r} - \frac{r\mathbf{u}}{c} \right] \left[1 - \frac{u^2}{c^2} \right] \qquad (3.39)$$

$$\mathbf{B} = \frac{q}{4\pi\varepsilon_0 c^2 s^3} [\mathbf{u}] \times \left[\mathbf{r} - \frac{r\mathbf{u}}{c} \right] \left[1 - \frac{u^2}{c^2} \right]$$

Since,

$$[\mathbf{u}] \times [\mathbf{u}] = 0$$

$$\mathbf{B} = \frac{q[\mathbf{u}] \times [\mathbf{r}]}{4\pi\varepsilon_0 c^2 s^3} \left[1 - \frac{u^2}{c^2} \right] \qquad (3.40)$$

where s is given by eqn (3.38). Eqns (3.39) and (3.40) relate $\mathbf{E}$ and $\mathbf{B}$ to the retarded position of the charge. These equations are not valid for accelerating charges.

The full expressions for the electric intensity and magnetic induction due to an accelerating charge, moving with a velocity comparable with the velocity of light, were developed from Gauss' law by Frisch and Wilets[2] using the transformations of the theory of special relativity and the principle of constant electric charge. Effectively, Frisch and Wilets assumed that the total flux of $\mathbf{E}$ from an accelerating charge was always equal to q/ε_0. Frisch and Wilets[2] showed that, in general, if $[\mathbf{a}]$ is the acceleration of the charge at its retarded position,

$$\mathbf{E} = \mathbf{E}_V + \mathbf{E}_A \qquad (3.41)$$

where,

$$\mathbf{E}_V = \frac{q}{4\pi\varepsilon_0 s^3} \left[\mathbf{r} - \frac{r\mathbf{u}}{c} \right] \left[1 - \frac{u^2}{c^2} \right] \qquad (3.42)$$

$$\mathbf{E}_A = \frac{q}{4\pi\varepsilon_0 s^3 c^2} \left\{ [\mathbf{r}] \times \left(\left[\mathbf{r} - \frac{r\mathbf{u}}{c} \right] \times [\mathbf{a}] \right) \right\} \qquad (3.43)$$

and,

$$\mathbf{B} = [\mathbf{r}] \times \mathbf{E}/[rc] = \mathbf{B}_V + \mathbf{B}_A \qquad (3.44)$$

where,

$$\mathbf{B}_V = \frac{q[\mathbf{u}] \times [\mathbf{r}]}{4\pi\varepsilon_0 c^2 s^3}\left[1 - \frac{u^2}{c^2}\right] \qquad (3.45)$$

$$\mathbf{B}_A = \frac{q[\mathbf{r}]}{4\pi\varepsilon_0 c^3 s^3 [r]} \times \left\{[\mathbf{r}] \times \left(\left[\mathbf{r} - \frac{r\mathbf{u}}{c}\right] \times [\mathbf{a}]\right)\right\} \qquad (3.46)$$

and

$$s = [r - \mathbf{r} \cdot \mathbf{u}/c] \qquad (3.47)$$

These equations can also be developed from conventional electro-magnetism in the way outlined in Section 5.8 of Chapter 5. All the terms in eqns (3.41) to (3.47), namely $[\mathbf{r}]$, $[\mathbf{u}]$ and $[\mathbf{a}]$, refer to the retarded position of the charge in *Figure 3.5(b)*. If $[\mathbf{a}]$ is zero, then $\mathbf{E}_A$ and $\mathbf{B}_A$ are both zero, and eqns (3.41) and (3.44) reduce to eqns (3.39) and (3.40) respectively, which are the same as eqns (3.24) and (3.27) respectively for the electric and magnetic fields of a 'point' charge moving with uniform velocity. Thus the first terms on the right-hand sides of eqns (3.41) and (3.44) namely $\mathbf{E}_V$ and $\mathbf{B}_V$, are similar to the fields of a 'point' charge moving with uniform velocity. For example, the contribution to the total electric intensity due to $\mathbf{E}_V$ diverges radially from the position the accelerating charge would have had, if it had carried on with uniform velocity $\mathbf{u}$. That is, the contribution to the electric field associated with $\mathbf{E}_V$ diverges radially from the 'projected' position of the charge in *Figure 3.5(b)*. If $[\mathbf{a}] = 0$ and $[\mathbf{u}] = 0$, eqns (3.41) and (3.44) reduce to Coulomb's law. If they are present, $\mathbf{E}_A$ and $\mathbf{B}_A$ predominate at large distances from the charge giving rise to the radiation fields. A fuller discussion of the radiation fields is given in Appendix 3. Our development of Maxwell's equations in Chapter 4 will be only for charges moving with uniform velocities. The reader should remember that this is a special case. The more general case of accelerating charges is discussed in Appendix 3(c).

REFERENCES

[1] PANOFSKY, W. K. H. and PHILLIPS, M. *Classical Electricity and Magnetism*, 2nd Ed. Ch. 19. Addison-Wesley, Reading, Mass. 1962
[2] FRISCH, D. H. and WILETS, L. *Amer. J. Phys.* 24 (1956) 574

PROBLEMS

The existence of magnetic monopoles has been suggested on theoretical grounds (for example, P.A.M. Dirac, *Phys. Rev.* **74** (1948) 817, but so far, they have not been observed experimentally. (K. W. Ford, *Scientific American* **209** (1963) No. 6, 122.) Even if they are observed experimentally, it is almost certain that they play no significant role in normal electromagnetic phenomena. However, hypothetical magnetic monopoles can be used to give the reader some interesting problems. There is no generally accepted definition of unit magnetic monopole strength in M.K.S. units. A widely used definition, in the past, was in terms of the relation

$$\mathbf{f} = \frac{(q_m)_1 (q_m)_2 \mathbf{r}}{\mu_0 4 \pi r^3} = (q_m)_2 \mathbf{H}_1 \tag{3.48}$$

for the force between two stationary magnetic monopoles of strength $(q_m)_1$ and $(q_m)_2$ a distance r apart in empty space. When defined in this way, q_m is measured in Webers. A plus sign denotes a north seeking monopole, that is a monopole from which the magnetic field lines diverge. Some people prefer to define unit magnetic monopole, in terms of the equation

$$\mathbf{f} = \frac{\mu_0 (q_m^*)_1 (q_m^*)_2 \mathbf{r}}{4 \pi r^3} = (q_m^*)_2 \mathbf{B}_1 \tag{3.49}$$

For 'point' magnetic monopoles in empty space, the theories developed using the two definitions of pole strength are equivalent, if

$$q_m = \mu_0 q_m^*; \quad \mathbf{H} = \mathbf{B}/\mu_0 \quad \text{and} \quad \mathbf{D} = \varepsilon_0 \mathbf{E} \tag{3.50}$$

In the problems, we shall generally use the definition of monopole strength in terms of eqn (3.49) since this gives the field vectors $\mathbf{E}$ and $\mathbf{B}$. When solving problems involving magnetic monopoles the principle of constant magnetic monopole strength should be taken as axiomatic, that is it should be assumed that q_m^* should be independent of the velocity of the monopole, and q_m^* should have the same numerical value in all inertial reference frames.

Problem 3.1—In Σ', a magnetic monopole of strength q_m^* is at rest on the y' axis at a distance y' from the origin. An electric charge $+q$ is moving with uniform velocity v along the *negative x'* axis. Show that the force on the stationary monopole q_m^* when the electric charge is at the origin of Σ' is

$$f_z' = q_m^* B_z', \quad \text{where} \quad B_z' = -qv/4 \pi \varepsilon_0 c^2 y'^2 \sqrt{1 - v^2/c^2}$$

[Hint: Use eqn (3.27)] Transform the force on the monopole to the inertial frame Σ, which moves with uniform velocity $-v$ relative to Σ' along their common x axis. In Σ, q is at rest and gives rise to an electric field only, so that the force on the monopole must be due to the electric field of the charge q. Show that the magnitude of the electric force on the moving monopole due to the stationary electric charge is consistent with the relation

$$\mathbf{f} = -q_m^*\mathbf{v} \times \mathbf{E}/c^2 (= -q_m\mathbf{v} \times \mathbf{D})$$

where $\mathbf{v}$ is the velocity of the monopole and $\mathbf{E}$ is the electric field due to the charge q measured in Σ. This result illustrates that there should be a force on a magnetic monopole moving in an electric field.

Problem 3.2—Show that for the conditions of the previous problem, the force on the moving charge q due to the stationary monopole when the charge is at the origin of Σ' is $f_z' = qvB'$, where $B' = \mu_0 q_m^*/4\pi y'^2$. Transform this force to Σ. In Σ the charge is at rest, and this force must be interpreted as an electric force, illustrating that a moving magnetic monopole gives rise to an electric field, which in the present circumstances is equal to

$$E_z = \mu_0 q_m^* v / 4\pi y^2 \sqrt{1 - v^2/c^2} \text{ in } \Sigma$$

Problem 3.3—Consider an electric charge q and a magnetic monopole q_m^* at rest on the y' axis of Σ'. There is no force between them. Show that, if this force is transformed to Σ, there is still no resultant force between them, though both are moving with uniform velocity v. Relative to Σ, the moving charge has a magnetic force $q\mathbf{v} \times \mathbf{B}_m$ acting on it, due to $\mathbf{B}_m$, the magnetic field due to the moving monopole. According to the expression for the Lorentz force, the total force on the charge is $q\mathbf{E}_m + q\mathbf{v} \times \mathbf{B}_m$. Show that if this force is to be zero, then the moving monopole must give an electric field $\mathbf{E}_m = -\mathbf{v} \times \mathbf{B}_m$ where $\mathbf{v}$ is the velocity of the monopole.

In Σ, there is a magnetic force $q_m^*\mathbf{B}$ on the moving monopole due to $\mathbf{B}$, the magnetic field of the moving charge. Show that, if the total force on the magnetic monopole is to be zero, then there must be an electric force on it, equal to $-q_m^*\mathbf{v} \times \mathbf{E}/c^2$, where $\mathbf{E}$ is the electric field due to the moving charge. These results suggest that the force on a magnetic monopole q_m^* moving with velocity $\mathbf{u}$ in an external electric field $\mathbf{E}$ and an external magnetic field $\mathbf{B}$ in empty space is

$$\mathbf{f} = q_m^*(\mathbf{B} - \mathbf{u} \times \mathbf{E}/c^2) \qquad (3.51)$$

or

$$\mathbf{f} = q_m(\mathbf{H} - \mathbf{u} \times \mathbf{D}) \qquad (3.52)$$

and that in empty space a magnetic monopole moving with a uniform velocity $\mathbf{u}$ gives an electric field $\mathbf{E}_m$, which is related to its magnetic field $\mathbf{B}_m$ by

$$\mathbf{E}_m = -\mathbf{u} \times \mathbf{B}_m \qquad (3.53)$$

or

$$\mathbf{D}_m = -\mathbf{u} \times \mathbf{H}_m/c^2 \qquad (3.54)$$

44

Problem 3.4—Consider a magnetic monopole $(q_m^*)_2$ moving with uniform velocity **v** along the x axis of an inertial frame Σ. Let it be at the origin at $t = 0$. Let a monopole $(q_m^*)_1$ move with velocity **u** and let it be at the point x, y, z at a distance **r** from the origin at $t = 0$. In Σ', $(q_m^*)_2$ is at rest. Use Coulomb's law in the form of eqn (3.49) for the force between magnetic monopoles to calculate the force on $(q_m^*)_1$ in Σ'. Transform this force back to Σ as was done for the case of moving electric charges in Section 3.1. Show that the force $\mathbf{f}_1$ on $(q_m^*)_1$ can be expressed in the form

$$\mathbf{f}_1 = (q_m^*)_1(\mathbf{B}_2 - \mathbf{u} \times \mathbf{E}_2/c^2) \tag{3.55}$$

where,

$$\mathbf{B}_2 = \frac{\mu_0(q_m^*)_2(1-\beta^2)\mathbf{r}}{4\pi r^3(1 - \beta^2 \sin^2 \theta)^{3/2}} \tag{3.56}$$

$$\mathbf{E}_2 = -\mathbf{v} \times \mathbf{B}_2 = -\frac{\mu_0(q_m^*)_2(1-\beta^2)\mathbf{v} \times \mathbf{r}}{4\pi r^3(1 - \beta^2 \sin^2\theta)^{3/2}} \tag{3.57}$$

where **v** is the velocity of $(q_m^*)_2$, $\beta = v/c$ and θ is the angle between **r** and **v**.

Problem 3.5—Relate the expression for the electric and magnetic fields of a moving magnetic monopole developed in Problem 3.4 to the retarded position of the monopole.

Problem 3.6—Sketch the magnetic and electric fields of a north seeking magnetic monopole moving with a velocity **u** comparable with the velocity of light. Illustrate how the lines of **B** diverge radially from the monopole and bunch in a direction perpendicular to the direction of motion. Illustrate how the electric field lines form closed circles concentric with the direction **u**. Show that the direction of **E** is given by a *left*-handed corkscrew rule. Compare and contrast these fields with those of a moving electric charge. Show that the direction of **E** is consistent with the Maxwell equation $\nabla \times \mathbf{E} = -\partial \mathbf{B}/\partial t$. (Hint: cf. Section 4.7.)

Problem 3.7—A proton is moving with a uniform velocity $0 \cdot 8c$ along the x axis of the inertial reference frame Σ. A second proton is moving with uniform velocity $0 \cdot 6c$ along the y axis. The second proton is at a position $y = 10^{-4}$ metres, when the proton, which is moving along the x axis, is at the origin. Calculate the magnitude and directions of the electric and the magnetic forces between the charges at this instant. The charge on a proton is $1 \cdot 60 \times 10^{-19}$ coulombs and

$$\varepsilon_0 = 8 \cdot 85 \times 10^{-12} \text{ farads/metre}$$

[*Ans*: $(f_1)_{\text{elec}} = 1 \cdot 5 \times 10^{-20}$ newtons; $(f_1)_{\text{mag}} = 0$

$(f_2)_{\text{elec}} = 3 \cdot 8 \times 10^{-20}$ newtons;

$(f_2)_{\text{mag}} = 1 \cdot 8 \times 10^{-20}$ newtons]

4

MAXWELL'S EQUATIONS VIA RELATIVITY

4.1. INTRODUCTION

In Chapter 3, the electric intensity $\mathbf{E}$ and the magnetic induction $\mathbf{B}$ of a 'point' charge q, such as an electron or a proton, moving with *uniform* velocity $\mathbf{u}$ were developed from Coulomb's law, by taking the principle of constant electric charge and the transformations of the theory of special relativity as axiomatic. The fields at a distance $\mathbf{r}$ from the charge are given by eqns (3.24) and (3.27), namely

$$\mathbf{E} = \frac{q\mathbf{r}(1-\beta^2)}{4\pi\varepsilon_0 r^3(1-\beta^2\sin^2\theta)^{3/2}} \qquad (4.1)$$

$$\mathbf{B} = \mathbf{u}\times\mathbf{E}/c^2 \qquad (4.2)$$

where θ is the angle between $\mathbf{r}$ and $\mathbf{u}$ and $\beta = u/c$. In eqns (4.1) and (4.2), the fields are related to the 'present' position of the charge, that is the position of the charge at the time the fields are determined at the field point. In this chapter, eqns (4.1) and (4.2) will be used to develop Maxwell's equations, which relate the values of $\mathbf{E}$ and $\mathbf{B}$ at the field point. It was illustrated in Section 3.3, that when a charge is accelerating, eqns (4.1) and (4.2) must be extended. In order to keep the discussion of Maxwell's equations in this chapter as simple as possible, the discussion is confined to systems of charges moving with uniform velocities. The more complicated case of accelerating charges is given in Appendix 3.

Originally Maxwell's equations were developed in the nineteenth century, from experiments based on matter in bulk form, before any precise atomic theory had been developed. It is now known that macroscopic charge and current distributions are built up of a finite number of microscopic (atomic) charges, namely electrons, protons and positive ions. The principle of superposition will be used to extend Maxwell's equations relating the fields of 'point' charges into a macroscopic theory for charge and current distributions of finite dimensions.

As an analogy, consider the mass density of a solid. When the macroscopic mass density of a solid varies with position, it is defined as

$$d = \Delta m/\Delta V \qquad (4.3)$$

where Δm is the mass of a small element of volume ΔV of the solid. In a macroscopic theory, the mass density is treated as a continuous function of position. Solids are built up of a discrete number of atoms. On the atomic scale there would be enormous fluctuations in mass density, the density being very high inside atomic nuclei and very much less outside nuclei. The diameters of nuclei are $\sim 10^{-14}$ metres whereas the diameters of atoms are $\sim 10^{-10}$ metres (or ~ 1 Å). In terms of atomic particles, eqn (4.3) for the mass density can be rewritten

$$d = \frac{\sum_{i=1}^{N} m_i}{\Delta V} \tag{4.4}$$

where m_i is the mass of the ith atom, the summation being over all N atoms in a small volume element ΔV of the solid. Now ΔV can be made small on the laboratory scale, say a 10^{-7} metre cube, and yet contain many atoms, $\sim 10^9$ for a 10^{-7} metre cube. Thus on the laboratory (macroscopic) scale d can be treated as a continuous function of position, since over distances of $\sim 10^{-7}$ metres, fluctuations due to the discrete structure of matter will average out. The macroscopic mass density is therefore an average mass density, obtained by averaging over a region of space, large on the atomic scale, but small on the laboratory scale. We will refer to such a procedure, as finding the *local space* average. This approach, using equations similar to eqn (4.4), will be used to define other macroscopic quantities, such as charge density ρ, the macroscopic electric intensity $\mathbf{E}$ and the macroscopic magnetic induction $\mathbf{B}$.

In practice macroscopic charge and current distributions are made up of very large numbers of electrons and positive ions. The macroscopic charge density ρ is defined as the electric charge per unit volume. If ρ varies from point to point, by analogy with eqn (4.4), the macroscopic charge density at a point x, y, z, is defined by the relation

$$\rho = \frac{\sum_{i=1}^{N} q_i}{\Delta V} \tag{4.5}$$

where the summation is over all the N microscopic (atomic) charges in the volume element ΔV at x, y, z. If the charges are moving, they are continually moving in and out of ΔV. The summation of the charges in ΔV must be carried out at a fixed time. If ρ varies with position, ΔV must be small on the laboratory scale, say a 10^{-7}

metre cube. In general, such a volume element will contain an enormous number of atomic charges, and effects associated with the discrete microscopic structure of macroscopic charge distributions will generally be averaged out on the laboratory scale and in a macroscopic theory, ρ is treated as a continuous function of position.

The *electric current density* $\mathbf{J}$ is defined as the current flow per unit area, the area being normal to the direction of current flow. If $\Delta \mathbf{I}$ is the current crossing an area ΔS, which is normal to $\Delta \mathbf{I}$, then

$$\mathbf{J} = \frac{\Delta \mathbf{I}}{\Delta S} \qquad (4.6)$$

This current flow is due to the motions of a large number of discrete atomic charges. In a simplified model in which the microscopic charges in the system have the same charge q and the same velocity $\mathbf{u}$, such that the macroscopic charge distribution moves with velocity $\mathbf{u}$, the current density is

$$\mathbf{J} = nq\mathbf{u} = \rho\mathbf{u} \qquad (4.7)$$

where n is the number of charges per unit volume. When dealing with metallic conductors, the current is carried mainly by moving conduction electrons which have a distribution of velocities. In addition, one has positive ions which are virtually at rest in a stationary conductor. In such cases,

$$\mathbf{J} = \Sigma n_i q_i \mathbf{u}_i \qquad (4.8)$$

where n_i is the number of charges of magnitude q_i per unit volume having velocity $\mathbf{u}_i$, the summation being over all possible velocities.

In order to extend the theory developed for a single 'point' charge into a macroscopic theory, we shall use the principle of superposition. It is assumed in classical electromagnetism, that, if one has a system of more than two charges, say q_1, q_2 and q_3, then the total force $\mathbf{f}$ on q_1 due to both q_2 and q_3 is the *vector* sum of $\mathbf{f}_{1,2}$ the force on q_1 due to q_2 and $\mathbf{f}_{1,3}$ the force on q_1 due to q_3, that is

$$\mathbf{f} = \mathbf{f}_{1,2} + \mathbf{f}_{1,3} \qquad (4.9)$$

For stationary charges

$$\mathbf{f}_{1,2} = q_1 \mathbf{E}_2 \qquad (4.10)$$

and

$$\mathbf{f}_{1,3} = q_1 \mathbf{E}_3 \qquad (4.11)$$

where $\mathbf{E}_2$ and $\mathbf{E}_3$ are the electric fields at the position of q_1 due to

q_2 and q_3 respectively. Adding eqns (4.10) and (4.11) vectorially,

$$\mathbf{f} = q_1(\mathbf{E}_2 + \mathbf{E}_3)$$

This can be represented by

$$\mathbf{f} = q_1\mathbf{E}$$

where,

$$\mathbf{E} = \mathbf{E}_2 + \mathbf{E}_3 \tag{4.12}$$

is the resultant electric field at the position of q_1 due to q_2 and q_3. According to the principle of superposition the resultant electric field at a point is obtained, by adding vectorially, the partial electric fields due to the various charges in the system.

Similarly, if q_1 has velocity $\mathbf{u}_1$ and q_2 and q_3 are both moving, the total magnetic force on q_1 is assumed to be

$$\mathbf{f}_{\text{magnetic}} = q_1\mathbf{u}_1 \times \mathbf{B}_2 + q_1\mathbf{u}_1 \times \mathbf{B}_3$$

$$= q_1\mathbf{u}_1 \times (\mathbf{B}_2 + \mathbf{B}_3) = q_1\mathbf{u}_1 \times \mathbf{B}$$

where, $\mathbf{B} = \mathbf{B}_2 + \mathbf{B}_3$ is the resultant magnetic induction field at the position of q_1 due to q_2 and q_3. Thus, according to the principle of superposition, the resultant magnetic induction field at a point is obtained by adding vectorially the partial magnetic induction fields due to the various charges in the system.

In atomic physics one is generally interested in the resultant *microscopic* resultant electric intensity $\mathbf{e}$ and the microscopic magnetic induction $\mathbf{b}$ near and inside atoms. The microscopic fields give the force on a moving 'point' charge placed at the field point, when substituted in the expression for the Lorentz force. The microscopic fields depend on the atomic structure of the material, and can vary enormously near and inside atoms. The case of electric fields inside dielectrics is discussed in Appendix 2(b). However, the scale of many phenomena is much larger than atomic dimensions. In such cases, the enormous fluctuations in the microscopic fields average out, and one is interested primarily in the local space averages of the microscopic fields, that is in what are known as the *macroscopic* fields. For example, the macroscopic electric intensity $\mathbf{E}$ can be defined as

$$\mathbf{E} = \frac{1}{\Delta V} \int \mathbf{e}\,dV$$

The volume ΔV over which the average of the microscopic electric intensity $\mathbf{e}$ is taken, when calculating the macroscopic electric intensity $\mathbf{E}$, must be large on the atomic scale, but kept small on the laboratory scale, say a 10^{-5} cm cube. Similarly, the macroscopic magnetic induction $\mathbf{B}$ can be defined in terms of the microscopic magnetic induction $\mathbf{b}$ by the equation

$$\mathbf{B} = \frac{1}{\Delta V} \int \mathbf{b}\,dV$$

where again ΔV is large on the atomic scale but small on the laboratory scale. Actually, due to the orbital motions of electrons in atoms, and fluctuations in conduction current on the atomic scale, the microscopic fields fluctuate rapidly in time periods of $\sim 10^{-16}$ sec, the period of rotation of orbital electrons. It is also necessary in macroscopic theories to average the microscopic fields over time periods long on the atomic scale, but short on the time scale of changes in macroscopic current and charge density. Since in Maxwell's equations it is the expressions for the curls and divergences of the macroscopic fields which appear, we shall find it convenient in this chapter to do the averaging of the microscopic fields by defining $\mathbf{V} \cdot \mathbf{E}$ and $\mathbf{V} \cdot \mathbf{B}$ using eqn (A1.10) of Appendix 1, keeping ΔV large on the atomic scale but small on the laboratory scale, and defining $\mathbf{V} \times \mathbf{E}$ and $\mathbf{V} \times \mathbf{B}$ using eqn (A1.12) of Appendix 1, keeping ΔS large on the atomic scale, but small on the laboratory scale. In a macroscopic theory, the macroscopic fields $\mathbf{E}$ and $\mathbf{B}$, and the macroscopic charge and current densities ρ and $\mathbf{J}$ are treated as a smooth continuous function of position, just as are other macroscopic quantities such as mass density, which is defined by eqn (4.4).

Throughout this chapter, we shall treat atomic charges such as electrons as classical point charges. In many phenomena on the atomic scale one has to allow for the 'wave nature' of the electron.

For field points in empty space, the microscopic and the macroscopic fields are equal. In this case the symbols $\mathbf{E}$ and $\mathbf{B}$ will be used for both the microscopic and macroscopic fields.

4.2. THE EQUATION $\mathbf{V} \cdot \mathbf{E} = \rho/\varepsilon_0$

Consider a 'point' charge q moving with *uniform* velocity $\mathbf{u}$ as shown in *Figure 4.1*. Let it be at O at the time t. Draw a spherical surface of radius r with centre at O. At the time t the electric field intensity due to q is given by

$$\mathbf{E} = \frac{q\mathbf{r}(1-\beta^2)}{4\pi\varepsilon_0 r^3(1-\beta^2\sin^2\theta)^{3/2}} \tag{4.1}$$

where r is the distance from q to the field point, $\beta = u/c$ and θ is the angle between $\mathbf{u}$ and $\mathbf{r}$. The electric field lines diverge radially from O and are everywhere normal to the spherical surface. Consider the element of the surface between θ and $\theta + d\theta$, as shown in *Figure 4.1.* The area of the element is $2\pi r^2 \sin \theta \, d\theta$. Hence, the electric flux through this area is

$$\delta\Psi = \mathbf{E} . \mathbf{n}dS$$

$$= \frac{q(1-\beta^2)}{4\pi\varepsilon_0 r^2(1-\beta^2 \sin^2 \theta)^{3/2}} 2\pi r^2 \sin \theta \, d\theta$$

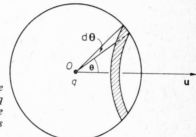

Figure 4.1. The calculation of the total flux of $\mathbf{E}$ *from a charge q moving with uniform velocity. The solid angle between* θ *and* $\theta + d\theta$ *is* $2\pi \sin \theta \, d\theta$

where $\mathbf{n}$ is a unit vector normal to dS. The total flux of $\mathbf{E}$ through the surface of the sphere is

$$\Psi = K \int_{\theta=0}^{\theta=\pi} \frac{\sin \theta \, d\theta}{(1-\beta^2 \sin^2 \theta)^{3/2}} = K \int_0^\pi \frac{\sin \theta \, d\theta}{[(1-\beta^2)+\beta^2 \cos^2 \theta]^{3/2}} \quad (4.13)$$

where

$$K = \frac{q(1-\beta^2)}{2\varepsilon_0} \quad (4.14)$$

Put

$$w = \beta \cos \theta \quad (4.15)$$

$$dw = -\beta \sin \theta \, d\theta$$

$$1-\beta^2 = a^2 \quad (4.16)$$

Hence,

$$I = \int \frac{\sin \theta \, d\theta}{(a^2+\beta^2 \cos^2 \theta)^{3/2}} = -\int \frac{dw}{\beta(a^2+w^2)^{3/2}}$$

51

Now,

$$\int \frac{dx}{(a^2+x^2)^{3/2}} = \frac{x}{a^2\sqrt{a^2+x^2}}$$

Hence,

$$I = -\frac{w}{\beta a^2\sqrt{a^2+w^2}}$$

Substituting for w from eqn (4.15) and for a^2 from eqn (4.16),

$$I = -\frac{\beta \cos\theta}{\beta(1-\beta^2)\sqrt{1-\beta^2\sin^2\theta}}$$

Substituting in eqn (4.13)

$$\Psi = -\frac{K}{(1-\beta^2)}\left[\frac{\cos\theta}{\sqrt{1-\beta^2\sin^2\theta}}\right]_{\theta=0}^{\theta=\pi} = \frac{2K}{(1-\beta^2)}$$

Substituting for K from eqn (4.14)

$$\Psi = \frac{2q(1-\beta^2)}{(1-\beta^2)2\varepsilon_0} = \frac{q}{\varepsilon_0}$$

Hence,

$$\int \mathbf{E}.\mathbf{n}dS = \frac{q}{\varepsilon_0} \tag{4.17}$$

where the integration is over the surface of the sphere shown in *Figure 4.1*. Thus the total electric flux from a charge moving with uniform velocity is the same as the electric flux from a stationary charge of the same magnitude, though the electric intensity varies with direction in the case of the moving charge. Since for a charge moving with uniform velocity $\mathbf{E}$ is proportional to $1/r^2$, the total electric flux is q/ε_0, whatever the radius of the sphere in *Figure 4.1*. Thus for an isolated 'point' charge, provided it has been moving with the same uniform velocity throughout its past history, the lines of $\mathbf{E}$ do not stop in empty space, but continue out radially from the present position of the charge to infinity. Wherever a Gaussian surface is drawn around the moving charge, and whatever the shape of the surface, the total flux of $\mathbf{E}$ from the surface is q/ε_0 as illustrated in *Figure 4.2(a)*. Thus for any surface, if a moving charge q is inside the Gaussian surface

$$\int \mathbf{E}.\mathbf{n}dS = q/\varepsilon_0 \tag{4.18}$$

If the charge is outside the surface, as shown in *Figure 4.2(b)*, as many lines of **E** leave the Gaussian surface as enter it, so that in this case

$$\int \mathbf{E} \cdot \mathbf{n} dS = 0 \qquad (4.18a)$$

Using eqn (A1.10) of Appendix 1, it follows from eqn (4.18a) that in the space outside the moving charge

$$\nabla \cdot \mathbf{E} = 0 \qquad (4.19)$$

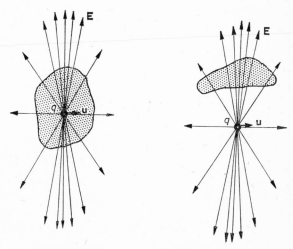

Figure 4.2. (a) The flux of **E** *through a Gaussian surface, due to a charge moving with uniform velocity, is equal to q/ε_0 if the charge is inside the surface, and (b) equal to zero if the charge is outside the Gaussian surface. The lines of* **E** *for an isolated 'point' charge, which is and always has been, moving with uniform velocity, continue radially outwards to infinity*

Now consider a system of 'point' charges moving with uniform velocities as shown in *Figure 4.3*. These charges build up a macroscopic charge and current distribution. Draw a small Gaussian surface of volume ΔV around the field point P, as shown in *Figure 4.3*. If the charge q_i is inside ΔV, then according to eqn (4.18)

$$\int_{\Delta V} \mathbf{E}_i \cdot \mathbf{n} dS = q_i / \varepsilon_0$$

where the integration is over the surface of ΔV. If q_j is outside ΔV, according to eqn (4.18a)

$$\int_{\Delta V} \mathbf{E}_j \cdot \mathbf{n} dS = 0$$

53

Adding for all the N charges in the complete system

$$\int_{\Delta V}(\mathbf{E}_1 \cdot \mathbf{n} + \mathbf{E}_2 \cdot \mathbf{n} + \ldots + \mathbf{E}_N \cdot \mathbf{n})\mathrm{d}S = \sum_{i=1}^{N'} q_i/\varepsilon_0 \qquad (4.20)$$

where N is the total number of charges in the complete system, and N' is the number of the charges inside the volume element ΔV in *Figure 4.3*. Now, using eqn (A1.2) of Appendix 1,

$$(\mathbf{E}_1 \cdot \mathbf{n} + \mathbf{E}_2 \cdot \mathbf{n} + \ldots + \mathbf{E}_N \cdot \mathbf{n}) = (\mathbf{E}_1 + \mathbf{E}_2 + \ldots + \mathbf{E}_N) \cdot \mathbf{n} = \mathbf{e} \cdot \mathbf{n}$$

where

$$\mathbf{e} = \mathbf{E}_1 + \mathbf{E}_2 + \ldots + \mathbf{E}_N$$

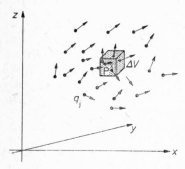

Figure 4.3. A system of 'point' charges moving with uniform velocities. The electric flux from the surface of ΔV is equal to $(1/\varepsilon_0)$ times the total charge inside ΔV. This leads to the relation $\nabla \cdot \mathbf{E} = \rho/\varepsilon_0$

is the total microscopic electric intensity at P due to all the charges in the complete system. Let ρ be the macroscopic charge density in the vicinity of P. Integrating eqn (4.5) over the volume ΔV of the Gaussian surface, we have

$$\sum_{i=1}^{N'} \frac{q_i}{\varepsilon_0} = \frac{1}{\varepsilon_0}\int_{\Delta V}\rho\mathrm{d}V$$

Eqn (4.20) can now be rewritten

$$\int_{\Delta V} \mathbf{e} \cdot \mathbf{n}\mathrm{d}S = \frac{1}{\varepsilon_0}\int_{\Delta V}\rho\Delta V \qquad (4.21)$$

If the Gaussian surface ΔV is made very small on the laboratory scale, but still large enough to contain many microscopic (atomic) charges, so that ρ can be treated as a continuous function of position, we have

$$\int_{\Delta V} \mathbf{e} \cdot \mathbf{n} dS = \frac{\rho \Delta V}{\varepsilon_0} \qquad (4.22)$$

If there are a large number of 'point' charges inside the volume element ΔV, in *Figure 4.3*, the local microscopic electric field $\mathbf{e}$ will vary enormously from point to point on the surface of ΔV. This would be particularly true if ΔV were inside a dielectric. The local microscopic field would be strong near atomic nuclei and weaker elsewhere. When dealing with phenomena on the atomic scale one would have to use the microscopic electric field $\mathbf{e}$, taking into account the atomic structure of the material. For phenomena whose scale is large on the atomic scale, one is generally interested in the macroscopic electric field (denoted $\mathbf{E}$) which is the local space average of the microscopic electric field $\mathbf{e}$. If the volume of the Gaussian surface ΔV, in *Figure 4.3*, is kept large on the atomic scale, fluctuations in the microscopic electric field $\mathbf{e}$ will average out, when evaluating $\int \mathbf{e} \cdot \mathbf{n} dS$ over the surface of ΔV, so that $\int \mathbf{e} \cdot \mathbf{n} dS$ should be equal to $\int \mathbf{E} \cdot \mathbf{n} dS$ evaluated over the surface of ΔV, where $\mathbf{E}$ is the local average or macroscopic electric field in the near vicinity of the field point P. Hence eqn (4.22) can be rewritten

$$\int_{\Delta V} \mathbf{E} \cdot \mathbf{n} dS = \frac{\rho \Delta V}{\varepsilon_0} \qquad (4.22a)$$

According to eqn (A1.10) of Appendix 1, the divergence of the vector $\mathbf{E}$ can be defined by the relation

$$\nabla \cdot \mathbf{E} = \underset{\Delta V \to 0}{\text{Limit}} \frac{\int \mathbf{E} \cdot \mathbf{n} dS}{\Delta V} \qquad (A1.10)$$

Substituting from eqn (4.22a), we obtain

$$\nabla \cdot \mathbf{E} = \frac{\rho}{\varepsilon_0} \qquad (4.23)$$

Eqn (4.23) is one of Maxwell's equations. It relates the macroscopic electric intensity $\mathbf{E}$ to the macroscopic charge density ρ. On the atomic scale, for example in a dielectric, there would be violent fluctuations in the local microscopic electric intensity $\mathbf{e}$ near and inside atoms. When the microscopic field $\mathbf{e}$ is averaged over a region of space, large on the atomic scale, to calculate the macroscopic field

E, the fluctuations on the atomic scale smooth out [cf. *Figure A2.2* of Appendix 2]. It is the divergence of the local space average or macroscopic electric field E which is $1/\varepsilon_0$ times the macroscopic charge density ρ defined by eqn (4.5). [On the atomic scale in the space *between* the electrons and ions in a solid, on the microscopic scale ρ would be zero and $\nabla \cdot \mathbf{e}$ would be zero, though, of course, for a Gaussian surface around a 'point' charge, eqn (4.18) would be valid.]

It is illustrated in Appendix 2 (a) that, in conventional approaches to electromagnetism, eqn (4.23) is developed from Coulomb's law for stationary charges. Yet in conventional approaches to electromagnetism it is *assumed* that eqn (4.23) is valid for the fields of *moving* charges *and* for the fields of *accelerating* charges. In this Section, eqn (4.23) was developed via relativity for charges moving with uniform velocities, but, even in this development, it was not shown that eqn (4.23) was valid for accelerating charges. It is clear that the use of eqn (4.23) for the fields of accelerating charges is an example of the application of Maxwell's equations in a context beyond the experimental evidence on which they are normally developed. Effectively, eqn (4.23) becomes one of the axioms of the theory and the validity of the use of eqn (4.23) in the case of moving and accelerating charges is that, predictions based on the use of eqn (4.23) are in agreement with the experimental results.

At this stage, the theory can be extended to field points inside *stationary* dielectrics. When a stationary dielectric is in an applied electric field, there is a separation of the positive and negative charges in atoms giving rise to induced atomic dipole moments. Polar molecules will tend to align themselves in the direction of the applied electric field, and any ions in crystals will tend to be displaced in the direction of the applied electric field. The macroscopic polarization P is defined as the resultant electric dipole moment per unit volume. By analogy with eqn (4.4) P can be defined by the relation

$$\mathbf{P} = \frac{1}{\Delta V} \sum_{i=1}^{N} \mathbf{p}_i \qquad (4.24)$$

where the vector summation is carried out over all the N atomic electric dipoles $\mathbf{p}_i$ inside a volume element ΔV, which is small on the laboratory scale but large on the atomic scale. In a macroscopic theory P is treated as a continuous function of position. It is shown in Appendix 2(b) that, for purposes of calculating the *macroscopic* (or local space average) electric field, a polarized dielectric can be replaced by a fictitious macroscopic surface charge distribution

$$\sigma' = \mathbf{P} \cdot \mathbf{n} \qquad (4.25)$$

56

and a fictitious macroscopic volume charge distribution

$$\rho' = -\nabla \cdot \mathbf{P} \qquad (4.26)$$

placed *in vacuo*. Thus for a field point inside a polarized dielectric, eqn (4.23) for the divergence of the macroscopic electric intensity becomes

$$\varepsilon_0 \nabla \cdot \mathbf{E} = \rho + \rho'$$

where ρ is the density of true charge placed on the dielectric, and ρ' is the fictitious volume charge density associated with the polarization of the dielectric. Using eqn (4.26),

$$\nabla \cdot (\varepsilon_0 \mathbf{E} + \mathbf{P}) = \rho \qquad (4.27)$$

Now the macroscopic quantity $\varepsilon_0 \mathbf{E} + \mathbf{P}$ appears so frequently in electromagnetic theory that it is convenient, *in the interests of simplicity*, to give it a special name, the electric displacement, and to give it a special symbol, namely $\mathbf{D}$, so that *by definition*

$$\mathbf{D} = \varepsilon_0 \mathbf{E} + \mathbf{P} \qquad (4.28)$$

Eqn (4.27) can be rewritten

$$\nabla \cdot \mathbf{D} = \rho \qquad (4.29)$$

One could just as well use eqn (4.27) and not introduce $\mathbf{D}$ at all. Eqn (4.29) is one of Maxwell's equations. It relates two macroscopic (that is local space average) quantities $\mathbf{D}$ and ρ. In conventional approaches to electromagnetism, eqn (4.29) is generally developed from electrostatics, in the way outlined in Appendix 2.

Eqn (4.28) is often rewritten in the form

$$\mathbf{D} = \varepsilon_r \varepsilon_0 \mathbf{E} \qquad (4.30)$$

where ε_r is the dielectric constant or relative permittivity. Eqn (4.30) is one of the constitutive equations.

4.3. RELATION BETWEEN THE SPATIAL AND TIME DERIVATIVES OF THE FIELDS OF A CHARGE MOVING WITH UNIFORM VELOCITY

The electromagnetic fields of a 'point' charge moving with uniform velocity are carried along convectively with the charge. For example, the lines of $\mathbf{E}$ always diverge radially from the 'present' position of the charge. Consider the case illustrated in *Figure 4.4(a)*. The 'point' charge q is moving along the x axis with uniform velocity $\mathbf{u}$,

For the conditions shown in *Figure 4.4(a)*, let an observer stationary at P measure an increase ΔE in the electric field intensity at P in a time interval Δt. If at a fixed time, that is considering the charge q as fixed in space, the observer at P moved a distance

$$\Delta x = -u\Delta t \qquad (4.31)$$

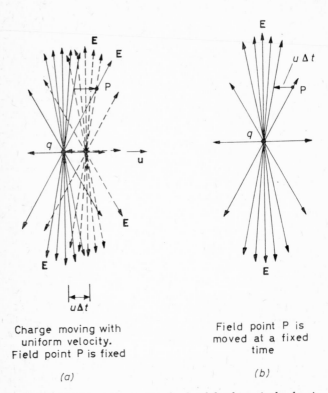

Charge moving with
uniform velocity.
Field point P is fixed

(a)

Field point P is
moved at a fixed
time

(b)

Figure 4.4. (a) The field point P is fixed and the change in the electric intensity **E** *in a time Δt due to the motion of the charge is measured. (b) The field point P is moved a distance $u\Delta t$ to the left at a fixed time. The same change in* **E** *is measured in both (a) and (b)*

to the left parallel to the x axis as shown in *Figure 4.4(b)*, he would measure the same increase ΔE in the field intensity. For the case shown in *Figure 4.4(a)* the x, y and z co-ordinates of the field point are constant, so that

$$\Delta \mathbf{E} = \left(\frac{\partial \mathbf{E}}{\partial t}\right)_{x,y,z} \Delta t \qquad (4.32)$$

58

For the conditions shown in *Figure 4.4(b)*, y, z and t are constant, so that

$$\Delta \mathbf{E} = \left(\frac{\partial \mathbf{E}}{\partial x}\right)_{y,z,t} \Delta x \qquad (4.33)$$

Equating the right hand sides of eqns (4.32) and (4.33) and using eqn (4.31),

$$\left(\frac{\partial \mathbf{E}}{\partial t}\right)_{x,y,z} = -u \left(\frac{\partial \mathbf{E}}{\partial x}\right)_{y,z,t} \qquad (4.34)$$

The magnetic field of a point charge moving with uniform velocity is also carried along convectively, so that by a similar argument

$$\left(\frac{\partial \mathbf{B}}{\partial t}\right)_{x,y,z} = -u \left(\frac{\partial \mathbf{B}}{\partial x}\right)_{y,z,t} \qquad (4.35)$$

In general, for the fields and potentials of a 'point' charge moving with *uniform* velocity $\mathbf{u}$ parallel to the x axis

$$\frac{\partial}{\partial t} = -u \frac{\partial}{\partial x} \qquad (4.36)$$

where $\partial/\partial t$ represents the rate of change with time at a fixed field point and $\partial/\partial x$ represents the rate of change with position, if the field point is changed at a fixed time, that is keeping the position of the charge fixed.

If $\mathbf{u}$ is not parallel to the x axis

$$\frac{\partial}{\partial t} = -\mathbf{u} \cdot \nabla \qquad (4.37)$$

4.4. THE EQUATION $\nabla \times \mathbf{E} = -(\partial \mathbf{B}/\partial t)$ FOR 'POINT' CHARGES MOVING WITH UNIFORM VELOCITIES

Consider an isolated 'point' charge q coulombs, moving with uniform velocity $\mathbf{u}$ along the x axis of an inertial reference frame Σ, as shown in *Figure 4.5*. Let q be at O, the origin of Σ, at the time t when the electric intensity $\mathbf{E}$ and the magnetic induction $\mathbf{B}$ are determined at a field point P, which has co-ordinates x, y, z and is at a radial distance $\mathbf{r}$ from O. Choose spherical polar co-ordinates, measuring the polar angle θ from the x axis, the direction of motion of the charge, as illustrated in *Figure 4.5*. Let $\mathbf{a}_r$, $\mathbf{a}_\theta$ and $\mathbf{a}_\phi$ be unit vectors in the directions of increasing r, θ and ϕ respectively as shown in *Figure 4.5*. According to eqns (3.25) and (3.26), the electric intensity $\mathbf{E}$ at P, when q is at the origin, is

$$\mathbf{E} = \mathbf{a}_r E_r \qquad (4.38)$$

where

$$E_r = \frac{q(1-\beta^2)}{4\pi\varepsilon_0 r^2(1-\beta^2 \sin^2 \theta)^{3/2}} \qquad (4.39)$$

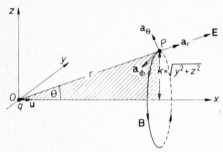

Figure 4.5. Spherical polar coordinates r, θ and ϕ are used to represent the electric and magnetic fields at P due to a charge q moving with uniform velocity $\mathbf{u}$. The polar angle θ is measured from the x axis, the direction of $\mathbf{u}$. The electric field diverges radially from the 'present' position of the charge and is in the direction of $\mathbf{a}_r$. The magnetic field is in the direction of $\mathbf{a}_\phi$

where $\beta = u/c$. Since $|\mathbf{u} \times \mathbf{r}| = ur \sin \theta$, the magnetic field at P, given by eqns (3.28) and (3.29), is,

$$\mathbf{B} = \mathbf{a}_\phi B_\phi \qquad (4.40)$$

where,

$$B_\phi = \frac{qu(1-\beta^2) \sin \theta}{4\pi\varepsilon_0 c^2 r^2(1-\beta^2 \sin^2 \theta)^{3/2}} \qquad (4.41)$$

It was illustrated in Section 3.2, that for a 'point' charge moving with uniform velocity the electric field lines bunch towards a direction perpendicular to $\mathbf{u}$, the velocity of the charge, as illustrated in *Figure 4.6*. Consider the line integral of $\mathbf{E}$, that is $\oint \mathbf{E} . \, d\mathbf{l}$, around the contour $ABCD$ in *Figure 4.6*. The arc AB is at a radial distance r from O and the arc DC is at a radial distance $r + \Delta r$ from O. Since the electric field is radial, there is no electric intensity along the arcs AB and CD, so that

$$\int_A^B \mathbf{E} . \, d\mathbf{l} = 0 \qquad (4.42)$$

and

$$\int_C^D \mathbf{E} . \, d\mathbf{l} = 0 \qquad (4.43)$$

The electric intensity at any point on the section BC is greater than the electric intensity at a point along AD at the same radial distance from O. Let the radial electric intensity along AD at a distance r from O be denoted by $E_r(r,\theta)$, where E_r is a function of r and θ. At the same radial distance from O, the electric intensity along BC is given by

$$E_r(r,\theta+\Delta\theta) = E_r(r,\theta)+\frac{\partial E_r(r,\theta)}{\partial \theta}\Delta\theta$$

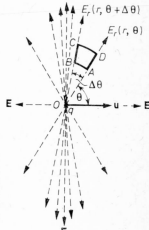

Figure 4.6. The calculation of the curl of the electric intensity $\mathbf{E}$ due to a charge q moving with uniform velocity. Since the electric intensity is greater along BC than AD, $\oint \mathbf{E} \cdot d\mathbf{l}$ evaluated around $ABCD$ is finite, so that $\nabla \times \mathbf{E}$ is finite. It is shown that $\nabla \times \mathbf{E} = -(\partial \mathbf{B}/\partial t)$. Both $\mathbf{E}$ and $\mathbf{B}$ arise from the moving charge

Hence,

$$\int_B^C \mathbf{E} \cdot d\mathbf{l}+ \int_D^A \mathbf{E} \cdot d\mathbf{l} = \int_B^C \mathbf{E} \cdot d\mathbf{l}- \int_A^D \mathbf{E} \cdot d\mathbf{l}$$

$$= \int_r^{r+\Delta r} \left[E_r(r,\theta)+\frac{\partial E_r(r,\theta)}{\partial \theta} \Delta\theta - E_r(r,\theta) \right] dr$$

$$= \frac{\partial E_r(r,\theta)}{\partial \theta} \Delta\theta\Delta r \qquad (4.44)$$

Adding eqns (4.42), (4.43) and (4.44)

$$\oint_{ABCD} \mathbf{E} \cdot d\mathbf{l} = \frac{\partial E_r}{\partial \theta}\Delta\theta\Delta r \qquad (4.45)$$

61

The curl of a vector can be defined by the relation

$$(\nabla \times \mathbf{E}) \cdot \mathbf{n} = \underset{\Delta S \to 0}{\text{Limit}} \frac{\oint \mathbf{E} \cdot d\mathbf{l}}{\Delta S} \qquad (A1.12)$$

Now for the area $ABCD$ in *Figure 4.6*, for integration from A to B to C to D

$$\mathbf{n}\Delta S = -\mathbf{a}_\phi r\Delta\theta\Delta r$$

Hence,

$$(\nabla \times \mathbf{E}) \cdot \mathbf{a}_\phi = -\frac{1}{r\Delta\theta\Delta r} \oint \mathbf{E} \cdot d\mathbf{l}$$

Substituting for $\oint \mathbf{E} \cdot d\mathbf{l}$ from eqn (4.45)

$$(\nabla \times \mathbf{E})_\phi = -\frac{1}{r}\frac{\partial E_r}{\partial \theta} \qquad (4.46)$$

Alternatively, we can start from the general expression for $\nabla \times \mathbf{E}$ in spherical polar co-ordinates, given by eqn (A1.29), of Appendix 1. Since $E_\theta = E_\phi = 0$, and $\partial E_r/\partial\phi$ is zero, we have

$$\nabla \times \mathbf{E} = \mathbf{a}_\phi \left(-\frac{1}{r}\frac{\partial E_r}{\partial \theta} \right) \qquad (4.47)$$

Differentiating eqn (4.39) for E_r partially with respect to θ, gives

$$\frac{\partial E_r}{\partial \theta} = \frac{q(1-\beta^2)}{4\pi\varepsilon_0 r^2} \frac{\partial}{\partial \theta}\left[\frac{1}{(1-\beta^2\sin^2\theta)^{3/2}} \right]$$

$$= \frac{q(1-\beta^2)}{4\pi\varepsilon_0 r^2}\left[\frac{3\beta^2 \sin\theta \cos\theta}{(1-\beta^2\sin^2\theta)^{5/2}} \right]$$

Using eqn (4.47)

$$\nabla \times \mathbf{E} = -\mathbf{a}_\phi \frac{1}{r}\frac{\partial E_r}{\partial \theta}$$

$$\nabla \times \mathbf{E} = -\mathbf{a}_\phi \frac{3q(1-\beta^2)\beta^2 \sin\theta \cos\theta}{4\pi\varepsilon_0 r^3(1-\beta^2\sin^2\theta)^{5/2}} \qquad (4.48)$$

This is the value of $\nabla \times \mathbf{E}$ at the point P in *Figure 4.5*, when the charge q is at the origin. We will now calculate $\partial\mathbf{B}/\partial t$ at the same field point P at the same time.

According to eqn (4.35), for a 'point' charge moving with uniform velocity $\mathbf{u}$ along the x axis, we have

$$\frac{\partial \mathbf{B}}{\partial t} = -u \frac{\partial \mathbf{B}}{\partial x} \tag{4.49}$$

Now

$$\mathbf{B} = \mathbf{a}_\phi B_\phi \tag{4.40}$$

where

$$B_\phi = \frac{qu(1-\beta^2)\sin\theta}{4\pi\varepsilon_0 c^2 r^2 (1-\beta^2 \sin^2\theta)^{3/2}} = \frac{K\sin\theta}{r^2(1-\beta^2\sin^2\theta)^{3/2}} \tag{4.50}$$

where

$$K = \frac{qu(1-\beta^2)}{4\pi\varepsilon_0 c^2}$$

In *Figure 4.5*, since P has co-ordinates x, y, z,

$$\sin\theta = \frac{\sqrt{y^2+z^2}}{r} = \frac{k}{r}$$

where,

$$k = \sqrt{y^2+z^2} = r\sin\theta$$

Substituting for $\sin\theta$, in eqn (4.50),

$$B_\phi = \frac{Kk}{r^3(1-\beta^2 k^2/r^2)^{3/2}}$$

or,

$$B_\phi = \frac{C}{(r^2-A)^{3/2}} \tag{4.51}$$

where,

$$C = Kk = \frac{qu(1-\beta^2)r\sin\theta}{4\pi\varepsilon_0 c^2} = \frac{qu(1-\beta^2)\sqrt{y^2+z^2}}{4\pi\varepsilon_0 c^2} \tag{4.52}$$

and

$$A = \beta^2 k^2 = \beta^2 r^2 \sin^2\theta = \beta^2(y^2+z^2) \tag{4.53}$$

Differentiating eqn (4.40) partially with respect to x

$$\frac{\partial \mathbf{B}}{\partial x} = \frac{\partial}{\partial x}\mathbf{a}_\phi B_\phi = \mathbf{a}_\phi \frac{\partial B_\phi}{\partial x} \tag{4.54}$$

63

Differentiating eqn (4.51) partially with respect to x remembering that $r = (x^2+y^2+z^2)^{1/2}$ contains x explicitly, but A and C do not, we find

$$\frac{\partial B_\phi}{\partial x} = -\frac{3Cr\partial r/\partial x}{(r^2-A)^{5/2}}$$

Now,

$$\frac{\partial r}{\partial x} = \frac{\partial}{\partial x}(x^2+y^2+z^2)^{1/2} = \frac{x}{(x^2+y^2+z^2)^{1/2}} = \cos\theta$$

Hence,

$$\frac{\partial B_\phi}{\partial x} = -\frac{3Cr\cos\theta}{(r^2-A)^{5/2}} = -\frac{3C\cos\theta}{r^4(1-A/r^2)^{5/2}}$$

Using eqn (4.54),

$$\frac{\partial \mathbf{B}}{\partial x} = \mathbf{a}_\phi \frac{\partial B_\phi}{\partial x} = -\mathbf{a}_\phi \frac{3C\cos\theta}{r^4(1-A/r^2)^{5/2}} \tag{4.55}$$

From eqns (4.49) and (4.55)

$$\frac{\partial \mathbf{B}}{\partial t} = -u\frac{\partial \mathbf{B}}{\partial x} = \mathbf{a}_\phi \frac{3uC\cos\theta}{r^4(1-A/r^2)^{5/2}}$$

Substituting for C from eqn (4.52) and for A from eqn (4.53),

$$-\frac{\partial \mathbf{B}}{\partial t} = -\mathbf{a}_\phi \frac{3q\beta^2(1-\beta^2)\sin\theta\cos\theta}{4\pi\varepsilon_0 r^3(1-\beta^2\sin^2\theta)^{5/2}} \tag{4.56}$$

Comparing eqns (4.48) and (4.56), it can be seen that the electric intensity $\mathbf{E}$ and the magnetic induction $\mathbf{B}$ at the point P in *Figure 4.5* are related by the equation

$$\mathbf{V} \times \mathbf{E} = -\frac{\partial \mathbf{B}}{\partial t} \tag{4.57}$$

This relation holds whatever the position of the 'point' charge in *Figures 4.5* and *4.6*. Eqn (4.57) is one of Maxwell's equations. If we know how $\mathbf{B}$ is varying with time at any fixed field point such as P, due to a 'point' charge moving with uniform velocity, then $\mathbf{V} \times \mathbf{E}$, at that field point, can be calculated using eqn (4.57). Conversely, if $\mathbf{V} \times \mathbf{E}$ is given at any point then $\partial \mathbf{B}/\partial t$ can be calculated. There is no need to enquire what the position of the charge is. Eqn (4.57) is a relation between the field quantities $\mathbf{E}$ and $\mathbf{B}$ at the field point, where the fields are determined. Eqn (4.57) was developed for the fields of a charge moving with uniform velocity. It is assumed in classical electromagnetism, that eqn (4.57) also holds for the fields of accelerating charges.

It will now be assumed that the contour $ABCD$ in *Figure 4.6* is made from a stationary coil of thin wire as illustrated in *Figure 4.7*. As the charge q approaches the coil from the left, as illustrated in *Figure 4.7*, the magnetic flux through the coil in the direction upwards towards the reader is increasing. At the same time, when the charge is approaching the coil, the electric field is greater along BC than along AD, giving a resultant $\oint \mathbf{E} \cdot d\mathbf{l}$, which is the e.m.f. which drives a conduction current around the stationary coil of wire in the

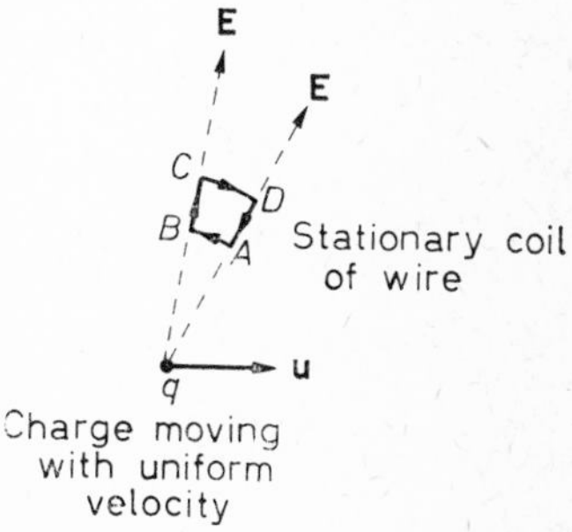

Figure 4.7. The moving charge q gives rise to an induced e.m.f. in the circuit ABCD. The e.m.f. is equal to the rate of change of the magnetic flux through ABCD. The direction of current flow is from A to B to C to D to A in agreement with Lenz' law. (The electric field due to the charge q is greater at AB than at CD, so that AB is at a different electric potential to CD and electric charge distributions will be induced on the surface of the conductor ABCD, which will vary as the position of the charge q varies. Cullwick[1] shows that the effect on the total induced e.m.f. is negligible)

direction A to B to C to D, as shown in *Figure 4.7*. According to the right-handed corkscrew rule, this current flow gives rise to an extra magnetic field which is downwards into the paper, away from the reader in *Figure 4.7*. It is in such a direction as 'to tend to oppose' the increase in magnetic flux through the coil $ABCD$ due to the 'point' charge, which is approaching the coil. This illustrates Lenz' law.

After the 'point' charge q has moved to the right of the loop $ABCD$ in *Figure 4.7*, the magnetic flux through the coil is still in a direction upwards towards the reader, but the magnetic flux through the coil is now decreasing with time as the charge moves away from the coil. In this case the electric field intensity due to the 'point' charge is less on the left-hand side of $ABCD$ than on the right-hand side, so that the e.m.f. and current flow in the coil are in the opposite direction to previously, namely from A to D to C to B to A. The magnetic field associated with this current flow is upwards towards the reader in such a direction as to 'oppose' the decrease in magnetic flux through the coil, as the charge moves away from the coil. This is again in agreement with Lenz' law.

In this example, it is not strictly correct to say that it is the varying magnetic field $\partial \mathbf{B}/\partial t$ which gives rise to the e.m.f. in the coil $ABCD$ in *Figure 4.7*. Both the electric field giving rise to the e.m.f. and the magnetic field passing through the coil are due to the same moving charge, which gives rise to both $\mathbf{E}$ and $\mathbf{B}$. What has been shown is that $\nabla \times \mathbf{E}$ is related to $\partial \mathbf{B}/\partial t$ by eqn (4.57). If we know one of the two, we can calculate the other. Both $\mathbf{E}$ and $\mathbf{B}$ are propagated from the charge with a speed c in empty space.

Integrating eqn (4.57) over the surface of any closed loop of any dimensions

$$\int (\nabla \times \mathbf{E}) \cdot \mathbf{n} dS = -\int \frac{\partial \mathbf{B}}{\partial t} \cdot \mathbf{n} dS$$

Using Stokes' theorem, eqn (A1.25),

$$\oint \mathbf{E} \cdot d\mathbf{l} = -\frac{\partial}{\partial t} \int \mathbf{B} \cdot \mathbf{n} dS \tag{4.58}$$

that is, the e.m.f. induced in any stationary coil, due to a moving 'point' charge, is equal to the rate of change of the total magnetic flux through the coil due to the charge.

Now consider an assembly of 'point' charges all moving with various *uniform* velocities, such that they build up a macroscopic charge and current distribution of the type illustrated previously in *Figure 4.3*. For any charge q_i moving with velocity $\mathbf{u}_i$, its electric field $\mathbf{E}_i$ and its magnetic field $\mathbf{B}_i$ at any field point P satisfy the equation

$$\nabla \times \mathbf{E}_i = -\frac{\partial \mathbf{B}_i}{\partial t}$$

Adding the contributions of all the 'point' charges in the complete system, we have

$$\nabla \times \mathbf{E}_1 + \nabla \times \mathbf{E}_2 + \ldots + \nabla \times \mathbf{E}_n = -\frac{\partial \mathbf{B}_1}{\partial t} - \frac{\partial \mathbf{B}_2}{\partial t} \ldots - \frac{\partial \mathbf{B}_n}{\partial t}$$

But, using eqn (A1.17)

$$\nabla \times \mathbf{E}_1 + \nabla \times \mathbf{E}_2 + \ldots + \nabla \times \mathbf{E}_n = \nabla \times (\mathbf{E}_1 + \mathbf{E}_2 \ldots \mathbf{E}_n) = \nabla \times \mathbf{E}$$

Also,

$$\frac{\partial \mathbf{B}_1}{\partial t} + \frac{\partial \mathbf{B}_2}{\partial t} + \ldots + \frac{\partial \mathbf{B}_n}{\partial t} = \frac{\partial}{\partial t} (\mathbf{B}_1 + \mathbf{B}_2 + \ldots \mathbf{B}_n) = \frac{\partial \mathbf{B}}{\partial t}$$

where $\mathbf{E}$ is the resultant electric intensity and $\mathbf{B}$ is the resultant magnetic induction at P due to all the charges in the complete system. Hence for a system of 'point' charges moving with uniform velocities such that they build up a macroscopic charge and current distribution of the type shown in *Figure 4.3*, we have

$$\nabla \times \mathbf{E} = -\frac{\partial \mathbf{B}}{\partial t} \tag{4.59}$$

Eqn (4.59) is one of Maxwell's equations. It is a relation between the resultant electric intensity $\mathbf{E}$ and the resultant magnetic induction $\mathbf{B}$ at any field point. It is independent of the precise positions of the 'point' charges in the system, and the macroscopic charge and current distribution they build up.

Strictly, eqn (4.59) was developed for the microscopic fields $\mathbf{e}$ and $\mathbf{b}$. It will be confirmed in Section 4.9 that eqn (4.59) does hold, without modification, for the macroscopic, or local average fields $\mathbf{E}$ and $\mathbf{B}$ due to a system of 'point' charges moving with uniform velocities. If eqn (4.59) is integrated over the surface of any closed loop

$$\int (\nabla \times \mathbf{E}) \cdot \mathbf{n}dS = -\int \frac{\partial \mathbf{B}}{\partial t} \cdot \mathbf{n}dS$$

Using Stokes' theorem, eqn (A1.25) of Appendix 1,

$$\oint \mathbf{E} \cdot d\mathbf{l} = -\frac{\partial}{\partial t} \int \mathbf{B} \cdot \mathbf{n}dS = -\frac{\partial \Phi}{\partial t} \tag{4.60}$$

where Φ is the total magnetic flux, through any surface bounded by the loop, due to all the moving charges in the system. The quantities $\mathbf{E}$ and $\mathbf{B}$ in eqn (4.60) are the values of the resultant electric and magnetic fields at the surface. According to eqn (4.60), the total e.m.f. in a stationary circuit is equal to the rate of change of the total flux of magnetic induction through the circuit. Eqn (4.60) is valid whatever the positions and magnitudes of the charge and current distributions in the system. If the rate of change of the flux of magnetic induction through the circuit is known, the induced e.m.f. can be calculated and vice versa. The electric and magnetic fields in eqns (4.59) and (4.60) arise from a common cause, namely the moving charges in the system. Eqn (4.60) is generally known as either Faraday's or Neumann's law of induction. The direction of the induced e.m.f. has been shown to be consistent with Lenz' law.

It is illustrated in Appendix 2(g) that, in conventional approaches to electromagnetism eqn (4.59) is generally developed from experiments with transformers carried out at low frequency, that is quasi-stationary conditions. By considering an example due to Sherwin[2], it is shown in Appendix 5(b) that, at low frequencies (that is quasi-stationary conditions), for air-cored transformers, the main contribution to the electric field in eqns (4.59) and (4.60) arises from the velocity dependent electric field $\mathbf{E}_V$ in eqn (3.41) for the total electric field of an accelerating charge. The contribution to the total electric field, due to the acceleration dependent or radiation field $\mathbf{E}_A$, given by eqn (3.43), is generally negligible for low frequency transformers. At the field point, eqn (3.42) for $\mathbf{E}_V$ is similar to eqn (4.39) which was developed from Coulomb's law in Chapter 3 (cf. eqn (3.21)). Hence the analysis and interpretation of the equation $\nabla \times \mathbf{E} = -\partial \mathbf{B}/\partial t$ developed earlier in this section is applicable almost directly to air-cored transformers, used at low frequency. In Appendix 5(b), the induced electric field in such cases, which is associated with $\mathbf{E}_V$, is traced to retardation effects. At low frequencies, the radiation fields in transformers are negligible. However, it is assumed in classical electromagnetism that eqn (4.59) is also valid for the electric and magnetic fields of accelerating charges. This goes beyond the experimental evidence on which eqn (4.59) is normally developed. It is found experimentally that eqn (4.59) does apply to the fields of accelerating charges. The case of iron-cored transformers is discussed in Appendix 2(g). It is found experimentally that eqn (4.59) also holds for iron-cored transformers.

4.5. THE EQUATION $\nabla \cdot \mathbf{B} = 0$

It was illustrated in Section 3.2, that for a charge q moving with uniform velocity $\mathbf{u}$, ($\beta = u/c$)

$$\mathbf{B} = \mathbf{u} \times \mathbf{E}/c^2 \qquad (4.61)$$

where

$$\mathbf{E} = \frac{q\mathbf{r}(1-\beta^2)}{4\pi\varepsilon_0 r^3(1-\beta^2 \sin^2\theta)^{3/2}}. \qquad (4.62)$$

The lines of $\mathbf{E}$ diverge radially from the 'present' position of the charge. Since $\mathbf{B}$ is perpendicular to both $\mathbf{u}$ and $\mathbf{E}$ the lines of $\mathbf{B}$, for an isolated 'point' charge moving with uniform velocity, form closed circles in a plane perpendicular to $\mathbf{u}$, the velocity of the charge, as illustrated in *Figures 3.4* and *4.8*. The lines of $\mathbf{B}$ for a single 'point' charge are continuous lines in spaces. Unlike the electric field lines they neither start nor end at the charge. Consider the volume element ΔV at the field point P in *Figure 4.8*. At P, the lines of $\mathbf{B}$ are upwards out of the paper. As many lines of $\mathbf{B}$ leave ΔV from the upper surfaces as enter it from the lower surfaces, so that for a single point charge,

$$\int \mathbf{B} \cdot \mathbf{n} dS = 0$$

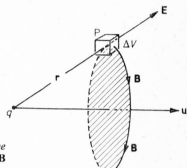

Figure 4.8. Calculation of $\nabla \cdot \mathbf{B}$ at the field point P. Since as many lines of $\mathbf{B}$ leave ΔV as enter it, $\nabla \cdot \mathbf{B}$ is zero

where the integration is over the surface of the volume element ΔV. Consider the system of moving charges shown earlier in *Figure 4.3*. Following the procedure of Section 4.2, adding for all the charges in the system, we have

$$\int (\mathbf{B}_1 \cdot \mathbf{n} dS + \ldots + \mathbf{B}_N \cdot \mathbf{n} dS) = \int (\mathbf{B}_1 + \mathbf{B}_2 + \ldots + \mathbf{B}_N) \cdot \mathbf{n} dS$$

$$= \int \mathbf{b} \cdot \mathbf{n} dS = 0 \tag{4.63}$$

where $\mathbf{b}$ is the resultant microscopic magnetic induction at the field point P in *Figure 4.3*. The divergence of the macroscopic magnetic induction $\mathbf{B}$ can be defined by eqn (A1.10) of Appendix 1,

$$\nabla \cdot \mathbf{B} = \underset{\Delta V \to 0}{\text{Limit}} \frac{\int \mathbf{B} \cdot \mathbf{n} dS}{\Delta V} \tag{4.64}$$

If the volume of the volume element ΔV in *Figure 4.3* is kept large on the atomic scale, fluctuations in the microscopic field $\mathbf{b}$ should average out, if $\int \mathbf{b} \cdot \mathbf{n} dS$ is evaluated over the surface of ΔV, so that $\int \mathbf{b} \cdot \mathbf{n} dS$ should be equal to $\int \mathbf{B} \cdot \mathbf{n} dS$, where $\mathbf{B}$ is the average or macroscopic magnetic induction in the vicinity of the field point P. Since $\int \mathbf{b} \cdot \mathbf{n} dS$ is zero, it follows that $\int \mathbf{B} \cdot \mathbf{n} dS$ is zero, so that from eqn (4.64)

$$\nabla \cdot \mathbf{B} = 0 \tag{4.65}$$

where $\mathbf{B}$ is the macroscopic magnetic induction due to a macroscopic charge distribution, built up from atomic charges such as electrons and ions. Eqn (4.65) is one of Maxwell's equations. According to eqn (4.65), if we have a small volume element ΔV, the net flux of $\mathbf{B}$ into or from the surface of the volume element is zero. For a fuller discussion, the reader is referred to Slepian[3].

It is interesting to note that the equation $\mathbf{V} . \mathbf{B} = 0$ is related to the equation $\mathbf{V} \times \mathbf{E} = -\partial \mathbf{B}/\partial t$. Since the divergence of the curl of any vector is zero,

$$\mathbf{V} . (\mathbf{V} \times \mathbf{E}) = -\mathbf{V} . \left(\frac{\partial \mathbf{B}}{\partial t}\right) = 0 \qquad (4.66)$$

Eqn (4.66) can be rewritten

$$\frac{\partial}{\partial t}(\mathbf{V} . \mathbf{B}) = 0$$

Integrating,

$$\mathbf{V} . \mathbf{B} = \text{constant} \qquad (4.67)$$

Thus if the divergence of the macroscopic magnetic induction is zero at one instant of time, it follows from eqn (4.67) that $\mathbf{V} . \mathbf{B}$ is zero at all times. In the general application of Maxwell's equations to electromagnetic phenomena, one can treat the equation $\mathbf{V} . \mathbf{B} = 0$, as an initial condition. It has been illustrated that for magnetic fields of charges moving with uniform velocities $\mathbf{V} . \mathbf{B}$. is zero. From time to time it has been suggested that magnetic monopoles may exist, or might be produced by high energy accelerators (e.g. Dirac[4]). If one did have magnetic monopoles, one might expect the lines of $\mathbf{B}$ to diverge from the magnetic monopoles in a way similar to the way electric field lines diverge from electric charges. For a system of such magnetic monopoles, $\mathbf{V} . \mathbf{B}$ would not be zero. It is left as an exercise for the reader (Problem 4.2) to show that in the presence of magnetic monopoles

$$\mathbf{V} . \mathbf{B} = \mu_0 \rho_m^* = \rho_m \qquad (4.68)$$

where ρ_m^*, the macroscopic density of magnetic monopoles, is defined by eqn (4.143) of Problem (4.2). If the existence of magnetic monopoles is ever established, it is probable that it will be in only very special circumstances that the equations of electromagnetism will have to be modified to allow for their presence. It is reasonably safe to conclude that magnetic monopoles play no significant role in normal terrestrial electromagnetic phenomena, and one writes

$$\mathbf{V} . \mathbf{B} = 0 \qquad (4.65)$$

Fundamental particles such as electrons, protons and neutrons do have intrinsic magnetic dipole moments. In classical electromagnetism it is assumed that as many lines of $\mathbf{B}$ return to such particles as leave them.

It is illustrated in Appendix 2(d), that, in conventional approaches to electromagnetism, eqn (4.65) is generally developed from the Biot–Savart law, eqn (A2.22) of Appendix 2 or (A4.10) of Appendix 4. The Biot–Savart law is normally developed on the basis of experiments with steady currents in complete circuits. It is shown in Appendix 4, that the Biot–Savart law arises from the velocity dependent magnetic field $\mathbf{B}_V$ in the expression for the magnetic field of an accelerating charge, eqn (3.44). The Biot–Savart law does not hold for the radiation fields. In classical electromagnetism it is *assumed* that eqn (4.65) is also valid for the magnetic fields of varying currents in incomplete circuits, and for the radiation fields of accelerating charges. This is another example of the use of one of Maxwell's equations in a wider experimental context than the experimental evidence on which it is normally developed, the Biot–Savart law in this instance.

4.6. DEVELOPMENT OF THE EQUATION $\varepsilon_0 c^2 \nabla \times \mathbf{B} = \mathbf{J} + \varepsilon_0 \partial \mathbf{E} / \partial t$ USING VECTOR ANALYSIS

In Section 4.4, the curl of the electric intensity $\mathbf{E}$ at a field point was related to the time rate of change of the magnetic induction $\mathbf{B}$ at the same field point. Vector analysis will now be used to develop a relation between $\nabla \times \mathbf{B}$ and $\partial \mathbf{E} / \partial t$ for a 'point' charge moving with uniform velocity, and for a macroscopic charge and current distribution made up of atomic charges such as electrons and ions moving with *uniform* velocities.

(a) 'Point' Charge Moving With Uniform Velocity

Consider the 'point' charge q moving with uniform velocity $\mathbf{u}$ along the x axis as shown in *Figure 4.9*. Let the charge be at the origin O at the time t when the fields $\mathbf{E}$ and $\mathbf{B}$ are determined at a field point P which has co-ordinates x, y, z, and is at a radial distance $\mathbf{r}$ from O, the 'present' position of the charge. According to eqn (4.2), at the field point P, the magnitude of the magnetic induction $\mathbf{B}$ is related to the electric field $\mathbf{E}$ by

$$\mathbf{B} = \frac{\mathbf{u} \times \mathbf{E}}{c^2} \tag{4.69}$$

Hence,

$$\mathbf{\nabla} \times \mathbf{B} = \mathbf{\nabla} \times (\mathbf{u} \times \mathbf{E}/c^2) = (1/c^2)\mathbf{\nabla} \times (\mathbf{u} \times \mathbf{E}) \qquad (4.70)$$

From vector analysis, for any two vectors $\mathbf{A}$ and $\mathbf{C}$,

$$\mathbf{\nabla} \times (\mathbf{A} \times \mathbf{C}) = (\mathbf{C} \cdot \mathbf{\nabla})\mathbf{A} - (\mathbf{A} \cdot \mathbf{\nabla})\mathbf{C} + \mathbf{A}(\mathbf{\nabla} \cdot \mathbf{C}) - \mathbf{C}(\mathbf{\nabla} \cdot \mathbf{A})$$
$$(A1.23)$$

Putting $\mathbf{A} = \mathbf{u}$ and $\mathbf{C} = \mathbf{E}$ in eqn (A1.23), we obtain

$$\mathbf{\nabla} \times (\mathbf{u} \times \mathbf{E}) = (\mathbf{E} \cdot \mathbf{\nabla})\mathbf{u} - (\mathbf{u} \cdot \mathbf{\nabla})\mathbf{E} + \mathbf{u}(\mathbf{\nabla} \cdot \mathbf{E}) - \mathbf{E}(\mathbf{\nabla} \cdot \mathbf{u}) \qquad (4.71)$$

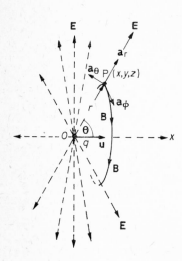

Figure 4.9. The 'point' charge q is moving with uniform velocity **u**. The magnetic field lines are closed circles. It is shown in the text that the curl of the magnetic induction **B** at the field point P is equal to $(1/c^2)$ times the time rate of change of the electric intensity **E** at the field point P, that is $\mathbf{\nabla} \times \mathbf{B} = (1/c^2)\partial\mathbf{E}/\partial t$. Both **E** and **B** are due to the same moving charge

The vector **u**, which is the velocity of the charge q, does not vary, if the position of the field point P is changed at a fixed time. Hence,

$$\frac{\partial \mathbf{u}}{\partial x} = \frac{\partial \mathbf{u}}{\partial y} = \frac{\partial \mathbf{u}}{\partial z} = 0,$$

so that

$$\mathbf{\nabla} \cdot \mathbf{u} = \frac{\partial u_x}{\partial x} + \frac{\partial u_y}{\partial y} + \frac{\partial u_z}{\partial z} = 0$$

and

$$(\mathbf{E} \cdot \mathbf{\nabla})\mathbf{u} = \left(E_x \frac{\partial}{\partial x} + E_y \frac{\partial}{\partial y} + E_z \frac{\partial}{\partial z} \right)\mathbf{u} = 0$$

For a charge moving with uniform velocity, according to eqn (4.37)

$$(\mathbf{u} \cdot \nabla)\mathbf{E} = -\frac{\partial \mathbf{E}}{\partial t}$$

Eqn (4.71) reduces to

$$\nabla \times \mathbf{B} = \frac{1}{c^2}\nabla \times (\mathbf{u} \times \mathbf{E}) = \frac{1}{c^2}\left\{ +\frac{\partial \mathbf{E}}{\partial t} + \mathbf{u}(\nabla \cdot \mathbf{E}) \right\} \qquad (4.72)$$

For a field point in empty space, according to eqn (4.19)

$$\nabla \cdot \mathbf{E} = 0$$

Substituting in eqn (4.72)

$$\nabla \times \mathbf{B} = \frac{1}{c^2}\nabla \times (\mathbf{u} \times \mathbf{E}) = \frac{1}{c^2}\frac{\partial \mathbf{E}}{\partial t} \qquad (4.73a)$$

Multiplying both sides by $\varepsilon_0 c^2$,

$$\varepsilon_0 c^2 \nabla \times \mathbf{B} = \varepsilon_0 \frac{\partial \mathbf{E}}{\partial t} \qquad (4.73b)$$

Eqn (4.73) is one of Maxwell's equations. It relates the curl of the magnetic induction **B** at a field point in empty space, due to a 'point' charge moving with *uniform* velocity to the time rate of change, at the same field point, of the electric intensity due to the same charge. Eqn (4.73) is a relation between the field quantities **E** and **B**. Both **E** and **B** arise from the same moving charge. Eqn (4.73) is valid at any field point whatever the position of the moving charge. The term $\varepsilon_0 \partial \mathbf{E}/\partial t$ will be called the Maxwell term, in honour of James Clerk Maxwell, who first introduced it. (It is generally known in the literature as the displacement 'current'.)

Some readers may prefer to develop eqn (4.73) using the more pictorial approach used in Section 4.6, when eqn (4.57) was derived. In *Figure 4.9* (as in *Figure 3.2*), from eqns (3.28) and (3.29)

$$\mathbf{B} = \mathbf{a}_\phi B_\phi = \mathbf{a}_\phi \frac{qu(1-\beta^2)\sin\theta}{4\pi\varepsilon_0 c^2 r^2(1-\beta^2\sin^2\theta)^{3/2}}$$

If at a fixed time, the field point P in *Figure 4.9* is moved in the direction of increasing **r** keeping θ and ϕ constant, the magnitude of B_ϕ varies; that is $\partial B_\phi/\partial r$ is finite. Hence, at the field point P in *Figure 4.9* $\nabla \times \mathbf{B}$ has a component in the direction of increasing θ. Similarly, if the field point P is moved in the direction of increasing θ, keeping r and ϕ fixed, the magnitude of B_ϕ varies, $\partial B_\phi/\partial\theta$ is finite, so that $\nabla \times \mathbf{B}$ has a component in the direction of increasing r at the field point P in *Figure 4.9*. According to eqn (A1.29) of Appendix 1, if $B_\theta = B_r = 0$

$$\nabla \times \mathbf{B} = \mathbf{a}_r \frac{1}{r \sin \theta} \left[\frac{\partial}{\partial \theta} (B_\phi \sin \theta) \right] - \mathbf{a}_\theta \frac{1}{r} \frac{\partial}{\partial r} (rB_\phi) \quad \text{(A1.29)}$$

Carrying out the differentiations, we find

$$\nabla \times \mathbf{B} = \frac{qu(1-\beta^2)}{4\pi\varepsilon_0 c^2} \left[\mathbf{a}_r \frac{2 \cos \theta [1 + \tfrac{1}{2}\beta^2 \sin^2 \theta]}{r^3 (1 - \beta^2 \sin^2 \theta)^{5/2}} \right.$$
$$\left. + \mathbf{a}_\theta \frac{\sin \theta}{r^3 (1 - \beta^2 \sin^2 \theta)^{3/2}} \right] \quad \text{(4.74)}$$

This is in agreement with our qualitative discussion, which showed that $\nabla \times \mathbf{B}$ has components in the directions of increasing r, and increasing θ. Now, according to eqns (3.25) and (3.26)

$$\mathbf{E} = \mathbf{a}_r E_r = \mathbf{a}_r \frac{q(1-\beta^2)}{4\pi\varepsilon_0 r^2 (1 - \beta^2 \sin^2 \theta)^{3/2}}$$

Let the field point P in *Figure 4.9* be fixed. In a time Δt the charge q moves a distance $u\Delta t$ along the x axis. The electric field then diverges radially from the new position of the charge. Hence both the magnitude and direction of the electric field at the field point P in *Figure 4.9* change with time. Hence, $\partial \mathbf{E}/\partial t$ has components in the directions of increasing r and increasing θ. Using eqn (4.37), for a charge moving with uniform velocity,

$$\frac{1}{c^2} \frac{\partial \mathbf{E}}{\partial t} = -\frac{1}{c^2} (\mathbf{u} \cdot \nabla) \mathbf{E}$$

In terms of the unit vectors $\mathbf{a}_r$, $\mathbf{a}_\theta$ and $\mathbf{a}_\phi$ appropriate to the field point P in *Figure 4.9*, the velocity of the charge can be expressed as follows,

$$\mathbf{u} = \mathbf{a}_r u \cos \theta - \mathbf{a}_\theta u \sin \theta$$

From eqn (A1.27), of Appendix 1,

$$\nabla = \mathbf{a}_r \frac{\partial}{\partial r} + \mathbf{a}_\theta \frac{1}{r} \frac{\partial}{\partial \theta} + \mathbf{a}_\phi \frac{1}{r \sin \theta} \frac{\partial}{\partial \phi}$$

Since,

$$\frac{\partial \mathbf{a}_r}{\partial r} = 0; \quad \frac{\partial \mathbf{a}_r}{\partial \theta} = \mathbf{a}_\theta; \quad \mathbf{E} = \mathbf{a}_r E_r;$$

$$\frac{1}{c^2}\frac{\partial \mathbf{E}}{\partial t} = -\frac{1}{c^2}(\mathbf{u} \cdot \nabla)\mathbf{E} = -\frac{u\cos\theta}{c^2}\frac{\partial}{\partial r}(\mathbf{a}_r E_r) + \frac{u\sin\theta}{c^2 r}\frac{\partial}{\partial \theta}(\mathbf{a}_r E_r)$$

$$= \mathbf{a}_r\left(-\frac{u\cos\theta}{c^2}\frac{\partial E_r}{\partial r} + \frac{u\sin\theta}{c^2 r}\frac{\partial E_r}{\partial \theta}\right) + \mathbf{a}_\theta\frac{u\sin\theta}{c^2 r}E_r$$

Evaluating $\partial E_r/\partial r$ and $\partial E_r/\partial \theta$, and substituting also for E_r, the reader can show that the above expression for $(1/c^2)\partial \mathbf{E}/\partial t$ reduces to the right-hand side of eqn (4.74) confirming that eqn (4.73a) is valid for the fields of a 'point' charge moving with uniform velocity in empty space.

By applying the principle of superposition, eqn (4.73b) will be extended in Section 4.9 to the case of the macroscopic electric and magnetic fields at fields points inside a macroscopic current distribution. As an introduction, the appropriate equation will be developed mathematically, using vector analysis, in Section 4.6(b). This will be followed by two illustrative examples in Sections 4.7 and 4.8, before proceeding to the full physical analysis in Section 4.9.

(b) Macroscopic Charge and Current Distribution

Consider a system of atomic charges such as electrons and ions, all moving with *uniform* velocities as shown in *Figure 4.10*. These charges build up a macroscopic charge distribution at the field point P. For any one of the point charges in the system, say q_i, its electric field $\mathbf{E}_i$ and magnetic field $\mathbf{B}_i$ at P, satisfy eqn (4.72) namely

$$c^2 \nabla \times \mathbf{B}_i = \partial \mathbf{E}_i/\partial t + \mathbf{u}_i(\nabla \cdot \mathbf{E}_i)$$

Adding, for all the N atomic charges in the *complete* system

$$c^2 \sum_{i=1}^{N} \nabla \times \mathbf{B}_i = \sum_{i=1}^{N} \frac{\partial \mathbf{E}_i}{\partial t} + \sum_{i=1}^{N} \mathbf{u}_i(\nabla \cdot \mathbf{E}_i) \qquad (4.75)$$

Now using eqn (A1.17) of Appendix 1,

$$\sum_{i=1}^{N} \nabla \times \mathbf{B}_i = \nabla \times (\mathbf{B}_1 + \mathbf{B}_2 + \dots \mathbf{B}_N) = \nabla \times \mathbf{b} \qquad (4.76)$$

where **b** is the resultant microscopic magnetic induction at P due to *all* the charges in the complete system. Similarly, if **e** is the resultant microscopic electric intensity at P due to *all* the charges in the system

$$\sum_{i=1}^{N} \frac{\partial \mathbf{E}_i}{\partial t} = \frac{\partial}{\partial t}(\mathbf{E}_1 + \mathbf{E}_2 \ldots + \mathbf{E}_N) = \frac{\partial \mathbf{e}}{\partial t} \qquad (4.77)$$

The simplest case would be if *all* the charges in the complete system

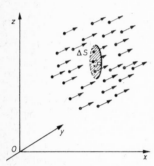

Figure 4.10. The system of 'point' charges moving with uniform velocities build up a macroscopic charge and current distribution. A practical example of a macroscopic current would be a conduction current in a metal. There is of the order of one free electron per atom in a metal, so that in a metal the average distance between neighbouring moving free electrons would be of the order of atomic dimensions ($\sim 10^{-10}$ metres). There would be large fluctuations in the separation of charges on the atomic scale giving large fluctuations in the microscopic magnetic induction **b**. If the surface ΔS used to calculate the curl of the macroscopic magnetic induction **B** at P is large on the atomic scale a finite number of 'point' charges cross ΔS per unit time, far more than can be shown in Figure 4.10. The fluctuations in the microscopic fields would also be more violent than Figure 4.10 would suggest

moved with the same uniform velocity **u**. (This is true for the problem discussed in Section 4.8.) In that case

$$\sum_{i=1}^{N} \mathbf{u}_i(\mathbf{\nabla} \cdot \mathbf{E}_i) = \mathbf{u} \sum_{i=1}^{N} \mathbf{\nabla} \cdot \mathbf{E}_i = \mathbf{u}\mathbf{\nabla} \cdot \left(\sum_{i=1}^{N} \mathbf{E}_i\right) = \mathbf{u}(\mathbf{\nabla} \cdot \mathbf{e})$$

$$(4.78)$$

where **e** is the resultant microscopic electric intensity at P due to all the charges in the system. If we consider a field point between the moving charges in *Figure 4.10*, then from eqn (4.19) the divergence of the microscopic electric field is zero. Substituting from eqns (4.76), (4.77) and (4.78) into eqn (4.75), we find for the microscopic fields,

$$\mathbf{\nabla} \times \mathbf{b} = \frac{1}{c^2}\frac{\partial \mathbf{e}}{\partial t}$$

However, if we are interested in the macroscopic or local space average electric and magnetic fields **E** and **B**, we cannot put $\mathbf{\nabla} \cdot \mathbf{E}$

76

equal to zero in eqn (4.78), but should try putting

$$\mathbf{V} \cdot \mathbf{E} = \rho/\varepsilon_0$$

where ρ is the macroscopic charge density at P. If the atomic charges in the vicinity of P move with the same uniform velocity $\mathbf{u}$, they give rise to a current density $\mathbf{J} = \rho\mathbf{u}$. Hence for macroscopic fields, we expect eqn (4.78) to become

$$\sum_{i=1}^{N} \mathbf{u}_i(\mathbf{V} \cdot \mathbf{E}_i) = \mathbf{u}(\mathbf{V} \cdot \mathbf{E}) = \mathbf{u}\rho/\varepsilon_0 = \mathbf{J}/\varepsilon_0 \qquad (4.78a)$$

This suggests that for field points inside a macroscopic current distribution, eqn (4.75) should become

$$c^2 \nabla \times \mathbf{B} = \frac{\partial E}{\partial t} + \frac{\mathbf{J}}{\varepsilon_0}$$

or

$$\varepsilon_0 c^2 \nabla \times \mathbf{B} = \varepsilon_0 \frac{\partial E}{\partial t} + \mathbf{J} \qquad (4.79)$$

where $\mathbf{E}$, $\mathbf{B}$ and $\mathbf{J}$ are the macroscopic electric intensity, the macroscopic magnetic induction and the macroscopic current density respectively. The term $\varepsilon_0 \partial E/\partial t$ in eqn (4.79) will be called the Maxwell term. (It is generally known in the literature as the displacement 'current'.) A more rigorous development of eqn (4.79) including the full physical interpretation of the terms and symbols in it, in particular the roles of the Maxwell term and the $\mathbf{J}$ term, will be developed in Section 4.9, after considering two illustrative examples in Sections 4.7 and 4.8.

Integrating eqn (4.79) over a finite surface S,

$$\varepsilon_0 c^2 \int_S (\nabla \times \mathbf{B}) \cdot \mathbf{n} dS = \varepsilon_0 \int_S \frac{\partial E}{\partial t} \cdot \mathbf{n} dS + \int_S \mathbf{J} \cdot \mathbf{n} dS$$

Using Stokes' theorem, eqn (A1.25), this can be rewritten as

$$\varepsilon_0 c^2 \oint \mathbf{B} \cdot d\mathbf{l} = \varepsilon_0 \frac{\partial}{\partial t} \int_S \mathbf{E} \cdot \mathbf{n} dS + \int_S \mathbf{J} \cdot \mathbf{n} dS \qquad (4.80)$$

According to eqn (4.80), $\varepsilon_0 c^2$ times (or $1/\mu_0$ times) the line integral of $\mathbf{B}$ around the boundary of any surface S is equal to ε_0 times the rate of change of the flux of $\mathbf{E}$ through the surface plus the total electric current flowing across the surface.

77

4.7. THE MAGNETIC FIELD OF A 'POINT' CHARGE MOVING WITH A UNIFORM VELOCITY VERY MUCH LESS THAN THE VELOCITY OF LIGHT

Consider a 'point' charge q moving with uniform velocity $\mathbf{u}$, which is $\ll c$, along the x axis as shown in *Figure 4.11*. According to eqn (3.27), if $u \ll c, \beta \ll 1$, the magnetic induction at a field point P

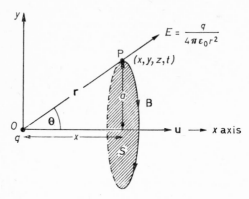

Figure 4.11. The magnetic induction at the field point P due to a 'point' charge moving with uniform velocity u≪c is calculated using Maxwell's equations

having co-ordinates x, y, z at the time t, when q is at the origin, is given to a good approximation by,

$$\mathbf{B} = \frac{q\mathbf{u} \times \mathbf{r}}{4\pi\varepsilon_0 c^2 r^3}$$

or

$$B = \frac{qu \sin \theta}{4\pi\varepsilon_0 c^2 r^2} \tag{4.81}$$

where $\mathbf{r}$ is the distance from q to the field point P, and θ is the angle between $\mathbf{r}$ and $\mathbf{u}$. Eqn (4.81) for the magnetic field of a moving 'point' charge with velocity $\ll c$ is generally known as the Biot–Savart law.

According to eqn (3.24), under the same conditions, $u \ll c$, the electric field intensity at P is given to a good approximation by

$$\mathbf{E} = \frac{q\mathbf{r}}{4\pi\varepsilon_0 r^3} \tag{4.82}$$

Maxwell's equations will now be used to calculate the electric and magnetic fields at P. If $u \ll c$, then $\mathbf{B}$ is very small and $\partial\mathbf{B}/\partial t$ is small, so that the equation $\nabla \times \mathbf{E} = -\partial\mathbf{B}/\partial t$ reduces to $\nabla \times \mathbf{E} \approx 0$. Integrating the equation

$$\nabla \cdot \mathbf{E} = \frac{\rho}{\varepsilon_0} \tag{4.23}$$

for a sphere of radius r, centre at O the position of the charge, we have

$$\int \nabla \cdot \mathbf{E} \, dV = \int \frac{\rho}{\varepsilon_0} \, dV$$

Using Gauss' theorem, eqn (A1.26)

$$\int \mathbf{E} \cdot \mathbf{n} \, dS = \int \frac{\rho}{\varepsilon_0} \, dV = \frac{q}{\varepsilon_0} \tag{4.83}$$

Since $\nabla \times \mathbf{E} \approx 0$, the magnitude of $\mathbf{E}$ does not vary with θ, as was the case in *Figure 3.3(b)*. Hence,

$$4\pi r^2 E_r = q/\varepsilon_0$$

or

$$E_r = \frac{q}{4\pi\varepsilon_0 r^2} \tag{4.84}$$

In this calculation, Maxwell's equations have been used to calculate the electric field of a charge moving with a uniform velocity $u \ll c$. The result is in agreement with eqn (4.82). Now, whatever the position of the charge in *Figure 4.11*, according to Maxwell's equations the magnetic field at the field point P in empty space is related to the electric field, at the same field point, by the equation

$$\nabla \times \mathbf{B} = \frac{1}{c^2} \frac{\partial\mathbf{E}}{\partial t} \tag{4.73}$$

According to eqn (4.65), $\nabla \cdot \mathbf{B}$ is zero. Hence, the lines of $\mathbf{B}$ are continuous. By symmetry considerations, the magnetic field must be symmetrical with respect to the x axis, the direction of $\mathbf{u}$, so that the lines of $\mathbf{B}$ are expected to form closed circles whose centres are on the x axis, as shown in *Figure 4.11*. Consider a circular disk shaped surface S of radius a, with centre on the x axis and with P on its circumference, as shown shaded in *Figure 4.11*. Integrating eqn (4.73) over the surface of the circular disk surface S,

$$\int (\nabla \times \mathbf{B}) \cdot \mathbf{n} \, dS = \frac{1}{c^2} \int \frac{\partial\mathbf{E}}{\partial t} \cdot \mathbf{n} \, dS \tag{4.85}$$

D*

Applying Stokes' theorem, eqn (A1.25),

$$c^2 \oint \mathbf{B} \cdot d\mathbf{l} = \frac{\partial}{\partial t} \int \mathbf{E} \cdot \mathbf{n} dS = \frac{\partial \Psi}{\partial t} \tag{4.86}$$

where

$$\Psi = \int \mathbf{E} \cdot \mathbf{n} dS$$

is the flux of $\mathbf{E}$ through the surface S. The line integral of $\mathbf{B}$ is evaluated in the clockwise direction, as seen from the left-hand side of the surface S shown in *Figure 4.11*. According to the definition of the operator $\nabla \times$ introduced in Appendix 1, the direction of $\mathbf{n}$ to be used in eqn (4.85) for each element of surface is given by the right-handed corkscrew rule. Hence in *Figure 4.11* the unit vector $\mathbf{n}$ leaves the surface S from the same side as the lines of $\mathbf{E}$ leave the surface. Hence, for the case shown in *Figure 4.11*, Ψ, the flux of $\mathbf{E}$ through the surface S is positive.

If $\mathbf{E}$ is given by eqn (4.84), the flux of $\mathbf{E}$ through the circular disk surface S in *Figure 4.11* due to the charge $+q$ is equal to $+q\Omega/4\pi\varepsilon_0$, where Ω is the solid angle subtended by the disk at O, the position of the charge. Now

$$\Omega = 2\pi(1 - \cos\theta)$$

(Reference: Pugh and Pugh[5].) Hence,

$$\Psi = \frac{q}{2\varepsilon_0}(1 - \cos\theta)$$

In *Figure 4.11*

$$\cos\theta = \frac{x}{\sqrt{a^2 + x^2}}$$

Hence,

$$\Psi = \frac{q}{2\varepsilon_0}\left(1 - \frac{x}{\sqrt{a^2 + x^2}}\right) \tag{4.87}$$

According to eqn (4.36), for a charge moving with *uniform* velocity along the x axis

$$\frac{\partial}{\partial t} = -u\frac{\partial}{\partial x}$$

Hence,

$$\frac{\partial \Psi}{\partial t} = -u \frac{\partial \Psi}{\partial x} = -u \frac{\partial}{\partial x}\left[\frac{q}{2\varepsilon_0}\left(1 - \frac{x}{\sqrt{a^2 + x^2}}\right)\right]$$

$$= \frac{qua^2}{2\varepsilon_0(a^2 + x^2)^{3/2}} = \frac{qua^2}{2\varepsilon_0 r^3}$$

Substituting in eqn (4.86)

$$\oint \mathbf{B} \cdot d\mathbf{l} = \frac{1}{c^2}\frac{\partial \Psi}{\partial t} = \frac{qua^2}{2\varepsilon_0 c^2 r^3}$$

By symmetry $\mathbf{B}$ has the same value along the circumference of the circular disk surface S. Since the circumference of the disk is $2\pi a$,

$$2\pi a B = \frac{qua^2}{2\varepsilon_0 c^2 r^3}$$

$$B = \frac{qua}{4\pi\varepsilon_0 c^2 r^3} = \frac{qu \sin\theta}{4\pi\varepsilon_0 c^2 r^2} \qquad (4.88)$$

This is in agreement with eqn (4.81). Thus one can either calculate $\mathbf{B}$ using the Biot–Savart law, if the position and velocity of the charge are given, or $\mathbf{B}$ can be calculated using Maxwell's equations. In the latter method, the electric field is first calculated at the field point P in *Figure 4.11*. Then, using eqn (4.73), $\nabla \times \mathbf{B}$ at the field point P can be calculated from the known value of $\partial \mathbf{E}/\partial t$ at the field point P. In the present example, $\mathbf{B}$ could then be calculated from the value of $\nabla \times \mathbf{B}$. The Maxwell term $\varepsilon_0 \partial \mathbf{E}/\partial t$ does not produce the magnetic field at P. In *Figure 4.11* both the electric and magnetic fields at P arise from the moving charge. These fields are propagated from the charge with a speed c in empty space. The electric and magnetic fields at P have a common cause, namely the moving charge. According to the equation $\varepsilon_0 c^2 \nabla \times \mathbf{B} = \varepsilon_0 \partial \mathbf{E}/\partial t$, if the charge q in *Figure 4.11* gives rise at P to an electric field which varies with time, then the same charge also gives rise to a magnetic field at P, the curl of the magnetic field being equal to $(1/c^2)\partial \mathbf{E}/\partial t$.

As the 'point' charge q approaches the circular disk shaped surface S from the left, as shown in *Figure 4.11*, the electric flux through the surface increases until, just before the charge reaches the surface S, the flux of $\mathbf{E}$ through the surface S approaches the value $q/2\varepsilon_0$, as illustrated in *Figure 4.12(a)*. After the 'point' charge passes through the surface S, the lines of E go through the surface in the opposite direction to previously as shown in *Figure 4.12(b)*. Just after q has passed through the surface, the flux of E through the surface S is $-q/2\varepsilon_0$. The variation of Ψ, the flux of E through the circular disk surface S, is shown for various values of x_0 in *Figure 4.13*, where x_0 is the distance of the charge from the centre of the disk shaped surface S shown in *Figure 4.11*, x_0 being negative before the charge crosses the surface S. It can be seen that there is a discontinuity of

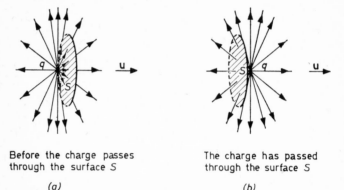

Before the charge passes
through the surface S

(a)

The charge has passed
through the surface S

(b)

Figure 4.12. (a) Just before the 'point' charge q crosses the circular disk shaped surface S, the flux of E through the surface S is equal to $+q/2\varepsilon_0$. The lines of E cross the surface S from left to right. (b) After the charge has passed through the surface S, the lines of E pass through the surface from right to left. There is a discontinuity in the flux of E, when the 'point' charge passes through the surface S, which according to eqn (4.17) is equal to $-q/\varepsilon_0$, whatever the speed of the charge. There is however no discontinuity in the magnitudes and directions of B and $\nabla \times B$ at a field point on the circumference of the surface S when the charge q crosses the surface S

$-q/\varepsilon_0$ in Ψ the electric flux through the surface S, when the 'point' charge crosses the surface S, and $\partial\Psi/\partial t$ is infinite when the point charge crosses the surface S in *Figure 4.11*.

It follows from the Biot–Savart law, eqn (4.81), that the magnitude and direction of the magnetic field at the field point P in *Figure 4.11* has almost the same value just before the charge q reaches the surface S, whilst the charge q coincides with the surface S and just after the charge q passes through the surface S. Hence the magnitude and direction of B (and hence the magnitude of $\oint B.dl$ evaluated around the circumference of the circular disk shaped surface S) should not change discontinuously when the point charge q crosses the surface S in *Figures 4.11* and *4.12*. Thus, when eqn (4.73) is applied the discontinuity of $-q/\varepsilon_0$ in Ψ, when the 'point' charge q crosses the surface S, cannot contribute to $c^2\oint \mathbf{B}.\mathbf{dl}$, even though this discontinuity contributes to $\partial\Psi/\partial t$. In a macroscopic theory, allowance must be made for the non-contribution to $\oint \mathbf{B}.\mathbf{dl}$ of the discontinuities in

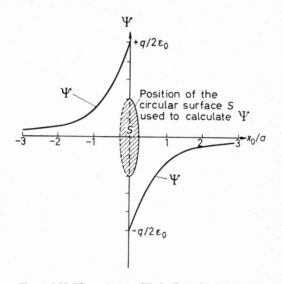

Figure 4.13. The variation of Ψ the flux of $\mathbf{E}$ through the circular disk surface S in Figure 4.11 with the position of the moving 'point' charge, x_0 is the distance of the charge from the surface S. There is a discontinuity of $-q/\varepsilon_0$ in Ψ when the 'point' charge crosses the surface (For a charge of finite size the variation of Ψ with x_0 is similar to that in Figure 4.15)

$\partial\Psi/\partial t$, when 'point' charges, such as conduction electrons, cross the surface around which $\oint \mathbf{B}.\mathbf{dl}$ is evaluated. In order to illustrate how the correction is made in a macroscopic theory, the above calculation will be repeated in Section 4.8 for the case of a charge of finite length.

Let the point charge q, which is moving with a uniform velocity $\ll c$ in *Figure 4.11*, pass outside the circular disk shaped surface S. At any instant, the electric flux Ψ through the surface S is equal to $q\Omega/4\pi\varepsilon_0$ where Ω is the solid angle subtended by the surface S at the position of the moving charge. As the charge approaches the surface

S from the left, Ω increases as the charge approaches the plane of the surface S, but if the charge does not pass through the surface S, Ω begins to decrease when the charge gets close to the plane of the surface S due to the decrease in the projection or normal area of the surface S as viewed from the charge, and Ω is zero when the charge is in the plane of the surface S. Then Ω begins to increase again as the charge moves further away from S, before decreasing again at large distances. This variation in Ω is continuous, so that, if the charge q in *Figure 4.11* does not pass through the surface S, there is no discontinuity of $- q/\varepsilon_0$ in Ψ, the flux of E through the surface S. It is only when the point charge q actually crosses the surface S that we have a discontinuity in Ψ, which is $-q/\varepsilon_0$, whatever the speed of the charge [cf. eqn (4.17)].

4.8. MAGNETIC FIELD OF A MOVING CHARGE OF FINITE LENGTH

Consider a thin line of charge of length l metres, having a charge λ coulombs/metre. Let the charge move with uniform velocity $\mathbf{u}$ (which is $\ll c$) in a direction parallel to its length along the x axis as shown in *Figure 4.14*. In this example, the field point P will be assumed to be on the y axis at a distance a from the origin as shown in *Figure 4.14*. Draw a circular disk shaped surface S of radius a, centre 0, the origin, in the plane $x = 0$ as shown in *Figure 4.14*. Let the middle point of the line of charge have co-ordinates $(x_0, 0, 0)$. Three positions of the charge will be considered (i) before the charge has reached the circular disk shaped surface S as shown in *Figure 4.14(a)* (that is $x_0 < -l/2$); (ii) whilst the charge is passing through the surface S (that is $-l/2 < x_0 < l/2$) as shown in *Figure 4.14(b)*; (iii) after the charge has passed completely through the surface S as shown in *Figure 4.14(c)*, (that is $x_0 > l/2$). In each case the magnetic field at P will be calculated using both the Biot–Savart law and Maxwell's equations.

Firstly, consider the case shown in *Figure 4.14(c)*. Consider the element of charge between x and $x + \Delta x$. According to eqn (4.81) for the Biot–Savart law, the contribution of this element of charge to the magnetic induction at P is

$$dB = \frac{qu \sin (\pi - \theta)}{4\pi\varepsilon_0 c^2 r^2} = \frac{\lambda u dx \cos \phi}{4\pi\varepsilon_0 c^2 r^2} \tag{4.89}$$

where r is the distance from the element of charge to the field point P, and θ and ϕ are as illustrated in *Figure 4.14(c)*. The direction of $d\mathbf{B}$ is given by the right-handed corkscrew rule, and it is upwards

84

from the paper in the $+z$ direction in *Figure 4.14(c)*. Now in *Figure 4.14(c)*

$$r = a \sec \phi$$

and

$$x = a \tan \phi$$

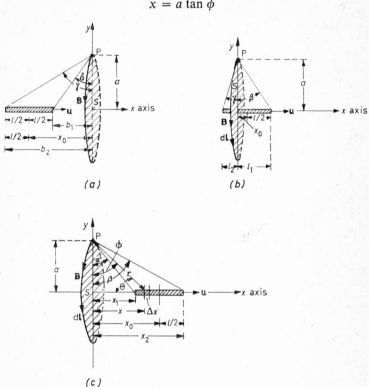

(a)

(b)

(c)

*Figure 4.14. The calculation of the magnetic field of a line of charge, moving with uniform velocity, using both the Biot–Savart law and Maxwell's equations. The flux of **E** through the circular disk shaped surface S having the field point P on its circumference is calculated for three positions of the charge: (a) before the charge reaches the surface, (b) whilst the charge is actually crossing the surface and (c) after the charge has passed completely through the surface*

Differentiating,

$$dx = a \sec^2 \phi d\phi$$

Substituting in eqn (4.89)

$$dB = \frac{\lambda u \cos \phi d\phi}{4\pi\varepsilon_0 c^2 a}$$

85

Integrating between the limits $\phi = \alpha$ and $\phi = \beta$,

$$B = \frac{\lambda u}{4\pi\varepsilon_0 c^2 a} \left[\sin \phi\right]_{\alpha}^{\beta} \qquad (4.90)$$

that is

$$B = \frac{\lambda u}{4\pi\varepsilon_0 c^2 a} \left[\sin \beta - \sin \alpha\right] \qquad (4.91)$$

The direction of **B** is upwards from the paper in the $+z$ direction in *Figure 4.14(c)*. Integrating around the circumference of the circular surface S in the clockwise direction as seen from the left-hand side of the disk in *Figure 4.14(c)*, that is in the direction of **B**, we have

$$\oint \mathbf{B} \cdot \mathbf{dl} = 2\pi a B = \frac{\lambda u}{2\varepsilon_0 c^2} \left[\sin \beta - \sin \alpha\right] \qquad (4.92)$$

The magnetic induction at P in *Figure 4.14(c)* will now be calculated using Maxwell's equations. From eqn (4.73) for a point in empty space

$$\mathbf{\nabla} \times (\mathbf{dB}) = \frac{1}{c^2} \frac{\partial}{\partial t} (\mathbf{dE})$$

Where d**B** and d**E** are the electric and magnetic fields due to the 'point' charge λdx between x and $x+dx$. Using the principle of superposition, summing the contributions of all the elements of charge in the line of charge, we obtain

$$\mathbf{\nabla} \times \mathbf{B} = \frac{1}{c^2} \frac{\partial \mathbf{E}}{\partial t} \qquad (4.93)$$

where **E** and **B** are the resultant electric and magnetic fields at P respectively due to the line of charge. Integrating eqn (4.93) over the surface of the circular disk surface S shown in *Figure 4.14(c)*,

$$\int (\mathbf{\nabla} \times \mathbf{B}) \cdot \mathbf{n} dS = \frac{1}{c^2} \int \frac{\partial \mathbf{E}}{\partial t} \cdot \mathbf{n} dS$$

Using Stokes' theorem

$$\oint \mathbf{B} \cdot \mathbf{dl} = \frac{1}{c^2} \frac{\partial}{\partial t} \int \mathbf{E} \cdot \mathbf{n} dS = \frac{1}{c^2} \frac{\partial \Psi}{\partial t} \qquad (4.94)$$

where Ψ is the flux of **E** through the circular disk S. The direction

MAGNETIC FIELD OF A MOVING CHARGE OF FINITE LENGTH

of dl in eqn (4.94) is taken to be the same as when developing eqn (4.92) that is in the clockwise direction as seen from the left-hand side of the disk surface S in *Figure 4.14(c)*. Under these conditions, for every element of surface, the unit vector $\mathbf{n}$ is in the $+x$ direction. For the case shown in *Figure 4.14(c)*, the lines of $\mathbf{E}$ go through the surface S in the opposite direction to $\mathbf{n}$, so that in this case, Ψ the flux of $\mathbf{E}$ through the disk, is negative. The magnitude of Ψ will now be calculated. According to eqn (4.87), the flux of $\mathbf{E}$ due to the element of charge between x and $x+dx$ is

$$d\Psi = -\frac{\lambda dx}{2\varepsilon_0}(1-\cos\theta) = -\frac{\lambda dx}{2\varepsilon_0}\left(1-\frac{x}{\sqrt{a^2+x^2}}\right)$$

Integrating to obtain the total flux of $\mathbf{E}$ through the disk, we have

$$\Psi = -\frac{\lambda}{2\varepsilon_0}\left[x-\sqrt{a^2+x^2}\right]_{x_1}^{x_2}$$

$$= -\frac{\lambda}{2\varepsilon_0}\left[x_2-\sqrt{a^2+x_2^2}-x_1+\sqrt{a^2+x_1^2}\right]$$

$$= -\frac{\lambda}{2\varepsilon_0}\left[l-\sqrt{a^2+x_2^2}+\sqrt{a^2+x_1^2}\right] \tag{4.95}$$

where x_1 and x_2 are illustrated in *Figure 4.14(c)*, and $l = x_2-x_1$. If x_0 is the co-ordinate of the middle point of the charge, $x_2 = x_0+l/2$ and $x_1 = x_0-l/2$. Thus eqn (4.95) can be rewritten

$$\Psi = -\frac{\lambda}{2\varepsilon_0}\left[l-\sqrt{a^2+(x_0+l/2)^2}+\sqrt{a^2+(x_0-l/2)^2}\right] \tag{4.96}$$

The variation of Ψ with x_0 for the region $x_0>l/2$ is represented by the section TU of the curve in *Figure 4.15*. In a time Δt there is the same change in the flux of $\mathbf{E}$ through the fixed surface S in *Figure 4.14* as, if at a fixed time, the position of the middle point of the charge changed from x_0 to $x_0+\Delta x_0$ keeping the field point fixed, where $\Delta x_0 = u\Delta t$. Hence

$$\frac{\partial\Psi}{\partial t} = u\frac{\partial\Psi}{\partial x_0} \tag{4.97}$$

For both $\partial/\partial t$ and $\partial/\partial x_0$ the field point remains fixed, Compared with eqn (4.36), there is no negative sign in eqn (4.97). In the present case dx_0 refers to a displacement of the charge, whereas, in eqn (4.36), dx refers to a change in the position of the field point. Moving the charge one way is equivalent to moving the field point in the opposite direction. Using eqns (4.96) and (4.97)

$$\frac{1}{c^2}\frac{\partial \Psi}{\partial t} = -\frac{\lambda u}{2\varepsilon_0 c^2}\frac{\partial}{\partial x_0}\left[l - \sqrt{a^2+(x_0+l/2)^2} + \sqrt{a^2+(x_0-l/2)^2}\right]$$

$$= \frac{\lambda u}{2\varepsilon_0 c^2}\left[\frac{(x_0+l/2)}{\sqrt{a^2+(x_0+l/2)^2}} - \frac{(x_0-l/2)}{\sqrt{a^2+(x_0-l/2)^2}}\right]$$

$$= \frac{\lambda u}{2\varepsilon_0 c^2}[\sin \beta - \sin \alpha]$$

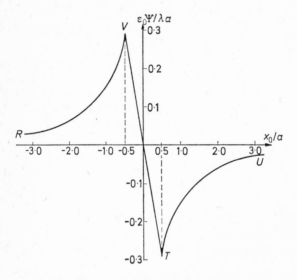

Figure 4.15. *The variation of* Ψ, *the flux of* **E** *through the circular disk shaped surface S in Figure 4.14 for various positions of the moving line of charge. (Actually* $\varepsilon_0 \Psi/\lambda a$ *is plotted against* x_0/a, *where* $\lambda = $ *charge/unit length, and a, the distance of the field point from the x axis, is equal to l the length of the charge)*

Using eqn (4.94)

$$\oint \mathbf{B} \cdot d\mathbf{l} = \frac{1}{c^2}\frac{\partial \Psi}{\partial t} = \frac{\lambda u}{2\varepsilon_0 c^2}(\sin \beta - \sin \alpha) \qquad (4.98)$$

or

$$B = \frac{\lambda u}{4\pi\varepsilon_0 c^2 a}(\sin \beta - \sin \alpha) \qquad (4.99)$$

Eqns (4.92) and (4.98) are the same; so are eqns (4.91) and (4.99). The direction of **B** is the same in both cases. This shows that for the case shown in *Figure 4.14(c)*, application of the Biot–Savart law to the moving charge, or applying Maxwell's equations calculating **B** from $\partial \mathbf{E}/\partial t$ give the same result.

Now consider the case shown in *Figure 4.14(a)* in which the moving charge distribution has not reached the circular disk surface S. The electric flux through the surface S in this case can be obtained from eqn (4.95) by changing x_1 into b_1 and x_2 into b_2 where b_1 and b_2 are as illustrated in *Figure 4.14(a)*. The sign of the flux must be changed (since for the case illustrated in *Figure 4.14(a)* the flux of **E** goes through the surface S in the opposite direction to the case shown in *Figure 4.14(c)*). For the case shown in *Figure 4.14(a)*, from eqn (4.95)

$$\Psi = +\frac{\lambda}{2\varepsilon_0}\left[l - \sqrt{a^2 + b_2^2} + \sqrt{a^2 + b_1^2}\right] \qquad (4.100)$$

In eqn (4.100) b_1 and b_2 are the magnitudes of the distances shown. Now in *Figure 4.14(a)*

$$b_1 = |x_0| - l/2 = -x_0 - l/2$$
$$b_2 = |x_0| + l/2 = -x_0 + l/2$$

where x_0, the co-ordinate of the mid point of the line of charge, is negative. Substituting in eqn (4.100)

$$\Psi = \frac{\lambda}{2\varepsilon_0}\left(l - \sqrt{a^2 + (x_0 - l/2)^2} + \sqrt{a^2 + (x_0 + l/2)^2}\right) \quad (4.101)$$

The variation of Ψ with x_0 for the region $x_0 < -l/2$ is represented by the portion RV of the curve in *Figure 4.15*. Using eqn (4.97)

$$\frac{1}{c^2}\frac{\partial \Psi}{\partial t} = \frac{u}{c^2}\frac{\partial \Psi}{\partial x_0}$$

$$= \frac{\lambda u}{2\varepsilon_0 c^2}\left[-\frac{(x_0 - l/2)}{\sqrt{a^2 + (x_0 - l/2)^2}} + \frac{(x_0 + l/2)}{\sqrt{a^2 + (x_0 + l/2)^2}}\right]$$

$$= \frac{\lambda u}{2\varepsilon_0 c^2}\left[\frac{(|x_0| + l/2)}{\sqrt{a^2 + (|x_0| + l/2)^2}} - \frac{(|x_0| - l/2)}{\sqrt{a^2 + (|x_0| - l/2)^2}}\right]$$

$$= \frac{\lambda u}{2\varepsilon_0 c^2}(\sin \gamma - \sin \delta)$$

89

Using eqn (4.94)

$$B = \frac{\lambda u}{4\pi\varepsilon_0 c^2 a}(\sin \gamma - \sin \delta) \qquad (4.102)$$

The magnetic induction can also be obtained from eqn (4.90) (which was obtained using the Biot–Savart law), by replacing α by $-\gamma$ and β by $-\delta$ giving

$$B = \frac{\lambda u}{4\pi\varepsilon_0 c^2 a}(\sin \gamma - \sin \delta)$$

in agreement with eqn (4.102). This shows that the use of the Biot–Savart law and Maxwell's equations give the same value for **B** in this case also.

Now consider the intermediate position shown in *Figure 4.14(b)*, when the charge is actually passing through the surface S. The calculation of **B** using the Biot–Savart law is straightforward. Putting $\alpha = -\gamma$ and $\beta = \beta$ in eqn (4.90), we obtain,

$$B = \frac{\lambda u}{4\pi\varepsilon_0 c^2 a}(\sin \beta + \sin \gamma)$$

and

$$\oint \mathbf{B}.\,d\mathbf{l} = \frac{\lambda u}{2\varepsilon_0 c^2}(\sin \beta + \sin \gamma) \qquad (4.103)$$

The use of Maxwell's equations should give the same value for **B** and for $\oint \mathbf{B} \cdot d\mathbf{l}$, evaluated around the circular disk.

The flux of **E** through the surface for the case illustrated in *Figure 4.14(b)* can be obtained from eqns (4.95) and (4.100) by splitting the charge into two portions, one of length l_1 on the right-hand side of the circular disk surface S and one of length l_2 on the left-hand side, as shown in *Figure 4.14(b)*. Putting $l = l_1$, $x_1 = 0$ and $x_2 = l_1$ in eqn (4.95), we have

$$\Psi_1 = -\frac{\lambda}{2\varepsilon_0}\left[l_1 - \sqrt{a^2 + l_1^2} + a\right]$$

Putting $l = l_2$, $b_1 = 0$ and $b_2 = l_2$ in eqn (4.100),

$$\Psi_2 = +\frac{\lambda}{2\varepsilon_0}\left[l_2 - \sqrt{a^2 + l_2^2} + a\right]$$

Adding Ψ_1 and Ψ_2 gives

$$\Psi = \Psi_1 + \Psi_2 = -\frac{\lambda}{2\varepsilon_0}\left[l_1 - \sqrt{a^2 + l_1^2} - l_2 + \sqrt{a^2 + l_2^2}\right]$$

Now $l_1 = x_0 + l/2$ and $l_2 = l/2 - x_0$, hence

$$\Psi = -\frac{\lambda}{2\varepsilon_0}\left[2x_0 - \sqrt{a^2 + (x_0 + l/2)^2} + \sqrt{a^2 + (l/2 - x_0)^2}\right]$$

(4.104)

The variation of Ψ with x_0 in the region $-l/2 < x_0 < l/2$ is represented by the section VT of the curve in *Figure 4.15*. It can be seen that Ψ decreases with increasing x_0 in this region and $\partial\Psi/\partial x_0$ is negative.

Using eqns (4.97) and (4.104)

$$\frac{1}{c^2}\frac{\partial\Psi}{\partial t} = \frac{u}{c^2}\frac{\partial\Psi}{\partial x_0}$$

$$= -\frac{\lambda u}{2\varepsilon_0 c^2}\left[2 - \frac{x_0 + l/2}{\sqrt{a^2 + (x_0 + l/2)^2}} - \frac{(l/2 - x_0)}{\sqrt{a^2 + (l/2 - x_0)^2}}\right]$$

$$= \frac{\lambda u}{2\varepsilon_0 c^2}\left[\sin\beta + \sin\gamma - 2\right]$$

Using eqn (4.94), substituting for $\partial\Psi/\partial t$,

$$c^2\oint\mathbf{B}\,.\,\mathbf{dl} = \frac{\partial\Psi}{\partial t}$$

(4.94)

$$c^2\oint\mathbf{B}\,.\,\mathbf{dl} = \frac{\lambda u}{2\varepsilon_0}\left[\sin\beta + \sin\gamma\right] - \frac{\lambda u}{\varepsilon_0}$$

(4.105)

giving

$$B = \frac{\lambda u}{4\pi\varepsilon_0 c^2 a}(\sin\beta + \sin\gamma) - \frac{\lambda u}{2\pi\varepsilon_0 c^2 a}$$

Thus in this case, the use of the Biot–Savart law and the equation $c^2\oint\mathbf{B}\,.\,\mathbf{dl} = \partial\Psi/\partial t$ do not give the same answer. In order to make eqn (4.105) consistent with eqn (4.103), one would have to add $\lambda u/\varepsilon_0$ to the right-hand side of eqn (4.105). This illustrates how eqn (4.94) must sometimes be extended, when one has a macroscopic current distribution.

Summarizing, Ψ, the flux of the electric intensity $\mathbf{E}$ through the circular disk shaped surface S in *Figure 4.14*, ($\Psi = \int\mathbf{E}\,.\,\mathbf{n}dS$), is shown for various values of x_0 in *Figure 4.15* for the special case when $l = a$. The values of Ψ are calculated in the limit $u \ll c$ using Coulomb's law. As the line of charge approaches the surface S from

91

the left in *Figure 4.14(a)*, Ψ, the flux of **E** through the surface, increases until $x_0 = -l/2$. In this region $\partial\Psi/\partial x_0$ is positive and the use of

$$c^2 \mathbf{V} \times \mathbf{B} = \frac{\partial \mathbf{E}}{\partial t} \tag{4.93}$$

in the integral form

$$c^2 \oint \mathbf{B} \cdot \mathbf{dl} = \frac{\partial\Psi}{\partial t} = \frac{\partial}{\partial t}\int \mathbf{E} \cdot \mathbf{n}dS \tag{4.94}$$

gives a value for the magnetic induction **B** in agreement with the Biot–Savart law. After the charge has passed completely through the surface ($x_0 > l/2$), the flux of **E** through the surface S is negative. In this region ($x_0 > l/2$), Ψ is getting less negative with increasing x_0, so that $\partial\Psi/\partial x_0$ is positive and the value of **B** obtained using eqn (4.94) and the Biot–Savart law are again the same.

When the charge is actually passing through the surface, that is $-l/2 < x_0 < l/2$ in *Figure 4.14(b)*, then Ψ is decreasing with increasing x_0 and $\partial\Psi/\partial x_0$ is negative. In this region the use of eqns (4.93) and (4.94) as they stand, do not give values of $\oint \mathbf{B} \cdot \mathbf{dl}$ consistent with the Biot–Savart law. In this region, the decrease in Ψ with increasing x_0 is due to the fact that, when an element of charge crosses the circular disk shaped surface S in *Figure 4.14(b)*. the direction in which the electric field, due to that element of charge, passes through the surface S changes direction leading to a small 'discontinuity' in the flux of **E** through the surface S of the type illustrated in *Figure 4.13*. These successive small 'discontinuities' on the atomic scale due to individual electrons' or ions' crossing the surface S build up the continuous decrease in Ψ in the region $-l/2 < x_0 < +l/2$, as shown in *Figure 4.15*. It was illustrated in Section 4.7 that, when eqn (4.94) is applied, the 'discontinuity' of $-q/\varepsilon_0$ in the flux of **E** through a surface, when a point charge q crosses that surface, does not contribute to $c^2 \oint \mathbf{B} \cdot \mathbf{dl}$, evaluated around the boundary of that surface. Thus the decrease in Ψ, the flux of **E** through the circular disk shaped surface S in *Figure 4.14(b)*, due to the effects of successive elements of charge's crossing the surface S, does not contribute to $c^2 \oint \mathbf{B} \cdot \mathbf{dl}$ when eqn (4.94) is applied, even though these successive 'discontinuities' in electric flux when elements of charge cross the surface S do contribute to $\partial\Psi/\partial t$. Let a small element of charge $\lambda\Delta x$ cross the circular disk shaped surface S in *Figure 4.14(b)* in a time Δt. Since a 'point' charge q gives a change of $-q/\varepsilon_0$ in Ψ, the electric flux through the surface S, when it crosses the surface S from left to right, the change in Ψ due to the charge $\lambda\Delta x$ crossing the surface S

in a time Δt is $-\lambda \Delta x / \varepsilon_0$. Hence the contribution of charge crossing the surface S in *Figure 4.14(b)* to $\partial \Psi / \partial t$ is therefore $-\lambda \Delta x / \varepsilon_0 \Delta t$ or $-\lambda u / \varepsilon_0$. This effect reduces the calculated value of $\partial \Psi / \partial t$, but makes no contribution to $c^2 \oint \mathbf{B} . d\mathbf{l}$ evaluated around the circumference of the surface S when eqn (4.94) is applied. If the contributions of charges' crossing the surface S are included in the calculated value of

$$\partial \Psi / \partial t [= (\partial / \partial t) \int \mathbf{E} . \mathbf{n} dS]$$

as was done in *Figure 4.15*, then, since these contributions to $\partial \Psi / \partial t$ do not contribute to $c^2 \oint \mathbf{B} . d\mathbf{l}$, when eqn (4.94) is applied, one must add $\lambda u / \varepsilon_0$ to the right-hand side of eqn (4.94) to compensate for the contributions of charges' crossing the surface to $\partial \Psi / \partial t$. Hence, we must have

$$c^2 \oint \mathbf{B} . d\mathbf{l} = \frac{\partial}{\partial t} \int \mathbf{E} . \mathbf{n} dS + \frac{\lambda u}{\varepsilon_0}$$

Eqns (4.103) and (4.105) are then the same.

In Section 4.6(b), using vector analysis, the equation

$$c^2 \nabla \times \mathbf{B} = \frac{\partial \mathbf{E}}{\partial t} + \frac{\mathbf{J}}{\varepsilon_0} \qquad (4.79)$$

was developed for a moving macroscopic charge distribution, from the formulae for the electric and magnetic fields of a 'point' charge moving with uniform velocity. Integrating eqn (4.79) over the surface of the circular disk shown in *Figure 4.14(b)*, and using Stokes' theorem, eqn (A1.25) of Appendix 1, we have

$$c^2 \int (\nabla \times \mathbf{B}) . \mathbf{n} dS = c^2 \oint \mathbf{B} . d\mathbf{l} = \frac{\partial}{\partial t} \int \mathbf{E} . \mathbf{n} dS + \frac{1}{\varepsilon_0} \int \mathbf{J} . \mathbf{n} dS \quad (4.106)$$

For the case shown in *Figure 4.14(b)*

$$\frac{1}{\varepsilon_0} \int \mathbf{J} . \mathbf{n} dS = \frac{\lambda u}{\varepsilon_0}$$

Thus in the present case the addition of the term $\mathbf{J}$ to the right-hand side of the equation $\varepsilon_0 c^2 \nabla \times \mathbf{B} = \varepsilon_0 \partial \mathbf{E} / \partial t$, gives a value for $\oint \mathbf{B} . d\mathbf{l}$ and hence for the magnetic field at the field point P in *Figure 4.14(b)* in agreement with the value calculated using the Biot–Savart law.

It is only when charge is actually crossing the surface S in *Figure 4.14(b)* that the term containing the current density $\mathbf{J}$ must be included in eqn (4.106). For example, if one removed a very thin slice of thickness Δs from the middle of the line of charge in *Figure 4.14*, as shown in *Figure 4.16(a)*, then the variation of Ψ with x_0

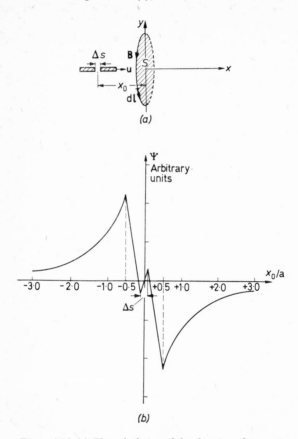

Figure 4.16. (a) The calculation of the electric and magnetic field of a line of charge with a complete break in the vicinity of its middle point. (b) The dependence of Ψ, the flux of $\mathbf{E}$ through the circular disk shaped surface S, on the position of the charge. It can be seen that $\partial \Psi / \partial x_0$ is negative only when charge is actually passing through the surface S

would be shown in *Figure 4.16(b)*. It can be seen that 'as the hole passes through the surface' that is $x_0 = 0$, then $\partial \Psi / \partial x_0$ is positive and the equation $c^2 \oint \mathbf{B} \cdot d\mathbf{l} = \partial \Psi / \partial t$ should give the correct value for the magnetic field. It can be seen that it is only when part of the charge is actually crossing the surface that $\partial \Psi/\partial x_0$ is negative in *Figures 4.15* and *4.16(b)*. The negative slopes are due to the fact that, when an element of charge passes through a surface, the direction in which its electric field passes through that surface changes, as illustrated in *Figure 4.12*. If the hole of length Δs is made very small, the effect of the absence of the charge $\lambda \Delta s$ on the magnetic field at P, calculated using the Biot–Savart law, should be negligible. However, the absence of $\lambda \Delta s$, when 'the hole passes through the surface' in *Figure 4.16(a)* changes *Figure 4.15* into *Figure 4.16(b)*, and $c^2 \oint \mathbf{B} \cdot \mathrm{d}\mathbf{l} = (\partial/\partial t) \int \mathbf{E} \cdot \mathbf{n} \mathrm{d}S$ should give a value of $\mathbf{B}$ in agreement with the Biot–Savart law. This again illustrates that one can obtain correct answers for $c^2 \oint \mathbf{B} \cdot \mathrm{d}\mathbf{l}$, evaluated around the boundary of a surface, when applying eqn (4.94), by removing the contribution to the change in the electric flux through the surface due to elements of charge's crossing the surface. The general case of eqns (4.79) and (4.93) will now be considered.

4.9. DISCUSSION OF THE EQUATION
$\varepsilon_0 c^2 \nabla \times \mathbf{B} = \varepsilon_0 \partial \mathbf{E}/\partial t + \mathbf{J}$

It was shown in Section 4.6(a) that at any field point, the electric and magnetic fields of *an isolated 'point' charge* moving with uniform velocity in empty space satisfy the relation

$$\varepsilon_0 c^2 \nabla \times \mathbf{B} = \varepsilon_0 \frac{\partial \mathbf{E}}{\partial t} \qquad (4.73)$$

This is a relation between field quantities, and is true whatever the position of the charge. It is not correct to say that it is the Maxwell term $\varepsilon_0 \partial \mathbf{E}/\partial t$ which gives rise to the magnetic field. The electric and magnetic fields have a common cause, since they are due to the same moving charge. When eqn (4.73) was used to calculate the magnetic field of a moving 'point' charge in Section 4.7, eqn (4.73) was integrated over a surface. There was a 'discontinuity' of $-q/\varepsilon_0$ in Ψ, the flux of $\mathbf{E}$ through this surface, when the 'point' charge crossed the surface. This 'discontinuity' in $\partial \Psi/\partial t$ did not contribute to $c^2 \oint \mathbf{B} \cdot \mathrm{d}\mathbf{l}$, when eqn (4.73) was applied, in the integrated form of eqn (4.86).

In Section 4.6(b) the equation

$$\varepsilon_0 c^2 \nabla \times \mathbf{B} = \varepsilon_0 \frac{\partial \mathbf{E}}{\partial t} + \mathbf{J} \qquad (4.79)$$

95

was developed using vector analysis for a field point inside a continuous charge distribution of the type shown in *Figure 4.10* (cf. page 76). In order to interpret eqn (4.79), we shall use the ideas developed in Sections 4.7 and 4.8 to develop eqn (4.79) for the macroscopic fields **E** and **B** from eqn (4.73) which is a relation between the fields of a single 'point' charge.

Consider a field point *P* inside the system of 'point' charges moving with *uniform* velocities as shown in *Figure 4.10*. (In a conductor we normally have stationary positive ions and moving conduction electrons. The mean separation of the conduction electrons is $\sim 10^{-10}$ metres in a metal.) If the *i*th moving charge gives rise to an electric intensity $\mathbf{E}_i$ and a magnetic induction $\mathbf{B}_i$ at the field point *P* in *Figure 4.10*, applying eqn (4.73), we have

$$c^2 \mathbf{V} \times \mathbf{B}_i = \frac{\partial \mathbf{E}_i}{\partial t}$$

Summing for all the *N* charges in the *complete* system, we have

$$c^2 \sum_{i=1}^{N} \mathbf{V} \times \mathbf{B}_i = \sum_{i=1}^{N} \frac{\partial \mathbf{E}_i}{\partial t} = \frac{\partial}{\partial t}\left(\sum_{i=1}^{N} \mathbf{E}_i \right)$$

Using eqn (A1.17) of Appendix 1,

$$c^2 \mathbf{V} \times \left(\sum_{i=1}^{N} \mathbf{B}_i \right) = \frac{\partial}{\partial t}\left(\sum_{i=1}^{N} \mathbf{E}_i \right)$$

or

$$c^2 \mathbf{V} \times \mathbf{b} = \frac{\partial \mathbf{e}}{\partial t} \tag{4.107}$$

where

$$\mathbf{b} = \sum_{i=1}^{N} \mathbf{B}_i; \quad \mathbf{e} = \sum_{i=1}^{N} \mathbf{E}_i$$

are the *resultant* microscopic fields at the field point *P* in *Figure 4.10*. As it stands, eqn (4.107) is a relation between the resultant microscopic electric and magnetic fields **e** and **b** at any field point in the empty space between the moving charges in *Figure 4.10*.

The macroscopic magnetic induction (denoted **B**), is the average of the microscopic magnetic induction **b** in the vicinity of the field point *P* in *Figure 4.10*. In a metal conductor the separation of individual conduction electrons is of the order of atomic dimensions. Hence the averaging of the local microscopic magnetic induction **b** must be carried out over regions of space large on the atomic scale, that is large compared with the distances between neighbouring moving charges in *Figure 4.10*. In practice when Maxwell's equations are applied to macroscopic phenomena, it is the expressions for the curls and divergences of the fields which are integrated over finite areas and volumes. It is convenient to define the curl of the *macroscopic* (or local space average) magnetic induction $\mathbf{B}$ in the vicinity of the field point P in *Figure 4.10*, by applying eqn (A1.12) of Appendix 1 to the surface ΔS in *Figure 4.10* giving

$$(\nabla \times \mathbf{B}) \cdot \mathbf{n} \Delta S = \underset{\Delta S \to 0}{\text{Limit}} \oint \mathbf{B} \cdot d\mathbf{l} \qquad (4.108a)$$

Though the surface ΔS in *Figure 4.10* is small on the laboratory scale, it is large on the atomic scale. If one evaluates the line integral of the microscopic magnetic induction $\mathbf{b}$ around the boundary of ΔS, that is, if one evaluates $\oint \mathbf{b} \cdot d\mathbf{l}$, since ΔS is large on the atomic scale, fluctuations in $\mathbf{b}$ on the atomic scale will average out in $\oint \mathbf{b} \cdot d\mathbf{l}$, which should equal $\oint \mathbf{B} \cdot d\mathbf{l}$ evaluated around the boundary of ΔS, where $\mathbf{B}$ is the local space average or macroscopic magnetic induction. Hence the eqn (4.108a) can be rewritten in the form

$$(\nabla \times \mathbf{B}) \cdot \mathbf{n} \Delta S = \underset{\Delta S \to 0}{\text{Limit}} \oint \mathbf{b} \cdot d\mathbf{l} \qquad (4.108b)$$

and used to define the *curl* of the macroscopic magnetic induction $\mathbf{B}$ in terms of the microscopic magnetic induction $\mathbf{b}$. The line integral $\oint \mathbf{b} \cdot d\mathbf{l}$ will now be evaluated. Integrating eqn (4.107) relating the microscopic fields $\mathbf{e}$ and $\mathbf{b}$ over the surface of the area ΔS in *Figure 4.10*, and applying Stokes' theorem, eqn (A1.25) of Appendix 1, we have

$$c^2 \int_{\Delta S} (\nabla \times \mathbf{b}) \cdot \mathbf{n} dS = c^2 \oint_{\Delta S} \mathbf{b} \cdot d\mathbf{l} = \frac{\partial}{\partial t} \int_{\Delta S} \mathbf{e} \cdot \mathbf{n} dS$$

If the area ΔS in *Figure 4.10* is large on the atomic scale, the fluctuations in the microscopic electric intensity $\mathbf{e}$ on the atomic scale will average out in $\int \mathbf{e} \cdot \mathbf{n} dS$ which can be replaced by $\mathbf{E} \cdot \mathbf{n} \Delta S$, where $\mathbf{E}$ is the space average or macroscopic electric intensity in the vicinity of ΔS. Hence, one *might*(?) expect the following relation to be valid for the surface ΔS,

$$c^2 \oint_{\Delta S} \mathbf{b} \cdot d\mathbf{l} = \frac{\partial}{\partial t}(\mathbf{E} \cdot \mathbf{n} \Delta S) = \frac{\partial \Psi}{\partial t} = \left(\frac{\partial \mathbf{E}}{\partial t}\right) \cdot \mathbf{n} \Delta S \qquad (4.109)$$

However, if the surface ΔS is inside a macroscopic current distribution (such as a conduction current in a metal), and ΔS is large on the atomic scale, then a large number of atomic charges cross ΔS per unit time, as shown in *Figure 4.10.* Each charge q gives a small 'discontinuity' of $-q/\varepsilon_0$ in the electric flux through ΔS, as it passes through the surface ΔS. (If the electrons and ions have finite dimensions these 'discontinuities' should be more like the continuous variation in *Figure 4.15* than the abrupt discontinuity in *Figure 4.13*.) It was illustrated in Section 4.7, how the 'discontinuities' of $-q/\varepsilon_0$ in Ψ, the electric flux through ΔS, when charges cross ΔS, contribute to $\partial\Psi/\partial t$ but not to $c^2\oint\mathbf{b}\,.\,\mathrm{dl}$, when eqn (4.94) or (4.109) is applied. If these 'discontinuities' are included in the calculated value of $\partial\Psi/\partial t$, then one must compensate for their inclusion when calculating $c^2\oint\mathbf{b}\,.\,\mathrm{dl}$.

Consider the 'point' charges crossing the surface ΔS in *Figure 4.10.* If the charges all have the same velocity $\mathbf{u}$ the number crossing ΔS per second is $N\Delta S\mathbf{u}\,.\,\mathbf{n}$, where $\mathbf{n}$ is a unit vector normal to ΔS, and N is the number of charges per metre^3. If each charge gives a change of $-q/\varepsilon_0$ in the electric flux through ΔS when it crosses ΔS, the change per second in the electric flux through ΔS due to charges' crossing the surface ΔS in *Figure 4.10* is $-qN\Delta S\mathbf{u}\,.\,\mathbf{n}/\varepsilon_0$, which is equal to $-\mathbf{J}\,.\,\mathbf{n}\Delta S/\varepsilon_0$, where $\mathbf{J} = qN\mathbf{u}$ is the macroscopic current density. If the contribution of $-\mathbf{J}\,.\,\mathbf{n}\Delta S/\varepsilon_0$ to the rate of change of electric flux through ΔS, due to charges' crossing ΔS, is included in the calculated value of $\partial\Psi/\partial t$, one can compensate for the non-contribution to $c^2\oint\mathbf{b}\,.\,\mathrm{dl}$ of the change of electric flux due to charges' crossing ΔS by adding $+\mathbf{J}\,.\,\mathbf{n}\Delta S/\varepsilon_0$ to $\partial\Psi/\partial t$ in eqn (4.109) giving

$$c^2\oint\mathbf{b}\,.\,\mathrm{dl} = \frac{\partial\Psi}{\partial t} + \mathbf{J}\,.\,\mathbf{n}\frac{\Delta S}{\varepsilon_0} = \left(\frac{\partial\mathbf{E}}{\partial t}\right)\,.\,\mathbf{n}\Delta S + \mathbf{J}\,.\,\mathbf{n}\frac{\Delta S}{\varepsilon_0}$$

Substituting in eqn (4.108b) gives

$$c^2(\nabla\times\mathbf{B})\,.\,\mathbf{n}\Delta S = \left(\frac{\partial\mathbf{E}}{\partial t} + \frac{\mathbf{J}}{\varepsilon_0}\right)\,.\,\mathbf{n}\Delta S$$

Hence,

$$\varepsilon_0 c^2\nabla\times\mathbf{B} = \varepsilon_0\frac{\partial\mathbf{E}}{\partial t} + \mathbf{J} \tag{4.110}$$

This is the same as eqn (4.79). All the quantities in eqn (4.110), namely $\mathbf{B}$, $\mathbf{E}$ and $\mathbf{J}$ are the macroscopic values, or the local space average values, at the field point P in *Figure 4.10.* The averages are taken over regions of space large on the atomic scale. The roles of the $\varepsilon_0\partial\mathbf{E}/\partial t$ and the $\mathbf{J}$ term in eqn (4.110) are different. The relation $\varepsilon_0 c^2\nabla\times\mathbf{b} = \varepsilon_0\partial\mathbf{e}/\partial t$ is the basic relation between the microscopic electric and magnetic fields in empty space, due to a system of moving 'point' charges. When eqn (4.108b) is used to calculate the curl of the *macroscopic* magnetic induction **B** *inside* a current distribution in terms of $\oint \mathbf{b} \cdot d\mathbf{l}$ evaluated around a surface ΔS large on the atomic scale, there are 'discontinuities' in Ψ the electric flux through ΔS due to charges' crossing the surface ΔS which contribute to $\partial \Psi/\partial t$ but not to $c^2 \oint \mathbf{b} \cdot d\mathbf{l}$, when eqn (4.109) is applied. To correct for this in a macroscopic theory the $\mathbf{J} \cdot \mathbf{n} \Delta S/\varepsilon_0$ term is added to the right-hand side of eqn (4.109), giving the **J** term on the right-hand side of eqn (4.110). It is left as an exercise for the reader to develop eqn (4.110) from eqn (4.73) for a field point inside a stationary conductor consisting of stationary positive ions, and negative electrons moving with uniform velocities. The case of accelerating charges is discussed in Appendix 3(c).

In eqn (4.110), **E** and **B** are the *resultant* macroscopic electric intensity and magnetic induction respectively, due to *all* the charges in the *complete* system, whereas the compensating term **J** depends only on the macroscopic current density at the field point where eqn (4.110) is applied. Integrating eqn (4.110) over a finite surface and applying Stokes' theorem, we had

$$\varepsilon_0 c^2 \oint \mathbf{B} \cdot d\mathbf{l} = \varepsilon_0 \frac{\partial}{\partial t} \int \mathbf{E} \cdot \mathbf{n} dS + \int \mathbf{J} \cdot \mathbf{n} dS \qquad (4.106)$$

In this way the line integral of the resultant macroscopic magnetic induction **B**, around the boundary of any finite surface S, can be related to the rate of change of the flux of the resultant macroscopic electric intensity **E** through the same surface and the total macro-scopic electric current crossing the same surface. The values of all the quantities in eqn (4.106) are their values at the surface S. Some of the charges in the system may be near the surface, others far away. Whatever the positions and velocities of the charges in the system, their resultant macroscopic electric and magnetic fields satisfy eqn (4.106). Since, in eqn (4.106), we are integrating over a finite surface, the 'discontinuities' in the flux of **E** through the surface S, due to charges' crossing the surface S, are important, and to compensate for these discontinuities the inclusion of the terms containing **J** in eqns (4.106) and (4.110) is essential. In fact, in the extreme case of a *steady* conduction current in a *complete* circuit, the rate of change of the *macroscopic* electric field is zero, that is $\partial \mathbf{E}/\partial t$ is zero, and one is left only with the **J** term, giving Ampère's circuital theorem,

$$\oint \mathbf{B} \cdot d\mathbf{l} = \frac{1}{\varepsilon_0 c^2} \int \mathbf{J} \cdot \mathbf{n} dS = \mu_0 \int \mathbf{J} \cdot \mathbf{n} dS$$

To illustrate the atomistic interpretation of Ampère's circuital theorem, we shall consider an example due to Vinti and Montgomery[20]. Consider an infinitely long line of 'point' charges q all a distance l apart, and all moving with the same uniform velocity u which is $\ll c$, as shown in *Figure 4.17(a)*. Consider one of the 'point' charges as it approaches very close to the surface S in *Figure 4.17(a)*. When the charge reaches the surface S, the other 'point' charges are

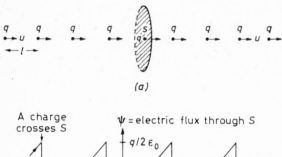

(a)

(b)

Figure 4.17. (a) An infinitely long line of 'point' charges q, a distance l apart, moving with uniform velocity u, give rise to a 'steady' macroscopic current I. (b) The variation of Ψ, the electric flux through the surface S with time. There is a discontinuity of $-q/\varepsilon_0$ in Ψ every time a 'point' charge crosses S. The mean value of Ψ is zero

distributed symmetrically on either side of the surface S, so that apart from the effect of the 'point' charge just reaching the surface S, the resultant electric flux through S due to all the other charges is zero. Hence just as a 'point' charge reaches the surface S in *Figure 4.17(a)*, the electric flux Ψ through the surface S is $+q/2\varepsilon_0$. Just after the charge crosses the surface S, the electric flux is $-q/2\varepsilon_0$. As the charges continue to move along, the electric flux then increases continuously until the next 'point' charge reaches the surface S, when there is another discontinuity of $-q/\varepsilon_0$ in Ψ. The variation of

Ψ with time is illustrated in *Figure 4.17(b)*. There is a 'saw tooth' type variation of Ψ with time. In the time interval between successive charges' crossing the surface S, $\partial \Psi/\partial t$ is positive and the equation

$$\oint \mathbf{B} \cdot d\mathbf{l} = \frac{1}{c^2} \frac{\partial \Psi}{\partial t} \qquad (4.94)$$

should give the correct value for $\oint \mathbf{B} \cdot d\mathbf{l}$. The successive discontinuities of $-q/\varepsilon_0$ in Ψ as successive charges cross the surface S do not contribute to $\oint \mathbf{B} \cdot d\mathbf{l}$, when eqn (4.94) is applied, even though the successive discontinuities of $-q/\varepsilon_0$ contribute to the time average values of Ψ and $\partial \Psi/\partial t$. If one takes away the contribution of these discontinuities to Ψ, we are left with a continuous positive increase in Ψ which can be used to calculate $\oint \mathbf{B} \cdot d\mathbf{l}$ using eqn (4.94).

Now the macroscopic electric intensity is generally defined as the local space *and time* average of the local microscopic electric field. Due to the effects of charges' crossing the surface S, the *time* average of Ψ and hence of $\partial \Psi/\partial t$ in *Figure 4.17(b)* is zero. In order to illustrate the time scale involved, consider a total current of one ampere, and let q equal $1 \cdot 6 \times 10^{-19}$ coulomb, the electronic charge. The current is the charge crossing the surface S per second, which for a current of one ampere is one coulomb. Hence, for a current of one ampere, the number of charges crossing the surface S in *Figure 4.17(a)* per second is $1/(1 \cdot 6 \times 10^{-19})$ and the time interval between successive charges' crossing the surface S is $1 \cdot 6 \times 10^{-19}$ sec. In such a time, light only travels a distance of $4 \cdot 8 \times 10^{-11}$ metres, which is of the order of atomic dimensions. The speed of light is the limiting speed for propagating forces. Hence for macroscopic phenomena, whose scales are much larger than atomic dimensions, it is sensible to take the time average of the microscopic electric field over time periods long compared with the atomic time scale (which is $\sim 1 \cdot 6 \times 10^{-19}$ sec in the present example), but short compared with the time variations of the macroscopic conduction current which are generally $> 10^{-10}$ sec, corresponding to frequencies $< 10^{10}$ Hertz. Thus in a *macroscopic* theory the flux of the macroscopic electric field through S would be zero in *Figure 4.17(a)* and (*b*), though on a *microscopic* scale one can relate $\oint \mathbf{B} \cdot d\mathbf{l}$ to the rate of change of the flux of the microscopic electric intensity through the surface S during the periods when no 'point' charges cross the surface S. For a current of I amperes, the number of charges crossing S per second is I/q. Since each charge crossing S gives a contribution of $-q/\varepsilon_0$ to the change in Ψ, the total decrease in Ψ per second due to charges' crossing S is $-I/\varepsilon_0$. These contributions to the rate of change of the macroscopic electric flux through S do not contribute to $\oint \mathbf{B} \cdot d\mathbf{l}$. To compensate for their inclusion in the macroscopic value of $\partial \Psi_{mac}/\partial t$, we can add an amount I/ε_0 to $\partial \Psi_{mac}/\partial t$. For a current 'steady' on the macroscopic time scale, since $\partial \Psi_{mac}/\partial t$ is zero, compensating for the inclusion of the discontinuities due to charges' crossing S, we have

$$\oint \mathbf{B} \cdot d\mathbf{l} = \frac{1}{c^2} \frac{\partial \Psi_{mac}}{\partial t} + \frac{I}{\varepsilon_0 c^2} \to \frac{I}{\varepsilon_0 c^2} \to \mu_0 I$$

This is Ampère's circuital theorem for the macroscopic fields. It has been illustrated that in this case also, on an atomistic level, the relation

$$c^2 \mathbf{\nabla} \times \mathbf{b} = \partial \mathbf{e}/\partial t$$

is the basic relation, where $\mathbf{e}$ is the microscopic electric field and $\mathbf{b}$ the microscopic magnetic field. The $\mathbf{J}$ term need only be introduced in a macroscopic theory when one averages the fields over regions of space and time to obtain the macroscopic fields. It is interesting to note that in conventional approaches to electromagnetism one starts with Ampère's circuital theorem (developed from the Biot–Savart law) and not from eqn (4.93).

The Maxwell term $\varepsilon_0 \partial \mathbf{E}/\partial t$ has sometimes been called the displacement 'current'. One is accustomed to associating many effects with electric currents which cannot be associated with the Maxwell term $\varepsilon_0 \partial \mathbf{E}/\partial t$. For example, one expects electric currents to give rise to a magnetic field according to the Biot–Savart law. One expects forces equal to $I d\mathbf{l} \times \mathbf{B}$ on current elements. If there are no forces on empty space, there cannot be a force proportional to $\varepsilon_0 (\partial \mathbf{E}/\partial t) \times \mathbf{B}$ on the 'displacement current' in empty space. In these respects the choice of the term 'current' to describe the Maxwell term $\varepsilon_0 \partial \mathbf{E}/\partial t$ is inappropriate in the context of classical electromagnetism. It was for this reason that the $\varepsilon_0 \partial \mathbf{E}/\partial t$ term in eqns (4.79) and (4.110) was referred to as the Maxwell term rather than the displacement 'current'. It is only in eqn (4.110) that the Maxwell term *appears* to behave like an electric current, but as was pointed out earlier in this section the roles of the term $\varepsilon_0 \partial \mathbf{E}/\partial t$ and the $\mathbf{J}$ term in eqn (4.110) are different, so that this comparison with an electric current is largely fortuitous.

When a dielectric is polarized, there is a relative displacement of positive and negative charges inside atoms, as illustrated in *Figure 4.18*. When the polarization $\mathbf{P}$ is varying with time, there is a continual movement of electric charge across the surface ΔS, as illustrated in *Figure 4.18*. It is left as an exercise for the reader to show, that when a dielectric is polarized, the equivalent macroscopic electric current density is $\partial \mathbf{P}/\partial t$. A simplified case is discussed in *Figure 4.18*. If the surface ΔS in *Figure 4.18* is used to define $\nabla \times \mathbf{B}$ for a field point inside a polarized dielectric, one must compensate for the 'discontinuities' in the electric flux through ΔS, when atomic charges cross ΔS when $\mathbf{P}$ is varying. Hence, inside a dielectric, one must add the term $\partial \mathbf{P}/\partial t$ to the $\mathbf{J}$ term in eqn (4.110) to compensate for the 'discontinuities' in the electric flux through ΔS, when charges cross ΔS. Hence inside a dielectric, we have for the macroscopic fields $\mathbf{E}$ and $\mathbf{B}$

$$\varepsilon_0 c^2 \nabla \times \mathbf{B} = \varepsilon_0 \frac{\partial \mathbf{E}}{\partial t} + \mathbf{J} + \frac{\partial \mathbf{P}}{\partial t} \qquad (4.111)$$

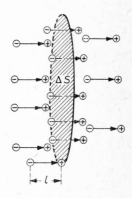

Figure 4.18. A simplified model of a polarized dielectric. It is assumed that each positive charge is displaced a distance l when the external electric field is applied, giving rise to a displacement of electric charge across the surface ΔS, which is perpendicular to the direction in which the positive charges are displaced. The number of positive charges crossing ΔS is equal to the number of positive charges within a distance l of ΔS, before the dielectric is polarized. The charge crossing ΔS is $Q = qNl\Delta S = Np\Delta S = P\Delta S$, where $p = ql$ is the dipole moment of each dipole, N is the number of dipoles per unit volume, and $P = Np$ is the polarization vector. When P is varying, e.g. due to variations in l, charge is crossing ΔS, the current being $\Delta Q/\Delta t = qN\Delta l\Delta S/\Delta t = (\Delta P/\Delta t)\Delta S$. Hence $J = \partial P/\partial t$

Using eqn (4.28), eqn (4.111) can be rewritten

$$\varepsilon_0 c^2 \nabla \times \mathbf{B} = \mathbf{J} + \frac{\partial \mathbf{D}}{\partial t} \qquad (4.112)$$

It should be stressed that in eqn (4.111) the roles of the $\varepsilon_0 \partial \mathbf{E}/\partial t$ and the $\partial \mathbf{P}/\partial t$ terms are different, the role of the $\partial \mathbf{P}/\partial t$ term being similar to the role of the $\mathbf{J}$ term. When the $\varepsilon_0 \partial \mathbf{E}/\partial t$ and the $\partial \mathbf{P}/\partial t$ terms are combined to form the $\partial \mathbf{D}/\partial t$ term in eqn (4.112), it tends to conceal their different roles in eqn (4.111) and tends to give the *illusion* that their roles are the same.

It is shown in Appendix 2(f), that for purpose of calculating the macroscopic (or local space average) magnetic induction due to a magnetized medium, the magnetized medium can be replaced by a *fictitious* macroscopic surface current $\mathbf{M} \times \mathbf{n}$ amperes per metre and a *fictitious* macroscopic volume current density

$$\mathbf{J}_m = \nabla \times \mathbf{M} \text{ amperes metre}^{-2} \qquad (4.113)$$

where $\mathbf{M}$, the magnetization vector, is the magnetic dipole moment per unit volume, defined in Appendix 2(f) by eqn (A2.40). The *macroscopic* magnetic induction $\mathbf{B}$ due to an isolated, stationary, uniformly magnetized, cylindrical bar magnet is shown in *Figure A2.4* of Appendix 2(f). The microscopic magnetic induction $\mathbf{b}$ varies rapidly near and inside atoms, due to the magnetic moments associated with electron orbital motions and intrinsic magnetic moments associated with electron spin and nuclear spin. When the microscopic magnetic field $\mathbf{b}$ is averaged over regions of space large compared with atomic dimensions, the irregularities on the atomic scale average out, and the lines of the *macroscopic* magnetic induction $\mathbf{B}$ are smooth and continuous, as shown in *Figure A2.4*. The line integral $\oint \mathbf{B} \cdot d\mathbf{l}$ and the curl of the macroscopic magnetic induction $\mathbf{B}$ in *Figure A2.4* are finite. It is shown in Appendix 2(f) that for a field point inside a stationary magnetized body, we have for the curl of the macroscopic magnetic induction $\mathbf{B}$

$$\varepsilon_0 c^2 \nabla \times \mathbf{B} = \nabla \times \mathbf{B}/\mu_0 = \mathbf{J}_m = \nabla \times \mathbf{M}$$

Hence for a field point inside a *stationary* magnetic dielectric in the presence of electric currents, eqn (4.111) becomes

$$\varepsilon_0 c^2 \nabla \times \mathbf{B} = \varepsilon_0 \frac{\partial \mathbf{E}}{\partial t} + \mathbf{J} + \frac{\partial \mathbf{P}}{\partial t} + \nabla \times \mathbf{M} \qquad (4.114)$$

The quantities $\mathbf{B}$, $\mathbf{E}$, $\mathbf{J}$, $\mathbf{P}$ and $\mathbf{M}$ are all *macroscopic* quantities, that is local space (and time) average quantities, the averaging being carried out over distances large on the atomic scale, but small on the laboratory scale.

In the interests of brevity, it is conventional to introduce the magnetizing force $\mathbf{H}$ *defined* as

$$\mathbf{H} = \varepsilon_0 c^2 \mathbf{B} - \mathbf{M} = \frac{\mathbf{B}}{\mu_0} - \mathbf{M} \qquad (4.115)$$

Using eqns (4.115) and (4.28), eqn (4.114) can be rewritten

$$\nabla \times \mathbf{H} = \mathbf{J} + \frac{\partial \mathbf{D}}{\partial t} \qquad (4.116)$$

Eqn (4.115) is sometimes rewritten in the form,

$$\mathbf{B} = \mu_0 \mathbf{H} + \mu_0 \mathbf{M} = \mu_r \mu_0 \mathbf{H} \qquad (4.117)$$

where μ_r is the relative permeability. Eqn (4.117) is one of the constitutive equations.

It is interesting to see why it is not necessary to add any extra terms to the equation

$$\mathbf{V} \times \mathbf{E} = -\partial \mathbf{B}/\partial t \qquad (4.59)$$

when there is a macroscopic current distribution at the field point. Consider a 'point' charge approaching the shaded surface S shown in *Figure 4.19(a)*. The lines of $\mathbf{B}$ through the surface, after the charge has crossed the surface S, are shown in *Figure 4.19(b)*. Since the lines of $\mathbf{B}$ form closed circles, which are in the same direction for field

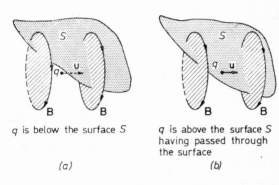

q is below the surface S

q is above the surface S
having passed through
the surface

(a)

(b)

Figure 4.19. *The charge q passes through the surface S. The magnetic field lines are upwards out of the surface both before and after the charge crosses the surface*

points situated in front of and behind the moving charge, as illustrated in *Figures 3.4* and *4.19*, there is no change in the direction of the flux of $\mathbf{B}$ through the surface when the charge crosses the surface. Since there is no discontinuity in the flux of $\mathbf{B}$ to compensate for when calculating $(\partial/\partial t) \int \mathbf{B} \cdot \mathbf{n} dS$ it is not necessary to add any extra terms to eqn (4.59). Since eqn (4.59) holds at all points for the microscopic fields it must also hold for the local space average or *macroscopic* electric and magnetic fields.

It has been suggested that magnetic monopoles may be produced in very high energy nuclear interactions. It has been postulated that in empty space, the lines of $\mathbf{B}$ would diverge from such monopoles in a way similar to the way the lines of $\mathbf{E}$ diverge from electric charges. If such a magnetic monopole crossed the shaded surface in *Figure 4.19*, then, when the monopole crossed the surface the flux of $\mathbf{B}$ through the surface would change from $+\mu_0 q_m^*/2$ to $-\mu_0 q_m^*/2$ where q_m^* is defined according to eqn (3.49). It is left as an exercise for the reader (Problem 4.2) to show that for a macroscopic distribution of moving magnetic monopoles, eqn (4.59) would have to be amended to

$$\mathbf{V} \times \mathbf{E} = -\frac{\partial \mathbf{B}}{\partial t} - \mathbf{J}_m = -\frac{\partial \mathbf{B}}{\partial t} - \mu_0 \mathbf{J}_m^* \qquad (4.118)$$

where $\mathbf{J}_m$ and $\mathbf{J}_m^*$ are the magnetic monopole current densities defined by eqns (4.144) of Problem 4.2. On the other hand, the electric field lines due to a magnetic monopole moving with uniform velocity form closed circles (Problem 3.4). There is no 'discontinuity' in the flux of $\mathbf{E}$ through a surface due to a moving magnetic monopole, when that monopole crosses the surface, and it is not necessary to add any extra terms to eqn (4.116) in the presence of magnetic monopoles. Since magnetic monopoles appear to play no significant role in normal electromagnetic phenomena, one generally writes $\mathbf{V} \cdot \mathbf{B} = 0$ and omits the term $\mathbf{J}_m^*$ in eqn (4.118).

An account of conventional approaches to eqn (4.116) is given in Appendix 2(h). Firstly, the equation $\mathbf{V} \times \mathbf{B} = \mu_0 \mathbf{J}$ is developed from the Biot–Savart law for *steady* macroscopic currents in *complete* circuits. By considering the charging of a capacitor at low frequencies, eqn (4.116) is then developed more by intelligent guesswork than anything else for quasi-stationary conditions, when the radiation electric and magnetic fields are generally negligible. When developed in this way, and in particular when it is applied to the radiation fields of accelerating charges, the validity of eqn (4.116) really depends on the fact that predictions based on eqn (4.116) are found to be in agreement with the experimental results, so that, in general, eqn (4.116) is checked *a posteriori*.

4.10. CALCULATION OF MAGNETIC FIELDS USING THE BIOT–SAVART LAW

It is shown in Appendix 4 that, if the velocity of the charges in a wire are $\ll c$, then the magnetic induction, for quasi-stationary conditions, is given to a good approximation by the Biot–Savart law, eqn (A4.10) of Appendix 4. If the conductor is of finite cross section, it can be divided into volume elements $\Delta V'$ at a position $\mathbf{r}'$ from the origin of a co-ordinate system, and $\mathbf{J}(\mathbf{r}')\Delta V'$ can be treated as a current element. It follows from eqn (A4.10) that the macroscopic magnetic induction at a field point at a position $\mathbf{r}$

from the origin can be written in the form

$$\mathbf{B} \simeq \int \frac{\mathbf{J}(\mathbf{r}') \times (\mathbf{r} - \mathbf{r}') \mathrm{d}V'}{4\pi\varepsilon_0 c^2 |\mathbf{r} - \mathbf{r}'|^3} \tag{4.119}$$

The integration must be over the whole of space. Only currents due to moving charges must be included in eqn (4.119). Some modern text books (such as *University Physics*, by Sears and Zemansky, 3rd Edition, page 719) still give the erroneous impression that one should include the Maxwell term $\varepsilon_0 \partial \mathbf{E}/\partial t$ in eqn (4.119). Such an approach is a relic of the nineteenth century ether theories. It has been illustrated in this chapter that the Maxwell term can be used to calculate the curl of the magnetic induction at a field point due to a system of moving charges, if the rate of change of the electric field at the same point due to the same system of charges is known. In itself, however, $\varepsilon_0 \partial \mathbf{E}/\partial t$ produces no magnetic field. It is associated with magnetic fields, only in as much as both the electric and magnetic fields in a system arise from a common cause, namely the moving charges in the system. It is only in eqn (4.114) that the Maxwell term *appears* to resemble an electric current. It so happens that, if one did calculate the electric field $\mathbf{E}(\mathbf{r}')$ of a point charge moving with a velocity $u \ll c$ in empty space, and then assumed that the displacement 'current' $\varepsilon_0 \partial \mathbf{E}/\partial t$ associated with this electric field gave rise to a magnetic field in other parts of space according to the Biot–Savart law, that is, if one calculated

$$\int \frac{\varepsilon_0 \dfrac{\partial \mathbf{E}(\mathbf{r}')}{\partial t} \times (\mathbf{r} - \mathbf{r}') \mathrm{d}V'}{4\pi\varepsilon_0 c^2 |\mathbf{r} - \mathbf{r}'|^3}$$

then, if the integral is evaluated over the whole of space, the integral would be zero. A short proof is given in Appendix 6. Using the principle of superposition the same must be true for a system of moving charges. Thus, if the integral in eqn (4.119) were evaluated over the whole of space, the addition of $\varepsilon_0(\partial \mathbf{E}/\partial t)$ to the current density $\mathbf{J}$ due to moving charges would not matter in practice, in the quasi-stationary conditions in which the Biot–Savart law, eqn (4.119), is applicable. However, it is wrong in principle to include the Maxwell term $\varepsilon_0(\partial \mathbf{E}/\partial t)$ in eqn (4.119).

When the polarization of a dielectric is varying with time, then atomic charges are being displaced giving rise to a current density $\partial \mathbf{P}/\partial t$ due to an actual movement of electric charge on the atomic scale, as illustrated in *Figure 4.18*. This movement of charge inside a dielectric gives rise to a magnetic field. Thus the term $\partial \mathbf{P}/\partial t$ must be

107

included in eqn (4.119). The macroscopic magnetic induction due to a magnetized medium can be calculated by replacing the magnetized medium by a fictitious volume current density $\mathbf{V}' \times \mathbf{M}$. Thus eqn (4.119) for the Biot–Savart law must be generalized to

$$\mathbf{B(r)} \simeq \int \frac{\left[\mathbf{J(r')} + \dfrac{\partial \mathbf{P(r')}}{\partial t} + \mathbf{V}' \times \mathbf{M(r')} \right] \times (\mathbf{r - r'}) dV'}{4\pi\varepsilon_0 c^2 \left| \mathbf{r - r'} \right|^3} \tag{4.120}$$

the integration being over all space.

Sometimes, it is more convenient to calculate magnetic fields using the vector potential $\mathbf{A}$ defined by the relation

$$\mathbf{B} = \mathbf{V} \times \mathbf{A}$$

For quasi-stationary conditions, eqn (4.120) can then be rewritten in the form

$$\mathbf{A(r)} = \frac{1}{4\pi\varepsilon_0 c^2} \int \frac{\left[\mathbf{J(r')} + \dfrac{\partial \mathbf{P(r')}}{\partial t} + \mathbf{V}' \times \mathbf{M(r')} \right] dV'}{\left| \mathbf{r - r'} \right|} \tag{4.121}$$

the integration being over the whole of space (Reference: Appendix 2(d)).

4.11. EXAMPLE: A CHARGING CAPACITOR

In order to illustrate the use of Maxwell's equations and the Biot–Savart law, the example illustrated in *Figure A2.6* of Appendix 2(h) will be considered in greater detail. It will be assumed that the plates of the capacitor are not infinite in area, so that there is a fringing electric field as shown in *Figure 4.20*. It will again be assumed that the plates are *in vacuo*, and that the frequency of the alternating current is low enough to satisfy the quasi-stationary approximation, so that the current I can be assumed to have the same value in all parts of the circuit.

Firstly, one can use the *Biot–Savart law* including all the moving charges in the system. In this method the magnetic field is related back to its sources, and the Biot–Savart law is equivalent to using eqn (3.31). At large distances from the capacitor, the magnetic induction calculated using the Biot–Savart law should approximate closely to

$$B = \frac{\mu_0 I}{2\pi r} \tag{4.122}$$

as the effects of the gap in the 'infinitely long wire' between the plates of the capacitor are then negligible. In eqn (4.122) r is the perpendicular distance from the field point to the wire. As one approaches the capacitor the effect of the missing part of the wire becomes more important and, for a capacitor of finite dimensions, the magnetic induction near the capacitor and between the plates of the capacitor should be a little less than the value given by eqn (4.122).

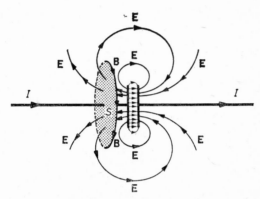

Figure 4.20. The charging of a capacitor. In this case the fringing electric field is important

The second approach is to use Maxwell's equation, eqn (4.110)

$$\varepsilon_0 c^2 \mathbf{V} \times \mathbf{B} = \mathbf{J} + \varepsilon_0 \frac{\partial \mathbf{E}}{\partial t} \qquad (4.123)$$

Integrating eqn (4.123) over the surface of the circular disk shaped surface S of radius r shown in *Figure 4.20*, we have

$$\varepsilon_0 c^2 \int_S (\mathbf{\nabla} \times \mathbf{B}) \cdot \mathbf{n} dS = \varepsilon_0 c^2 \oint_S \mathbf{B} \cdot d\mathbf{l} = \int_S \mathbf{J} \cdot \mathbf{n} dS + \varepsilon_0 \int_S \frac{\partial \mathbf{E}}{\partial t} \cdot \mathbf{n} dS \qquad (4.124)$$

By symmetry, $\mathbf{B}$ has the same value around the circumference of S, so that

$$2\pi r \varepsilon_0 c^2 B = I + \varepsilon_0 \frac{\partial \Psi}{\partial t}$$

or,

$$B = \frac{1}{2\pi r \varepsilon_0 c^2}\left[I + \varepsilon_0 \frac{\partial \Psi}{\partial t}\right] = \frac{\mu_0}{2\pi r}\left[I + \varepsilon_0 \frac{\partial \Psi}{\partial t}\right]$$

where $\Psi = \int \mathbf{E} \cdot \mathbf{n} dS$ is the flux of $\mathbf{E}$ through S due to the fringing field of the capacitor. As the current I charges up the capacitor, the magnitudes of $\mathbf{E}$ and $|\Psi|$ are increasing. However, Ψ is negative for the position of S shown in *Figure 4.20*, since $\mathbf{E}$ goes through the surface S in the opposite direction to the direction in which a right-handed corkscrew would advance, if rotated in the direction $\oint \mathbf{B} \cdot \mathbf{dl}$ is evaluated. Thus, when the capacitor is charging up $\partial \Psi / \partial t$ is negative, and

$$B < \frac{\mu_0 I}{2\pi r} \qquad (4.125)$$

This is in agreement with the Biot–Savart law. The nearer the surface S is to the capacitor, the bigger the numerical values of $|\Psi|$ and $|\partial \Psi / \partial t|$, and the smaller B becomes.

If the surface S in *Figure 4.20* is between the plates of the capacitor, $\mathbf{J}$ is zero everywhere on the surface S. However, in this case the lines of $\mathbf{E}$, between the plates of the capacitor, go through S in the positive direction, so that Ψ and $\partial \Psi / \partial t$ are positive, and the direction of B is the same as previously. It is shown in Appendix 2(h) that, when the fringing field is negligible, according to eqn (A2.52) of Appendix 2(h), $\partial \Psi / \partial t = A \, dE/dt$ is equal to I/ε_0 and $B = \mu_0 I / 2\pi r$. When the surface S is between the plates of the capacitor in *Figure 4.20*, the effect of the fringing electric field is to reduce Ψ, so that eqn (4.125) is still valid.

When the surface S in *Figure 4.20* is on the right-hand side of the capacitor, $\mathbf{J}$ is finite, but now Ψ and $\partial \Psi / \partial t$ are negative again, so that eqn (4.125) is again satisfied.

All the quantities in eqn (4.124), namely $\mathbf{B}, \mathbf{J}$ and $\mathbf{E}$, are their values on the surface S. The magnetic field is not related back to its sources when Maxwell's equations are used, in contrast to when the Biot–Savart law is used. In the latter case, the magnetic field must be related back to its sources, that is the moving charges in the system.

4.12. EXAMPLE: HEAVISIDE'S 'RATIONAL' CURRENT ELEMENT[6]

Consider a thin current element of length dl insulated except at its ends. The current element is immersed in a conducting liquid of infinite extent, as shown in *Figure 4.21(a)*. A steady current I is made to flow along the element. The current flows out from one end of the current element into the liquid and back into the current element from the liquid at the other end of the current element. One can

110

treat one end of the current element as a point source of current from which the current flows out radially into the liquid, as shown dotted in *Figure 4.21(a)*. The other end of the current element is a point sink of current, into which current flows radially from the liquid. The resultant lines of current flow in the liquid have the same shape as the electric field lines of an electric dipole. The problem can be solved using either the Biot–Savard law or Maxwell's equations.

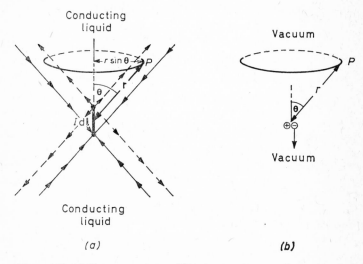

(a)   (b)

*Figure 4.21. (a) The current element I*d**l** is immersed in a conducting liquid of infinite extent. The current element is insulated except at its ends. Electric current flows into the current element at one end and out from the current element into the liquid at the other end. (b) A simplified model of an isolated current element in empty space. It consists of a stationary positive ion and an electron moving with uniform velocity*

Biot–Savart Law

Firstly, consider the current element Idl. According to the Biot–Savart law, it gives rise to a magnetic induction

$$dB \simeq \frac{Idl}{4\pi\varepsilon_0 c^2 r^2} \sin\theta \qquad (4.126)$$

at the field point P, which is at a distance **r**, θ from the current element.

Now consider the spherically symmetric radial flow of current from one end of the current element. Symmetry considerations rule out all possible directions for the magnetic field due to this current flow, except a magnetic field radially outwards from that end of the current element. But if $\mathbf{V} \cdot \mathbf{B}$ is to be zero, one cannot have such a radial magnetic field so that the magnetic field due to the flow of current from the current element into the liquid must be zero. Similarly, the current flow from the liquid into the other end of the element gives no resultant magnetic field. Hence the resultant magnetic field at P due to the complete current system is given by eqn (4.126).

Maxwell's Equations

Now from eqn (4.110)

$$\varepsilon_0 c^2 \mathbf{V} \times \mathbf{B} = \varepsilon_0 \frac{\partial \mathbf{E}}{\partial t} + \mathbf{J}$$

When the current is steady, $\partial \mathbf{E}/\partial t$ is zero so that

$$\varepsilon_0 c^2 \mathbf{V} \times \mathbf{B} = \mathbf{J} \qquad (4.127)$$

Consider a circular disk shaped surface having the field point P on its circumference as shown in *Figure 4.21(a)*. The surface is symmetric with respect to d**l**, the centre of the surface being along the direction of d**l**, and all points on the circumference of the surface are at a distance r, θ from d**l**. Integrating eqn (4.127) over this surface and using Stokes' theorem, we have

$$\varepsilon_0 c^2 \int (\mathbf{V} \times \mathbf{B}) \cdot \mathbf{n} dS = \varepsilon_0 c^2 \oint \mathbf{B} \cdot d\mathbf{l} = \int \mathbf{J} \cdot \mathbf{n} dS$$

By symmetry **B** must have the same value along all points of the circumference of the surface, so that

$$2\pi r \sin \theta \varepsilon_0 c^2 B = \int \mathbf{J} \cdot \mathbf{n} dS \qquad (4.128)$$

The electric field of an electric dipole of dipole moment **p** is given by

$$\mathbf{E} = \frac{p}{4\pi\varepsilon_0 r^3} (\mathbf{a}_r 2 \cos \theta + \mathbf{a}_\theta \sin \theta)$$

By analogy

$$\mathbf{J} = \frac{Id l}{4\pi r^3} [\mathbf{a}_r 2 \cos \theta + \mathbf{a}_\theta \sin \theta] \qquad (4.129)$$

where $\mathbf{a}_r$ and $\mathbf{a}_\theta$ are unit vectors in the directions of increasing r and θ respectively. The net flow of current through the circular disk surface in *Figure 4.21(a)* is the same as through a spherical cap surface of radius r subtended by the surface. Consider an element ΔS

of this spherical surface between θ and $\theta+d\theta$. The total flow of current through this surface is equal to

$$\mathbf{J} \cdot \mathbf{n}\Delta S = 2\pi r^2 \sin \theta d\theta J_r = 2\pi r^2 \sin \theta d\theta \frac{Idl}{4\pi r^3} 2 \cos \theta$$

$$= \frac{Idl}{2r} \sin 2\theta \, d\theta$$

Hence the total flow of current through the disk shaped surface in *Figure 4.21(a)* is given by

$$\int \mathbf{J} \cdot \mathbf{n}dS = \frac{Idl}{2r} \int_0^\theta \sin 2\theta \, d\theta = \frac{Idl}{2r}\left[-\frac{\cos 2\theta}{2}\right]_0^\theta$$

$$= \frac{Idl}{2r} \sin^2 \theta$$

Substituting in eqn (4.128)

$$2\pi r\varepsilon_0 c^2 B \sin \theta = \frac{Idl}{2r} \sin^2 \theta$$

$$B = \frac{Idl \sin \theta}{4\pi\varepsilon_0 c^2 r^2} \qquad (4.130)$$

This is in agreement with eqn (4.126), which was developed using the Biot–Savart law. Thus the use of the Biot–Savart law or Maxwell's equations gives the same result. When Maxwell's equations are used, one need only know the current actually flowing through the circular disk shaped surface in *Figure 4.21(a)*. One does not need to know the value of **J** elsewhere, when applying Maxwell's equations, though, of course, the current flowing through the surface depends on the conditions elsewhere. However, if **J** is given on the surface $\oint \mathbf{B} \cdot \mathbf{dl}$ can be evaluated.

It is interesting to note that according to the Biot–Savart law the current flow in the conducting liquid does not contribute to the magnetic field at P in *Figure 4.21(a)*. However, this current flow must be included when applying Maxwell's equations. If this current flow in the liquid were not present, the experimental conditions would be different, as illustrated in *Figure 4.21(b)* where Idl is represented by a stationary positive ion and a moving electron in empty space. This new problem is not a steady state problem as the electric flux through the circular disk surface due to the moving electron in *Figure 4.21(b)* is varying with time. In this case, since no current crosses the surface,

$$\varepsilon_0 c^2 \oint \mathbf{B} \cdot d\mathbf{l} = \varepsilon_0 \int \frac{\partial \mathbf{E}}{\partial t} \cdot \mathbf{n} dS$$

where $\mathbf{E}$ is the electric field due to the moving electron. This is the same problem as discussed in Section 4.7 where it was shown that the application of Maxwell's equations gives the same value for the magnetic field of a moving 'point' charge (electron) as the Biot–Savart law.

4.13. A CRITIQUE OF MAXWELL'S EQUATIONS

Taking Coulomb's law, the principle of constant charge and the transformations of the theory of special relativity as axiomatic, it was shown in Chapter 3 that, for a 'point' charge q moving with uniform velocity $\mathbf{u}$ in empty space, the electric intensity $\mathbf{E}$ and magnetic induction $\mathbf{B}$ are given by eqns (3.24) and (3.27), namely

$$\mathbf{E} = \frac{q\mathbf{r}(1-\beta^2)}{4\pi\varepsilon_0 r^3 (1-\beta^2 \sin^2 \theta)^{3/2}} \tag{4.1}$$

$$\mathbf{B} = \mathbf{u} \times \mathbf{E}/c^2 \tag{4.2}$$

where $\mathbf{r}$ is a vector from the 'present' position of the charge to the field point, θ is the angle between $\mathbf{r}$ and $\mathbf{u}$ and $\beta = u/c$. The vectors $\mathbf{E}$ and $\mathbf{B}$ are defined in terms of the Lorentz force acting on a moving test charge placed at the field point. In this chapter it was shown that at any field point in empty space, eqns (4.1) and (4.2) for the electric and magnetic fields of a 'point' charge moving with uniform velocity satisfy the relations

$$\nabla \cdot \mathbf{E} = 0 \tag{4.131}$$

$$\nabla \cdot \mathbf{B} = 0 \tag{4.132}$$

$$\nabla \times \mathbf{E} = -\frac{\partial \mathbf{B}}{\partial t} \tag{4.133}$$

$$\varepsilon_0 c^2 \nabla \times \mathbf{B} = \varepsilon_0 \frac{\partial \mathbf{E}}{\partial t} \tag{4.134}$$

These equations are known as Maxwell's equations. They are relations between fields quantities at the field point. Both the electric and magnetic fields are propagated from the 'retarded' position of the charge with a speed c in empty space. Eqns (4.131), (4.132), (4.133) and (4.134) were developed for the fields of a charge moving with uniform velocity. In classical electromagnetism it is assumed that these equations are also valid for the fields of an accelerating charge (cf. Appendix 3(c)). By considering an assembly of 'point' charges, such as positive ions and electrons, moving with uniform velocities, such that they build up macroscopic charge and current distributions, it was shown that if $\varepsilon_r = 1$ and $\mu_r = 1$, the *macroscopic* fields **E** and **B** satisfy the equations

$$\mathbf{V} \cdot \mathbf{E} = \rho/\varepsilon_0 \tag{4.135}$$

$$\mathbf{V} \cdot \mathbf{B} = 0 \tag{4.136}$$

$$\mathbf{V} \times \mathbf{E} = -\frac{\partial \mathbf{B}}{\partial t} \tag{4.137}$$

$$\varepsilon_0 c^2 \mathbf{V} \times \mathbf{B} = \varepsilon_0 \frac{\partial \mathbf{E}}{\partial t} + \mathbf{J} \tag{4.138}$$

where ρ is the macroscopic charge density defined by eqn (4.5) and **J** the macroscopic current density is defined by eqn (4.8). The macroscopic electric intensity **E** and the macroscopic magnetic induction **B** are obtained by averaging the microscopic electric intensity **e** and the microscopic magnetic induction **b** respectively over regions of space large on the atomic scale. In this way the large fluctuations in the microscopic fields on the atomic scale average out. The divergences and curls in eqns (4.135)–(4.138) say how these local space average or macroscopic electric fields **E** and **B** vary, when the position of the field point is varied at a fixed time. The time derivatives depend on how the values of the macroscopic fields vary with time at a fixed field point.

The equations for the macroscopic fields **E** and **B** should be applied to phenomena whose scale is very large on the atomic scale. When Maxwell's equations are applied to such macroscopic phenomena, it is the expressions for the curls and divergences of the macroscopic fields and not the macroscopic fields themselves which are integrated over surfaces and volumes large on the atomic scale. It is therefore convenient to do the averaging of the microscopic fields **e** and **b** by using eqns (A1.12) and (A1.10) respectively to define the curls and divergences of the macroscopic fields. For example, the curl of the macroscopic electric intensity, $\mathbf{V} \times \mathbf{E}$, can be defined as

$$(\mathbf{V} \times \mathbf{E}) \cdot \mathbf{n} = \underset{\Delta S \to 0}{\text{Limit}} \frac{\oint \mathbf{e} \cdot \mathbf{dl}}{\Delta S} \qquad (A1.12)$$

and $\mathbf{V} \cdot \mathbf{E}$ can be defined as

$$\mathbf{V} \cdot \mathbf{E} = \underset{\Delta V \to 0}{\text{Limit}} \frac{\int \mathbf{e} \cdot \mathbf{n} dS}{\Delta V} \qquad (A1.10)$$

Similar definitions are used for $\mathbf{V} \times \mathbf{B}$ and $\mathbf{V} \cdot \mathbf{B}$. In eqn (A1.10), ΔV is assumed to be very small on the laboratory scale, but large enough on the atomic scale to contain many atomic charges, say a 10^{-5} cm cube. In eqn (A1.12), it is assumed that ΔS is small on the laboratory scale but large on the atomic scale. On the atomic scale there will be large fluctuations in the local microscopic electric and microscopic magnetic fields $\mathbf{e}$ and $\mathbf{b}$ due to atomic structure. If ΔS and ΔV are kept large on the atomic scale, these iregularities will average out in the integrals in eqns (A1.10) and (A1.12) and these equations can be used to determine the curls and divergences of the macroscopic (or local space average) fields $\mathbf{E}$ and $\mathbf{B}$. It is illustrated in Appendix 3(c) how eqns (4.135), (4.136), (4.137) and (4.138) also hold for systems of accelerating charges.

The development of Maxwell's equations via relativity can be extended in the conventional way to systems containing *stationary* dielectrics, magnetic materials and electrical conductors. At a field point where the macroscopic polarization vector, defined by eqn (4.24), is $\mathbf{P}$ and the macroscopic magnetization vector, defined by eqn (A2.40) of Appendix 2, is $\mathbf{M}$, Maxwell's equations take the form

$$\mathbf{V} \cdot (\varepsilon_0 \mathbf{E} + \mathbf{P}) = \rho / \varepsilon_0 \qquad (4.139)$$

$$\mathbf{V} \cdot \mathbf{B} = 0 \qquad (4.140)$$

$$\mathbf{V} \times \mathbf{E} = -\frac{\partial \mathbf{B}}{\partial t} \qquad (4.141)$$

$$\mathbf{V} \times \left(\frac{\mathbf{B}}{\mu_0} - \mathbf{M} \right) = \mathbf{J} + \varepsilon_0 \frac{\partial \mathbf{E}}{\partial t} + \frac{\partial \mathbf{P}}{\partial t} \qquad (4.142)$$

All the quantities in the above equations are *macroscopic* quantities, $\mathbf{E}$ and $\mathbf{B}$ being the macroscopic fields. The properties of the materials come in via the macroscopic quantities $\mathbf{J}, \mathbf{P}$ and $\mathbf{M}$. The equations refer to one point of space at one instant of time. It is conventional,

116

but by no means necessary, to introduce the abbreviations $\mathbf{D} = \varepsilon_0\mathbf{E}+\mathbf{P}$ and $\mathbf{H} = \mathbf{B}/\mu_0-\mathbf{M}$, in which case eqns (4.139) and (4.142) can be rewritten in the shorter form

$$\nabla . \mathbf{D} = \rho \tag{4.139a}$$

$$\nabla \times \mathbf{H} = \mathbf{J}+\frac{\partial \mathbf{D}}{\partial t} \tag{4.142a}$$

To solve problems one must also be given the constitutive equations, which are generally written in the form

$$\mathbf{D} = \varepsilon_r\varepsilon_0\mathbf{E}; \quad \mathbf{B} = \mu_r\mu_0\mathbf{H}; \quad \mathbf{J} = \sigma\mathbf{E}$$

The dielectric constant ε_r, the relative permeability μ_r and the electrical conductivity σ depend on the properties of the material medium present at the field point. In a macroscopic theory, ε_r, μ_r and σ are treated as smooth continuous functions of position. Between Maxwell's equations, eqns (4.139a), (4.140), (4.141) and (4.142a), and the three constitutive equations there are 17 component equations, one each from eqns (4.139a) and (4.140) and three each from the others. However, there are only 16 scalar component variables, three each for $\mathbf{E}, \mathbf{D}, \mathbf{B}, \mathbf{H}$ and $\mathbf{J}$ and one for ρ. It was shown in Section 4.5, by taking the divergence of both sides of eqn (4.141), that $\partial(\nabla . \mathbf{B})/\partial t$ is zero. Hence eqn (4.140) is an initial condition. This reduces the number of equations to 16.

If in the definitions of $\nabla \times \mathbf{E}$, $\nabla . \mathbf{E}$, $\nabla \times \mathbf{B}$ and $\nabla . \mathbf{B}$, one made ΔS and ΔV small on the atomic scale, there would be little likelihood of an electron or atomic nucleus being inside ΔV or crossing ΔS, so that both $\mathbf{J}$ and ρ would be zero. Eqns (4.135), (4.136), (4.137) and (4.138) would then reduce, in the empty space between electrons and nuclei, to

$$\nabla . \mathbf{e} = 0$$

$$\nabla . \mathbf{b} = 0$$

$$\nabla \times \mathbf{e} = -\frac{\partial \mathbf{b}}{\partial t}$$

$$\nabla \times \mathbf{b} = \frac{1}{c^2}\frac{\partial \mathbf{e}}{\partial t}$$

where $\mathbf{e}$ and $\mathbf{b}$ are the microscopic electric intensity and magnetic induction respectively. The macroscopic polarization and magnetization vectors are not appropriate on the microscopic scale, and one would have to allow for the positions of individual atomic charges and atomic electric and magnetic dipoles in the vicinity of the field point. Consequently no conciseness is gained by introducing the abbreviations **D** and **H** in a microscopic theory.

The choice of which set of Maxwell's equations to use depends on the scale of the phenomenon under investigation. A typical case in which the use of Maxwell's equations is useful is the theory of the reflection of electromagnetic waves at the surface of a dielectric. In this case, one does not want to relate the electric intensity **E** and the magnetic induction **B**, associated with the radiation, back to the source of the radiation. One merely wants to find out what happens, at the boundary of the dielectric, to the electric and magnetic field vectors associated with the radiation. Maxwell's equations are relations between the macroscopic electric intensity **E** and the macroscopic magnetic induction **B** at the boundary, valid irrespective of the position of the source of the radiation. For visible radiation the wavelength is $\sim 5,000\text{Å}$ which is much greater than atomic dimensions, which are $\sim 1\text{Å}$. There is no need to allow for the discrete structure of matter and for the microscopic fluctuations in electric and magnetic fields when discussing the reflection of visible radiation, since, over distances of $\sim 5,000\text{Å}$, the irregularities in the fields and dielectric constants associated with atomic structure will average out. Maxwell's equations for the macroscopic fields give the appropriate boundary conditions, and in these boundary conditions the macroscopic (or local space average) values for **E** and **B** and the macroscopic values for dielectric constant and electrical conductivity can be used, treating the dielectric as a continuous medium. However, if radiation in the x-ray region is used, the wavelength is $\sim 1\text{Å}$, which is comparable with atomic dimensions. The use of the macroscopic dielectric constant and electrical conductivity is no longer appropriate. To calculate the scattering of x-rays, one must use a microscopic theory taking into account the crystal structure of the scattering material.

In Appendix 2 a brief review is given of the experimental evidence on which Maxwell's equations are normally developed in conventional approaches. For example, it is illustrated in Appendix 2(b) how the equation $\mathbf{\nabla} . \mathbf{D} = \rho$ is normally developed for electrostatic charge distributions. In the conventional approaches to electromagnetism, it is then assumed, generally without further comment, that this equation holds for the fields of moving and accelerating charges. This illustrates how Maxwell's equations are often used in a wider context, such as for accelerating charges, than the experimental evidence, on the basis of which they are normally developed.

One does not usually measure the radiation fields near antennae to 'establish' Maxwell's equations. The conventional approach is to develop the equation $\mathbf{V} . \mathbf{D} = \rho$ from electrostatics. The equations $\mathbf{V} . \mathbf{B} = 0$ and $\varepsilon_0 c^2 \, \mathbf{V} \times \mathbf{B} = \mathbf{J}$ are generally developed from the Biot–Savart law, which is valid only in the quasi-stationary approximation, when the radiation fields are generally negligible. The equation $\varepsilon_0 c^2 \mathbf{V} \times \mathbf{B} = \mathbf{J}$ is then generally extended by adding the Maxwell term $\varepsilon_0(\partial \mathbf{E}/\partial t)$ to the right-hand side, more by intelligent guesswork than anything else in an example where the radiation fields are again generally negligible. The equation $\mathbf{V} \times \mathbf{E} = -\partial \mathbf{B}/\partial t$ is developed from experiments on electromagnetic induction in conditions when the radiation fields are again generally negligible. After developing Maxwell's equations in this way for special cases, Maxwell's equations are then generally taken as axiomatic, and it is assumed that they are valid in all possible cases, including systems of accelerating charges moving with velocities comparable with the velocity of light. In this approach, the validity of Maxwell's equations does not depend on whether or not each individual law has been 'established' by experiment for all possible experimental conditions, but on whether or not the predictions of the theory as a whole are in agreement with the experimental results. In the general case of accelerating charges, even in the conventional approaches, Maxwell's equations are checked *a posteriori*. Hertz realized that Maxwell's equations could be applied in a wider context than the experimental evidence on which Maxwell developed them. Speaking to the German Association for the Advancement of Natural Science and Medicine in 1889, Hertz[7] said

It is impossible to study this wonderful theory without feeling as if the mathematical equations had an independent life and intelligence of their own, as if they were wiser than ourselves, indeed wiser than their discoverer, as if they gave forth more than had been put into them. And this is not altogether impossible; it may happen when the equations prove to be more correct than their discoverers could with certainty have known.

In his book on *Electric Waves*, Hertz[8] wrote

To the question, 'what is Maxwell's theory?' I know of no shorter or more definite answer than the following. Maxwell's theory is Maxwell's system of equations.

In the same book Hertz adds,

If we wish to add more colour to the theory, there is nothing to prevent us from supplementing all this and aiding our powers of imagination by concrete representations of the various conceptions as to the nature of electric polarization, the electric current, etc. But scientific accuracy requires of us that we should in no wise confuse the simple and homely figure as it is presented to us by nature, with the gay garment which we use to clothe it. Of our own free will we can make no change whatever in the form of the one, but the cut and colour of the other we can choose as we please.

In the present text, electric and magnetic fields were sometimes represented as continuous lines in space. The number of field lines in diagrams was limited to give a measure of field strength in terms of the closeness of adjoining field lines. In motional e.m.f.s it is conventional to talk in terms of cutting magnetic field lines. These pictures help our powers of imagination. They should be interpreted in the light of the above quotation from Hertz. Maxwell's theory is a set of equations which enable us to calculate the field vectors **E**, **B**, **D** and **H**. When these are known, they can be used to make predictions. For example, the force on a moving test charge can be calculated. The electromagnetic energy of the system can be calculated. The Poynting vector can be used to calculate the rate of energy flow. In the context of classical electromagnetism, the fields themselves are 'observed' when they 'interact' with fundamental particles. To quote Feynman, Leighton and Sands[9]

> The only sensible question is what is the most *convenient* way to look at electrical effects. Some people prefer to represent them as the inter-action at a distance of charges, and to use a complicated law. Others love the field lines . . .
> The best way is to use the abstract field idea. That it is abstract is unfortunate, but necessary. The attempts to try to represent the electric field as the motion of some kind of gear wheels, or in terms of lines, or of stresses in some kind of material have used up more effort of physicists than it would have taken simply to get the right answers about electro-dynamics.

The interpretation of Maxwell's equations as an axiomatic system to be checked *a posteriori* is similar to what is becoming the conventional approach to quantum mechanics. In elementary courses Schrödinger's equations can be 'developed' for a single particle from the 'wave' nature of the electron. However, when considering systems of more than one particle, one must start from a postulative basis. For example, Sherwin[10] lists five postulates, which enable one to set up the differential equations to determine the wave function for any system, and rules for calculating physical quantities such as position, linear momentum, angular momentum etc. from the wave function. One does not try and check each individual postulate of quantum mechanics separately. The theory as a whole is used to make predictions, and the degree of acceptance of the theory depends on whether or not the predictions of the theory as a whole are in agreement with the experimental results, that is, the theory is checked *a posteriori*.

In this monograph, only classical electromagnetism has been considered and all quantum effects were ignored. In order to account for quantum effects, the field equations must be quantized (Reference: Heitler[11]). Maxwell's equations and the equations for the potentials ϕ and **A** developed from Maxwell's equations are used as the starting point for the quantum theory of radiation. Maxwell's equations can, however, be used to interpret physical optics (Reference: Born and Wolf[12]). The reason can be illustrated by the following quotation from Feynman, Leighton and Sands[13]

> When we have the wave function of a single photon, it is the amplitude to find a photon somewhere. Although we haven't ever written it down there is an equation for the photon wave function analogous to the Schrödinger equation for the electron. The photon equation is just the same as Maxwell's equations for the electromagnetic field, and the wave function is the same as the vector potential **A**. The wave function turns out to be just the vector potential. The quantum physics is the same thing as the classical physics because photons are non-interacting Bose particles and many of them can be in the same state—as you know they *like* to be in the same state. The moment you have billions in the same state (that is, in the same electromagnetic wave), you can measure the wave function, which is the vector potential directly. Of course, it worked historically the other way. The first observations were on situations with many photons in the same state, and so we were able to discover the correct equation for a single photon by observing directly with our hands on a macroscopic level the nature of wave function.

Thus Maxwell's equations or the equations for the potentials ϕ and **A** can be used to interpret physical optics. Once ϕ and **A** or **E** and **H** are calculated from the charge and current distributions, the energy density of photons in empty space can be estimated from the expression $(\varepsilon_0 E^2/2 + \mu_0 H^2/2)$. The electric and magnetic radiation fields can be associated with photons (or quanta) as illustrated, for example, by experiments such as the photo-disintegration of a deuteron by a high energy γ-ray. (At photon energies a few keV above threshold, the photodisintegration of the deuteron is due mainly to electric dipole absorption. The proton arising from the dis-integration is emitted preferentially in the direction of the electric vector of the incident photon, Wilkinson[14].) The interactions between charges can be interpreted in terms of the exchange of virtual photons.

In this chapter, Maxwell's equations were developed via relativity. A brief outline was also given of conventional approaches to Maxwell's equations. The important thing is not how Maxwell's equations are developed, but the fact that in Maxwell's equations we have a theory which is consistent with the theory of special relativity and is in good agreement with the experimental results, provided all quantum effects can be neglected.

In Chapter 4 we have referred to electric and magnetic fields, seemingly as if they were separate independent entities. It was shown in Chapter 3, that the electric and magnetic fields represent two aspects of the forces between moving charges. The electric field depends on that part of the force on a test charge, which is independent of the velocity of the test charge, whereas the magnetic field depends on that part of the force on a test charge which depends on the velocity of the test charge. Both the expressions for the electric and magnetic field of a charge moving with uniform velocity were developed from Coulomb's law in Chapter 3. These expressions were then used to develop Maxwell's equations. This approach illustrates the essential unity of classical electromagnetism. This unity is not as apparent when Maxwell's equations are developed in the conventional way from a number of diverse phenomena.

A discussion of how various macroscopic electromagnetic phenomena are related to the electric and magnetic fields of accelerating 'point' charges is given in Appendix 3(d) [cf. eqns (A3.28) and (A3.29)].

A discussion of the relativistic invariance of Maxwell's equations, and a discussion of the electrodynamics of moving media is given in Chapter 6.

REFERENCES

[1] CULLWICK, E. G. *Electromagnetism and Relativity*, 2nd Ed., p. 207. Longmans Green, 1959
[2] SHERWIN, C. W. *Basic Concepts in Physics*, Ch. 5, Holt, Rinehart and Winston, New York, 1961
[3] SLEPIAN, J. *Amer. J. Phys.* **19** (1951) 87
[4] DIRAC, P. A. M. *Phys. Rev.* **74** (1948) 817
[5] PUGH, E. M. and PUGH, E. W. *Principles of Electricity and Magnetism*, p. 52. Addison-Wesley, Reading, Mass., 1960
[6] MOULLIN, E. B. *The Principles of Electromagnetism*, p. 22. Clarendon Press, Oxford, 1932

[7] HERTZ, H. Address to Sixty Second Meeting of German Assoc. for Advancement of Natural Science and Medicine, Heidelberg, 1889, Reprinted: *Autobiography of Science*, Editors, F. R. Moulton and J. J. Schifferes, p. 63. John Murray, London, 1963

[8] HERTZ, H. *Electric Waves*, pp. 21 and 28. Macmillan, London, 1893

[9] FEYNMAN, R. P., LEIGHTON, R. B. and SANDS, M. *The Feynman Lectures on Physics*, Vol. II, pp. 1–9. Addison-Wesley, Reading Mass., 1964

[10] SHERWIN, C. W. *Introduction to Quantum Mechanics*, Holt, Rinehart and Winston, New York, 1959

[11] HEITLER, W. *Quantum Theory of Radiation*, 2nd Ed. Clarendon Press, Oxford, 1954

[12] BORN, M. and WOLF, W. *Principles of Optics*, Pergamon Press, London, 1959

[13] FEYNMAN, R. P., LEIGHTON, R. B. and SANDS, M. *The Feynman Lectures on Physics*, Vol. III, pp. 21–6. Addison-Wesley, Reading, Mass., 1965

[14] WILKINSON, D. H. *Phil Mag.* Series 7, **43** (1952) 659

[15] WHITMER, R. M. *Amer. J. Phys.* **33** (1965) 481

[16] TRICKER, R. A. R. *Early Electrodynamics, the First Law of Circulation.* Pergamon Press, Oxford, 1965

[17] RASETTI, F. *Phys. Rev.* **66** (1944) 1

[18] FEYNMAN, R. P, LEIGHTON, R. B. and SANDS, M., *The Feynman Lectures on Physics*, Vol. II, pp. 23–2. Addison-Wesley, Reading, Mass., 1964

[19] FRENCH, A. P. and TESSMAN, J. R. *Amer. J. Phys.* **31** (1963) 201

[20] VINTI, J. P. and MONTGOMERY, D. J. X. *Amer. J. Phys.* **17** (1949) 298

PROBLEMS

Problem 4.1—Using eqns (3.56) and (3.57) show that the electric and magnetic fields of an isolated magnetic monopole moving with uniform velocity in empty space satisfy Maxwell's equations in the form

$$\mathbf{V} \cdot \mathbf{E} = 0$$

$$\mathbf{V} \cdot \mathbf{B} = 0$$

$$\mathbf{V} \times \mathbf{E} = -\frac{\partial \mathbf{B}}{\partial t}$$

$$\varepsilon_0 c^2 \mathbf{V} \times \mathbf{B} = \varepsilon_0 \frac{\partial \mathbf{E}}{\partial t}$$

Problem 4.2—Assume that magnetic monopole strength is defined by eqns (3.48) and (3.49) respectively

$$\mathbf{f} = \frac{(q_m)_1 (q_m)_2 \mathbf{r}}{\mu_0 4\pi r^3} = \frac{\mu_0 (q_m^*)_1 (q_m^*)_2 \mathbf{r}}{4\pi r^3}$$

Consider a system of magnetic monopoles moving with uniform velocities, such that they build up macroscopic magnetic pole distributions of magnetic charge density ρ_m (or ρ_m^*) and magnetic current density J_m (or J_m^*), defined in analogous ways to eqns (4.5) and (4.7) as

$$\rho_m = \sum_{i=1}^{N} (q_m)_i/\Delta V \left\{ \text{or} \quad \rho_m^* = \sum_{i=1}^{N} (q_m^*)_i/\Delta V \right\} \quad (4.143)$$

$$\mathbf{J}_m = nq_m\mathbf{u} \quad \{\text{or} \quad \mathbf{J}_m^* = nq_m^*\mathbf{u}\} \quad (4.144)$$

Using eqns (3.56) and (3.57) and the methods developed for moving electric charges in this chapter, show that

$$\mathbf{V} \cdot \mathbf{E} = 0$$

$$\mathbf{V} \cdot \mathbf{B} = \rho_m = \mu_0\rho_m^*$$

$$\mathbf{V} \times \mathbf{E} = -\frac{\partial \mathbf{B}}{\partial t} - \mathbf{J}_m = -\frac{\partial \mathbf{B}}{\partial t} - \mu_0\mathbf{J}_m^*$$

$$\varepsilon_0 c^2 \mathbf{V} \times \mathbf{B} = \varepsilon_0 \frac{\partial \mathbf{E}}{\partial t}$$

Problem 4.3—Show that for a system of moving electric charges and moving magnetic monopoles in otherwise empty space, Maxwell's equations take the form

$$\mathbf{V} \cdot \mathbf{E} = \rho/\varepsilon_0$$

$$\mathbf{V} \cdot \mathbf{B} = \rho_m = \mu_0\rho_m^*$$

$$\mathbf{V} \times \mathbf{E} = -\frac{\partial \mathbf{B}}{\partial t} - \mathbf{J}_m = -\frac{\partial \mathbf{B}}{\partial t} - \mu_0\mathbf{J}_m^*$$

$$\varepsilon_0 c^2 \mathbf{V} \times \mathbf{B} = \varepsilon_0\frac{\partial \mathbf{E}}{\partial t} + \mathbf{J}$$

where ρ and $\mathbf{J}$ are the volume density of electric charge and the electric current density respectively.

Problem 4.4—By a method similar to Section 4.7 use Maxwell's equations to develop the electric field of a magnetic monopole, in the limit when the velocity of the monopole is very much less than the velocity of light.

Problem 4.5—Repeat the calculation of Section 4.8 for a line of magnetic charge. Plot how Φ, the flux of $\mathbf{B}$ through the circular disk shaped surface in *Figure 4.14*, varies with x_0. Compare and contrast the variation of the flux of $\mathbf{B}$ due to a line of magnetic monopoles to the variation of the flux of $\mathbf{E}$ from a line of electric charge under identical conditions.

124

PROBLEMS

Problem 4.6—Consider a distribution of magnetic monopoles moving with uniform velocities. Show that in the limit when the velocities of the monopoles are very much less than the velocity of light, the electric field is

$$\mathbf{E}(\mathbf{r}) \simeq -\int \frac{\mathbf{J}_m(\mathbf{r}') \times (\mathbf{r}-\mathbf{r}') \mathrm{d}V'}{4\pi \left| \mathbf{r}-\mathbf{r}' \right|^3}$$

$$\simeq -\mu_0 \int \frac{\mathbf{J}_m^*(\mathbf{r}') \times (\mathbf{r}-\mathbf{r}') \mathrm{d}V'}{4\pi \left| \mathbf{r}-\mathbf{r}' \right|^3}$$

where $\mathbf{r}$ is the position vector of the field point and $\mathbf{r}'$ is the position vector of the source point.

Problem 4.7—If the magnetization $\mathbf{M}(\mathbf{r}')$ of a magnetic material is varying with time, show that it gives rise to an electric field given by

$$\mathbf{E}(\mathbf{r}) = -\mu_0 \int \frac{\dfrac{\partial \mathbf{M}(\mathbf{r}')}{\partial t} \times (\mathbf{r}-\mathbf{r}') \mathrm{d}V'}{4\pi \left| \mathbf{r}-\mathbf{r}' \right|^3}$$

(Hint: Use the magnetic pole model of atomic magnetic dipoles. By analogy with *Figure 4.18*, the magnetic current is $\mathbf{J}_m^* = \partial \mathbf{M}/\partial t$. Use the magnetic analogue of the Biot–Savart law, problem 4.6. For a derivation in terms of the Amperian model, the reader is referred to Whitmer[15].)

Problem 4.8—A solenoid of radius a, n_1 turns per metre carrying a current I_1 lies on the x axis and extends from $x = -\infty$ to the origin. Show that, if $x \gg a$, the magnetic induction at any point along the $+x$ axis is $\mu_0 n_1 I_1 \pi a^2/4\pi x^2$.

A second solenoid of radius b, n_2 turns per metre carrying a current I_2 lies on the x axis from $x = +r$ to $x = +\infty$. Show that, if $r \gg a$ and $r \gg b$, then the force between the two solenoids can be written in the form

$$f_x = -\frac{\mu_0}{4\pi} \frac{q_1^* q_2^*}{r^2}$$

where $q_1^* = \pi a^2 n_1 I_1$ and $q_2^* = \pi b^2 n_2 I_2$. Discuss the significance of the result in the interpretation of the forces between bar magnets. (Hint: Use the normal formula for the field on the axis of a solenoid to calculate the magnetic field. Calculate the total flux through the second solenoid and use

$$f = I_1 I_2 \frac{\partial M_{12}}{\partial r}$$

The case when the second solenoid is not along the x axis was discussed by Ampère. Reference: Tricker[16].)

Problem 4.9—The Amperian model is sometimes extended to atomic magnetic dipoles which are pictured as little current loops. This picture is reasonable for electron orbital motion, interpreted in terms of the Bohr Theory. How valid is the model in quantum mechanics and how valid is it for electron spin? In your answer discuss the interpretation of diamagnetism in terms of the various models of atomic magnetic dipoles.

Problem 4.10—Show that in an isolated permanent bar magnet the vectors **B** and **H** are approximately in opposite directions. If a *high* energy cosmic ray μ-meson of charge q velocity **u** passes through such a bar magnet, is the *average* magnetic force on the μ meson in the direction of $q\mathbf{u}\times\mathbf{B}$ or of $q\mathbf{u}\times\mathbf{H}$? (Hint: The experiments of Rasetti[17] show that if $u \sim c$ the average force is in the direction of $q\mathbf{u}\times\mathbf{B}$.)

If a magnetic monopole of strength q_m^* crosses a permanent magnet, is the average magnetic force in the direction of $q_m^*\mathbf{B}$ or the direction of $q_m^*\mathbf{H}$? (Hint: When in doubt wait for the experiment to be performed.)

Problem 4.11—The frequency of the e.m.f. giving rise to the current flow in *Figures 4.19* and *A2.6* increases steadily up to radio frequencies. Discuss how the electric and magnetic fields between the plates change with frequency. (Reference: Feynman, Leighton and Sands[18].)

Problem 4.12—The coil in a moving coil galvanometer is rotating in a steady uniform magnetic field. Hence, $\partial\mathbf{B}/\partial t$ is zero at the position of the coil. How then do you account for the electromagnetic damping of a galvanometer? (Hint: Is this a transformer or a motional e.m.f.? Cf. Appendix 2(g).)

Problem 4.13—Let S be a surface separating two media having dielectric constants $(\varepsilon_r)_1$ and $(\varepsilon_r)_2$, relative permeabilities $(\mu_r)_1$ and $(\mu_r)_2$ and electrical conductivities σ_1 and σ_2 respectively. Let **n** be a unit vector normal to S directed from medium 1 to medium 2. Using Maxwell's equations, show that the boundary conditions are:

(a) Tangential components of **E** are continuous, that is

$$\mathbf{n}\times(\mathbf{E}_2-\mathbf{E}_1) = 0$$

(b) The discontinuity in the normal component of **D** equals Ω, the true charge per unit area on the surface

$$\mathbf{n}\cdot(\mathbf{D}_2-\mathbf{D}_1) = \Omega$$

(c) The tangential components of the vector **H** are continuous, that is, provided neither σ_1 or σ_2 is infinite

$$\mathbf{n}\times(\mathbf{H}_2-\mathbf{H}_1) = 0$$

(d) The normal component of **B** is continuous

$$\mathbf{n}\cdot(\mathbf{B}_2-\mathbf{B}_1) = 0$$

126

Discuss what happens in cases (c) and (d) if σ_1 or σ_2 is infinite (a super-conductor). Discuss also the effects you would expect if there were magnetic monopoles on the surface S.

Problem 4.14—An alternating conduction current $I_0 \sin 2\pi ft$ flows along a thin long wire of conductivity σ and relative dielectric coefficient ε_r. Show that the displacement 'current' in the wire is

$\{2\pi f \varepsilon_r \varepsilon_0 I_0 / \sigma\} \cos 2\pi ft$. Show, if $\sigma = 10^7$ mhos per metre,

$\varepsilon_r = 1$, $\varepsilon_0 = 8 \cdot 85 \times 10^{-12}$ farads per metre, then $\partial D/\partial t$ is not significant at frequencies below optical frequencies.

(Hint: Use $\mathbf{J} = \sigma \mathbf{E} = \sigma \mathbf{D}/\varepsilon_r \varepsilon_0$ to calculate $\partial D/\partial t$.)

Problem 4.15—A large isolated parallel plate capacitor with air between the plates is charged such that there is a potential difference between the plates. An x-ray source is used to produce uniform ionization between the plates of the capacitor. Under the influence of the electric field between the plates, an electric current will flow between the plates leading to the slow discharge of the capacitor. Assuming plates of *infinite* area, calculate the magnetic field between the plates of the capacitor due to the current flow. (Hint: It is zero by both the Biot–Savart law and using Maxwell's equations. For example, using Maxwell's equations

$$\varepsilon_0 c^2 \oint \mathbf{B} \cdot d\mathbf{l} = \int \left(\varepsilon_0 \frac{\partial \mathbf{E}}{\partial t} + \mathbf{J} \right) \cdot \mathbf{n} dS$$

where in this case $\mathbf{J}$ gives rise to a change in the charge on the plates which reduces $\mathbf{E}$ such that

$$J = -\varepsilon_0 \frac{\partial E}{\partial t}$$

Reference: French and Tessman[19].)

Problem 4.16—Compare and contrast the roles of the vectors $\mathbf{E}$ and $\mathbf{B}$ in the interpretation of Physical Optics in terms of classical electromagnetism, with the role of the wave function in Quantum Mechanics.

5

THE VECTOR AND SCALAR POTENTIAL
VIA RELATIVITY

5.1. INTRODUCTION

It is sometimes more convenient to solve problems in classical electromagnetism by introducing the scalar potential ϕ and the vector potential $\mathbf{A}$. The full definitions of ϕ and $\mathbf{A}$ and the differential equations for ϕ and $\mathbf{A}$ will be developed from Maxwell's equations in Section 5.7. For the present, it will be sufficient to assume that ϕ and $\mathbf{A}$ are related to the electric intensity $\mathbf{E}$ and the magnetic induction $\mathbf{B}$ by the equations

$$\mathbf{E} = -\nabla\phi - \frac{\partial\mathbf{A}}{\partial t} \qquad (5.1)$$

$$\mathbf{B} = \nabla\times\mathbf{A} \qquad (5.2)$$

5.2. VECTOR AND SCALAR POTENTIALS OF A 'POINT' CHARGE MOVING WITH UNIFORM VELOCITY

Consider a 'point' charge q moving with uniform velocity $\mathbf{u}$ along the x axis of the inertial frame Σ as shown in *Figure 5.1*. Let q be at the origin at $t = 0$, when the electric and magnetic fields are determined at the field point P, which has co-ordinates $x, y, z, t = 0$, and which is at a distance $\mathbf{r}_0$ from the origin. The expressions for the electric and magnetic fields of a moving 'point' charge q moving with uniform velocity were developed in Chapter 3, taking the transformations of the theory of special relativity, the principle of constant electric charge and Coulomb's law as axiomatic. For the case illustrated in *Figure 5.1* from eqns (3.19) and (3.20),

$$\mathbf{E} = g\mathbf{r}_0 \qquad (5.3)$$

and

$$\mathbf{B} = \frac{\mathbf{u}}{c^2}\times\mathbf{E} \qquad (5.4)$$

where according to eqn (3.10)

$$g = \frac{q\left(1 - \dfrac{u^2}{c^2}\right)}{4\pi\varepsilon_0 s^3} \tag{5.5}$$

According to eqn (3.7)

$$s = \sqrt{x^2 + \left(1 - \frac{u^2}{c^2}\right)(y^2 + z^2)} \tag{5.6}$$

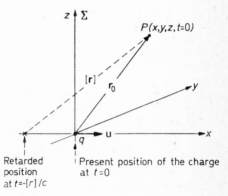

Figure 5.1. Calculation of the scalar and vector potentials at the field point P due to a charge q moving with uniform velocity **u**. *Note* $\mathbf{r}_0$ *is the distance from the 'present' position of the charge to the field point P, whereas* [**r**] *is the distance from the retarded position of the charge to P*

It will now be shown that, if it is assumed the scalar and vector potentials of a moving 'point' charge are given by

$$\phi = \frac{q}{4\pi\varepsilon_0 s} \tag{5.7}$$

$$\mathbf{A} = \frac{q\mathbf{u}}{4\pi\varepsilon_0 c^2 s} \quad \begin{cases} A_x = \dfrac{qu}{4\pi\varepsilon_0 c^2 s} \\[2mm] A_y = 0 \\[2mm] A_z = 0 \end{cases} \tag{5.8}$$

then the application of

$$\mathbf{E} = -\nabla\phi - \frac{\partial \mathbf{A}}{\partial t}$$

and

$$\mathbf{B} = \nabla \times \mathbf{A}$$

129

gives eqns (5.3) and (5.4) for the electric and magnetic fields of a 'point' charge moving with uniform velocity.

The x component of eqn (5.1) is

$$E_x = -\frac{\partial \phi}{\partial x} - \frac{\partial A_x}{\partial t} = -\frac{\partial}{\partial x}\left(\frac{q}{4\pi\varepsilon_0 s}\right) - \frac{\partial}{\partial t}\left(\frac{qu}{4\pi\varepsilon_0 c^2 s}\right)$$

Since the fields of a charge moving with uniform velocity are carried along convectively with the charge, according to eqn (4.37).

$$\frac{\partial}{\partial t} = -\mathbf{u} \cdot \nabla = -u\frac{\partial}{\partial x}$$

Hence,

$$E_x = -\frac{\partial}{\partial x}\left(\frac{q}{4\pi\varepsilon_0 s}\right) + u\frac{\partial}{\partial x}\left(\frac{qu}{4\pi\varepsilon_0 c^2 s}\right) \qquad (5.9)$$

Since u is a constant,

$$\frac{\partial}{\partial x}\left(\frac{1}{s}\right) = \frac{\partial}{\partial x}\left(\frac{1}{\sqrt{x^2+(1-u^2/c^2)(y^2+z^2)}}\right) = -\frac{x}{s^3}$$

Substituting in eqn (5.9)

$$E_x = \frac{qx}{4\pi\varepsilon_0 s^3} - \frac{qu^2 x}{4\pi\varepsilon_0 c^2 s^3} = \frac{q(1-u^2/c^2)x}{4\pi\varepsilon_0 s^3}$$

that is,

$$E_x = gx \qquad (5.10)$$

where g is given by eqn (5.5)

Now,

$$E_y = -\frac{\partial \phi}{\partial y} - \frac{\partial A_y}{\partial t}$$

Since,

$$A_y = 0,$$

$$E_y = -\frac{\partial}{\partial y}\left(\frac{q}{4\pi\varepsilon_0 s}\right)$$

Now,

$$\frac{\partial}{\partial y}\left(\frac{1}{s}\right) = \frac{\partial}{\partial y}\left(\frac{1}{\sqrt{x^2+(1-u^2/c^2)(y^2+z^2)}}\right)$$

$$\frac{\partial}{\partial y}\left(\frac{1}{s}\right) = -\frac{(1-u^2/c^2)y}{s^3}$$

130

Hence,

$$E_y = \frac{q(1-u^2/c^2)y}{4\pi\varepsilon_0 s^3} = gy \qquad (5.11)$$

Similarly,

$$E_z = \frac{q(1-u^2/c^2)z}{4\pi\varepsilon_0 s^3} = gz \qquad (5.12)$$

Combining eqns (5.10), (5.11) and (5.12)

$$\mathbf{E} = g\mathbf{r}_0$$

This is in agreement with eqn (5.3).

Now, $\mathbf{B} = \mathbf{V}\times\mathbf{A}$, where according to eqn (5.8) $\mathbf{A}$ has the component A_x only. Hence

$$B_x = \frac{\partial A_z}{\partial y} - \frac{\partial A_y}{\partial z} = 0 \qquad (5.13)$$

$$B_y = \frac{\partial A_x}{\partial z} - \frac{\partial A_z}{\partial x}$$

$$= \frac{\partial}{\partial z}\left(\frac{qu}{4\pi\varepsilon_0 c^2 s}\right)$$

Now,

$$\frac{\partial}{\partial z}\left(\frac{1}{s}\right) = -\frac{(1-u^2/c^2)z}{s^3}$$

Hence,

$$B_y = -\frac{qu(1-u^2/c^2)z}{4\pi\varepsilon_0 c^2 s^3} = -\frac{guz}{c^2}$$

or, using eqn (5.12)

$$B_y = -\frac{uE_z}{c^2} \qquad (5.14)$$

Now,

$$B_z = \frac{\partial A_y}{\partial x} - \frac{\partial A_x}{\partial y} = -\frac{qu}{4\pi\varepsilon_0 c^2}\frac{\partial}{\partial y}\left(\frac{1}{s}\right)$$

$$B_z = \frac{qu(1-u^2/c^2)y}{4\pi\varepsilon_0 c^2 s^3} = \frac{uE_y}{c^2} \qquad (5.15)$$

131

Since $\mathbf{u}$ has an x component only, eqns (5.13), (5.14) and (5.15) can be combined into the vector equation.

$$\mathbf{B} = \frac{\mathbf{u}}{c^2} \times \mathbf{E}$$

This is in agreement with eqn (5.4). Thus if it is assumed that ϕ and $\mathbf{A}$ are given by eqns (5.7) and (5.8), then application of the equations $\mathbf{E} = -\nabla\phi - \partial\mathbf{A}/\partial t$ and $\mathbf{B} = \nabla \times \mathbf{A}$ gives the expressions for the electric and magnetic fields of a 'point' charge moving with uniform velocity developed in Chapter 3.

From eqn (3.8)

$$s = r_0\sqrt{1 - (u^2/c^2)\sin^2\theta}$$

Eqns (5.7) and (5.8) can be rewritten in the form

$$\phi = \frac{q}{4\pi\varepsilon_0 r_0\sqrt{1 - (u^2/c^2)\sin^2\theta}} \tag{5.16}$$

$$\mathbf{A} = \frac{q\mathbf{u}}{4\pi\varepsilon_0 c^2 r_0\sqrt{1 - (u^2/c^2)\sin^2\theta}} \tag{5.17}$$

where $\mathbf{r}_0$ is a vector from the *present* position of the charge to the field point and θ is the angle between $\mathbf{r}_0$ and $\mathbf{u}$.

5.3. THE LORENTZ CONDITION

For the case of the charge q moving with uniform velocity $\mathbf{u}$ along the x axis, as illustrated in *Figure 5.1*, from eqn (5.8), we have at $t = 0$

$$A_x = \frac{qu}{4\pi\varepsilon_0 c^2 s}; \quad A_y = 0; \quad A_z = 0 \tag{5.8}$$

Hence

$$\nabla \cdot \mathbf{A} = \frac{\partial A_x}{\partial x} + \frac{\partial A_y}{\partial y} + \frac{\partial A_z}{\partial z} = \frac{qu}{4\pi\varepsilon_0 c^2}\frac{\partial}{\partial x}\left(\frac{1}{s}\right) \tag{5.18}$$

The scalar potential is given by eqn (5.7), namely

$$\phi = \frac{q}{4\pi\varepsilon_0 s} \tag{5.7}$$

132

For a charge moving with uniform velocity u along the x axis, from eqn (4.37)

$$\frac{\partial}{\partial t} = -\mathbf{u} \cdot \mathbf{\nabla} = -u \frac{\partial}{\partial x}$$

Hence,

$$\frac{\partial \phi}{\partial t} = -\frac{qu}{4\pi\varepsilon_0} \frac{\partial}{\partial x} \left(\frac{1}{s}\right)$$

$$\frac{1}{c^2} \frac{\partial \phi}{\partial t} = -\frac{qu}{4\pi\varepsilon_0 c^2} \frac{\partial}{\partial x} \left(\frac{1}{s}\right) \tag{5.19}$$

Adding eqns (5.18) and (5.19) gives

$$\mathbf{\nabla} \cdot \mathbf{A} + \frac{1}{c^2} \frac{\partial \phi}{\partial t} = 0 \tag{5.20}$$

This relation is known as the Lorentz condition. Thus the scalar potential ϕ and the vector potential $\mathbf{A}$ of a 'point' charge moving with uniform velocity, given by eqns (5.7) and (5.8), satisfy the Lorentz condition. Our development of eqns (5.7) and (5.8) was based on the transformations of the theory of special relativity. This illustrates that the Lorentz condition is consistent with the theory of special relativity.

5.4. THE SCALAR AND VECTOR POTENTIALS OF A 'POINT' CHARGE RELATED TO ITS RETARDED POSITION

If the fields $\mathbf{E}$ and $\mathbf{B}$ are propagated with speed c in empty space, then the potentials ϕ and $\mathbf{A}$ are also propagated from their source with a velocity c in empty space. If a 'point' charge is accelerating, the potentials ϕ and $\mathbf{A}$ must be related to the retarded position of the charge, that is the position the charge had at the time the fields 'left' it at a time $[r]/c$ before the fields are determined at the field point, as illustrated in *Figure 3.5*. It was shown in Section 3.3, eqn (3.38), that

$$s = \left[r - \frac{\mathbf{r} \cdot \mathbf{u}}{c} \right] \tag{5.21}$$

where $[\mathbf{r}]$ is the distance from the retarded position of the charge to the field point, at the time $[r]/c$ before the fields are determined at the field point.

Hence, eqns (5.7) and (5.8) can be rewritten

$$\phi = \frac{q}{4\pi\varepsilon_0[r - \mathbf{r} \cdot \mathbf{u}/c]} \qquad (5.22)$$

$$\mathbf{A} = \frac{q}{4\pi\varepsilon_0 c^2}\left[\frac{\mathbf{u}}{r - \mathbf{r} \cdot \mathbf{u}/c}\right] \qquad (5.23)$$

All the quantities inside the square brackets in eqns (5.22) and (5.23) refer to the retarded position of the charge. Eqns (5.22) and (5.23) were developed for charges moving with uniform velocity $\mathbf{u}$ taking the transformations of the theory of special relativity, the principle of constant electric charge and Coulomb's law as axioms. The potentials for the more general case of an accelerating charge should reduce to eqns (5.22) and (5.23), if the acceleration of the charge is zero. It can be shown, using classical electromagnetic theory in the way outlined in Section 5.8, that, in fact, the scalar and vector potentials of an accelerating 'point' charge are also given by eqns (5.22) and (5.23) respectively. These expressions for the potentials of a moving 'point' charge are known as the Liénard–Wiechert potentials.

5.5. VECTOR AND SCALAR POTENTIALS OF A DISTRIBUTION OF 'POINT' CHARGES

Eqns (5.22) and (5.23) were developed for a single isolated 'point' charge such as an electron, a proton or a positive ion. Eqns (5.22) and (5.23) relate the potentials to the retarded position of the charge. The principle of superposition can be used to calculate the total scalar and vector potential of a distribution of 'point' charges, which are moving with uniform velocities. Consider the distribution of moving 'point' charges shown in *Figure 5.2*. Let the total scalar potential ϕ be calculated at the field point P at the time t.

According to eqn (5.22) the contribution of the charge q_i to the scalar potential at P at the time t is

$$\phi_i = \frac{[q_i]}{4\pi\varepsilon_0[r_i - \mathbf{r}_i \cdot \mathbf{u}_i/c]} \qquad (5.24)$$

where $\mathbf{u}_i$ is the velocity of the charge q_i at its retarded position at the time $t - [r_i]/c$, where $[r_i]$ is the distance from the retarded position of q_i to the field point P. The total scalar potential is obtained by

134

summing the individual scalar potentials of the N individual 'point' charges in the complete system. Using eqn (5.24) we have

$$\phi_{\text{total}} = \sum_{i=1}^{N} \frac{[q_i]}{4\pi\varepsilon_0[r_i - \mathbf{r}_i \cdot \mathbf{u}_i/c]} \tag{5.25}$$

Similarly, using eqn (5.23)

$$\mathbf{A}_{\text{total}} = \sum_{i=1}^{N} \frac{[q_i\mathbf{u}_i]}{4\pi\varepsilon_0 c^2[r_i - \mathbf{r}_i \cdot \mathbf{u}_i/c]} \tag{5.26}$$

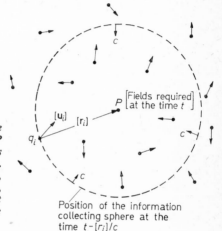

Figure 5.2. The calculation of the scalar and vector potentials at P due to a system of 'point' charges moving with uniform velocities. The 'information collecting sphere', shown dotted in Figure 5.2, collapses with speed c to arrive at the field point P at the time t the potentials at P are required

Position of the information collecting sphere at the time $t - [r_i]/c$

where in eqn (5.26) the summation is the vector addition of the vector potentials of the N 'point' charges in the complete system. Now the retarded times for the various 'point' charges in eqns (5.25) and (5.26) are different, since the various charges are at different distances from the field point P. For *purposes of exposition*, it is convenient, when summing the contributions of individual charges to eqns (5.25) and (5.26), to think in terms of an *imaginary* 'information collecting sphere' with centre at the field point P, and which collapses with velocity c in empty space, such that the 'information collecting sphere' arrives at P at the time t the potentials at P are required. This imaginary 'information collecting sphere' would pass the various charges at the appropriate retarded times. It will be *imagined* that the 'information collecting sphere' collects information about all the charges it passes as it collapses to arrive at P at the time t. The data on the charges, their velocities and positions at their

F   135

retarded positions can be used to calculate ϕ and $\mathbf{A}$ in eqns (5.25) and (5.26) respectively.

It was shown in Section 5.3 that for a single point charge, according to the Lorentz condition, eqn (5.20)

$$\nabla \cdot \mathbf{A}_i + \frac{1}{c^2}\frac{\partial \phi_i}{\partial t} = 0$$

Now,

$$\nabla \cdot \mathbf{A}_1 + \nabla \cdot \mathbf{A}_2 + \ldots = \nabla \cdot (\mathbf{A}_1 + \mathbf{A}_2 + \ldots) = \nabla \cdot \mathbf{A}$$

$$\frac{\partial \phi_1}{\partial t} + \frac{\partial \phi_2}{\partial t} + \ldots = \frac{\partial}{\partial t}(\phi_1 + \phi_2 \ldots) = \frac{\partial \phi}{\partial t}$$

Hence, for the total potentials ϕ and $\mathbf{A}$ of a distribution of 'point' charges moving with uniform velocities

$$\nabla \cdot \mathbf{A} + \frac{1}{c^2}\frac{\partial \phi}{\partial t} = 0 \qquad\qquad (5.27)$$

This relation, the Lorentz condition, relates the total potentials ϕ and $\mathbf{A}$ at a point in space at one instant of time. It is not necessary to relate ϕ and $\mathbf{A}$ back to their sources to apply eqn (5.27). It applies whatever the positions of the charges. It is a relation between field quantities.

5.6. THE RETARDED POTENTIALS

Having developed the potentials for a system of microscopic 'point' charges moving with uniform velocities, the potentials at a field point in empty space outside a macroscopic charge and current system will now be developed, by treating a continuous charge distribution as a collection of an extremely large number of moving 'point' charges such as electrons and ions.

Consider a macroscopic distribution of charge made up of n microscopic charges per cubic metre of magnitude q coulombs each, all moving with the same uniform velocity $\mathbf{u}$ so that the macroscopic charge distribution is moving with velocity $\mathbf{u}$. The potentials ϕ and $\mathbf{A}$ will be calculated at the field point P in empty space at the time t, as shown in *Figure 5.3*. Consider the surface element δS of the 'information collecting sphere' which is crossing the macroscopic charge distribution shown in *Figure 5.3*. The 'information collecting sphere' is at a distance $[r]$ from the field point P at the time $t - [r]/c$.

Let the element δS of the 'information collecting sphere' be at $CBB'C'$ at the time $t-[r]/c$ as shown in *Figure 5.3*. Let it be at the position $EFF'E'$ at a time δt later, where

$$\delta t = \delta r/c \qquad (5.28)$$

and where $\delta r = CE$.

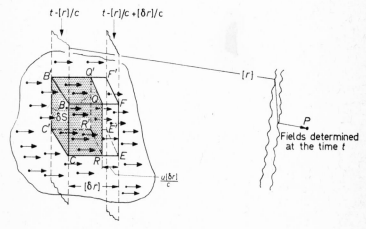

Figure 5.3. *The calculation of the scalar and vector potentials of a system of 'point' charges moving with the same uniform velocity* **u** *towards the field point P. The moving 'point' charges build up a macroscopic charge and current distribution*

Let $\delta V'$ be the volume element swept out by δS in the time δt so that

$$\delta V' = \delta S \delta r \qquad (5.29)$$

The positions of the charges in $\delta V'$ at the time $t-[r]/c$ are shown by dots in *Figure 5.3*. The charge density $[\rho]$ can be determined at the retarded time $t-[r]/c$ by counting the number of charges, that is dots, in $\delta V'$ at the fixed time $t-[r]/c$. Using eqn (4.5)

$$[\rho] = \frac{\sum_{i=1}^{N} q_i}{\delta V'} = \frac{Nq}{\delta V'} = nq \qquad (5.30)$$

where N is the total number of charges each of magnitude q, counted in $\delta V'$ at the fixed time $t-[r]/c$, and n is the number of charges per

137

unit volume at the time $t - [r]/c$. It is being assumed, initially, that all the individual microscopic charges in $\delta V'$ are moving directly towards P with the same uniform velocity $\mathbf{u}$, as shown in *Figure 5.3*, so that the macroscopic charge distribution has a velocity $\mathbf{u}$ towards P. The displacements of the charges in a time $\delta t = \delta r/c$ are represented by arrows in *Figure 5.3*. In the time δt the 'information collecting sphere' takes to go from $CBB'C'$ to $EFF'E'$, the charges move a distance $u\delta t$ towards P. Those charges which are at a distance less than $u\delta t$ from the surface $EFF'E'$ at a time $t - [r]/c$ when the 'information collecting sphere' is at $CBB'C'$ will have left the volume element $\delta V'$ through the surface $EFF'E'$, by the time the element δS of the 'information collecting sphere' reaches $EFF'E'$ at the time $t - [r]/c + \delta t$. Hence, the number of charges counted by the area δS of the 'information collecting sphere' in the volume element $\delta V'$ is equal to the number of charges which at the time $t - [r]/c$ were in the volume between $CBB'C'$ and $RQQ'R'$ in *Figure 5.3*, where

$$RE = u\delta t = \frac{u\delta r}{c}$$

that is in a volume $\delta S(\delta r - u\delta r/c) = \delta V'(1 - u/c)$. Since n is the number of atomic charges per unit volume counted at the fixed time $t - [r]/c$, the number of charges actually passed by the surface δS inside $\delta V'$ in the time δt is

$$N_0 = n\delta V'[1 - u/c] \qquad (5.31)$$

If $[\mathbf{u}]$ is not parallel to $[\mathbf{r}]$, then, in eqn (5.31), $[u/c]$ must be replaced by the component of $[\mathbf{u}/c]$ parallel to $[\mathbf{r}]$ that is by $[\mathbf{u} \cdot \mathbf{r}/rc]$ giving

$$N_0 = n\delta V'\left[1 - \frac{\mathbf{r} \cdot \mathbf{u}}{rc}\right] \qquad (5.32)$$

This is the generalization of eqn (5.31). Each one of these charges gives a contribution to the total scalar potential at P given by eqn (5.22). It is being assumed that all the charges have the same magnitude q, the same velocity $\mathbf{u}$ and that on the laboratory scale they are all at approximately the same distance $[r]$ from the field point. Hence the contribution to the scalar potential due to all the N_0 charges passed by the 'information collecting sphere' inside $\delta V'$ is

$$\delta\phi = \frac{qN_0}{4\pi\varepsilon_0[r - \mathbf{r} \cdot \mathbf{u}/c]} = \frac{qN_0}{4\pi\varepsilon_0[r][1 - \mathbf{r} \cdot \mathbf{u}/rc]}$$

Substituting for N_0 from eqn (5.32), gives

$$\delta\phi = \frac{qn\delta V'[1-\mathbf{r}\cdot\mathbf{u}/rc]}{4\pi\varepsilon_0[r][1-\mathbf{r}\cdot\mathbf{u}/rc]}$$

According to eqn (5.30)

$$nq = [\rho]$$

Hence,

$$\delta\phi = \frac{[\rho]\delta V'}{4\pi\varepsilon_0[r]} \tag{5.33}$$

The charge density $[\rho]$ is obtained by counting the number of point charges of magnitude q per unit volume at the fixed time $t-[r]/c$ and multiplying by q.

For an extended distribution of charge, every volume element of the charge distribution contributes to the scalar potential according to eqn (5.33), so that the total scalar potential at P in *Figure 5.3*, at the time t is given by

$$\phi = \frac{1}{4\pi\varepsilon_0}\int\frac{[\rho]\mathrm{d}V'}{[r]} \tag{5.34}$$

In eqn (5.34) the contributions due to the different volume elements $\delta V'$ of the macroscopic charge distribution are evaluated at different retarded times $t-[r]/c$ depending on the distance $[r]$ of the volume element $\delta V'$ from the field point P at the appropriate retarded time $t-[r]/c$.

Similarly, applying eqns (5.26), (5.31) and (5.30) to calculate the vector potential $\mathbf{A}$, we have

$$\mathbf{A} = \frac{1}{4\pi\varepsilon_0 c^2}\int\frac{[\rho\mathbf{u}]\mathrm{d}V'}{[r]} \tag{5.35}$$

Eqns (5.34) and (5.35) are the well-known equations for the retarded potentials.

It was shown that in Section 5.5 that Lorentz condition

$$\nabla\cdot\mathbf{A}+\frac{1}{c^2}\frac{\partial\phi}{\partial t} = 0 \tag{5.27}$$

holds for the potentials at a point in empty space due to a macroscopic distribution of charge, made up of a finite number of microscopic charges moving with uniform velocities.

139

5.7. DIFFERENTIAL EQUATIONS FOR ϕ AND A

An outline will now be given of the conventional developments of the differential equations for ϕ and **A** from Maxwell's equations. In conventional approaches to electromagnetism Maxwell's equations are developed initially for the macroscopic fields of macroscopic charge and current distributions in the way outlined in Appendix 2. According to Maxwell's equations $\mathbf{V} \cdot \mathbf{B}$ is always zero. Since the divergence of the curl of any vector is zero, we can always put

$$\mathbf{B} = \mathbf{V} \times \mathbf{A} \qquad (5.36)$$

where **A** is the vector potential. The divergence of **A** has yet to be specified.

The scalar potential ϕ is introduced into electrostatics, where it is defined by the relation

$$\mathbf{E} = -\mathbf{V}\phi \qquad (5.37)$$

However, since $\mathbf{V} \times \mathbf{E}$ is not zero for varying currents, eqn (5.37) must be extended. From Maxwell's equations

$$\mathbf{V} \times \mathbf{E} = -\frac{\partial \mathbf{B}}{\partial t} = -\frac{\partial}{\partial t}(\mathbf{V} \times \mathbf{A})$$

that is

$$\mathbf{V} \times \left(\mathbf{E} + \frac{\partial \mathbf{A}}{\partial t}\right) = 0$$

Since the curl of the gradient of any scalar function is zero, we can try putting

$$\mathbf{E} + \frac{\partial \mathbf{A}}{\partial t} = -\mathbf{V}\phi \quad \text{or} \quad \mathbf{E} = -\mathbf{V}\phi - \frac{\partial \mathbf{A}}{\partial t} \qquad (5.38)$$

Integrating around any closed loop, since $\oint \mathbf{V}\phi \cdot d\mathbf{l}$ is zero,

$$\oint \mathbf{E} \cdot d\mathbf{l} = -\oint \frac{\partial \mathbf{A}}{\partial t} \cdot d\mathbf{l} = -\frac{\partial}{\partial t} \oint \mathbf{A} \cdot d\mathbf{l}$$

Applying Stokes' theorem, eqn (A1.25) of Appendix 1, to $\oint \mathbf{A} \cdot d\mathbf{l}$ and using $\mathbf{B} = \mathbf{V} \times \mathbf{A}$, we obtain

$$\oint \mathbf{E} \cdot d\mathbf{l} = -\frac{\partial}{\partial t} \int (\mathbf{V} \times \mathbf{A}) \cdot \mathbf{n}dS = -\frac{\partial}{\partial t} \int \mathbf{B} \cdot \mathbf{n}dS$$

This is Faraday's (or Neumann's) law of induction. The term $-\partial A/\partial t$ in eqn (5.38) represents the contribution of electromagnetic induction to the electric field at a point.

From Maxwell's equations, if $\varepsilon_r = 1$ and $\mu_r = 1$, we have for the *macroscopic* fields

$$\varepsilon_0 c^2 \nabla \times \mathbf{B} = \varepsilon_0 \frac{\partial \mathbf{E}}{\partial t} + \mathbf{J}$$

Substituting for $\mathbf{E}$ and $\mathbf{B}$ from eqns (5.38) and (5.36) respectively,

$$\nabla \times (\nabla \times \mathbf{A}) = \frac{1}{c^2}\frac{\partial}{\partial t}\left(-\nabla\phi - \frac{\partial \mathbf{A}}{\partial t}\right) + \frac{\mathbf{J}}{\varepsilon_0 c^2}$$

But, from eqn (A1.24) of Appendix 1,

$$\nabla \times (\nabla \times \mathbf{A}) = \nabla(\nabla \cdot \mathbf{A}) - \nabla^2 \mathbf{A} \qquad (A1.24)$$

Hence,

$$\nabla(\nabla \cdot \mathbf{A}) - \nabla^2 \mathbf{A} = \frac{1}{c^2}\frac{\partial}{\partial t}\left(-\nabla\phi - \frac{\partial \mathbf{A}}{\partial t}\right) + \frac{\mathbf{J}}{\varepsilon_0 c^2}$$

Rearranging

$$\nabla^2 \mathbf{A} - \frac{1}{c^2}\frac{\partial^2 \mathbf{A}}{\partial t^2} - \nabla\left(\nabla \cdot \mathbf{A} + \frac{1}{c^2}\frac{\partial \phi}{\partial t}\right) = -\frac{\mathbf{J}}{\varepsilon_0 c^2} \qquad (5.39)$$

The divergence of $\mathbf{A}$ has still to be specified. The most popular choice is to specify $\nabla \cdot \mathbf{A}$ such that

$$\nabla \cdot \mathbf{A} + \frac{1}{c^2}\frac{\partial \phi}{\partial t} = 0 \qquad (5.40)$$

This is the *Lorentz condition*, cf. eqn (5.27). This choice for $\nabla \cdot \mathbf{A}$ gives equations which are relativistically invariant. This choice is sometimes called the covariant gauge. Using the Lorentz condition, eqn (5.39) becomes

$$\nabla^2 \mathbf{A} - \frac{1}{c^2}\frac{\partial^2 \mathbf{A}}{\partial t^2} = -\frac{\mathbf{J}}{\varepsilon_0 c^2} = -\mu_0 \mathbf{J} \qquad (5.41)$$

From Maxwell's equations, if $\varepsilon_r = 1$,

$$\nabla \cdot \mathbf{E} = \rho/\varepsilon_0$$

Substituting for $\mathbf{E}$ from eqn (5.38), we obtain

$$\nabla \cdot (\nabla\phi + \partial \mathbf{A}/\partial t) = -\rho/\varepsilon_0 \qquad (5.42)$$

But $\mathbf{V} \cdot (\nabla\phi) = \nabla^2\phi$ and from the Lorentz condition

$$\mathbf{V} \cdot \left(\frac{\partial\mathbf{A}}{\partial t}\right) = \frac{\partial}{\partial t}(\mathbf{V} \cdot \mathbf{A}) = \frac{\partial}{\partial t}\left(-\frac{1}{c^2}\frac{\partial\phi}{\partial t}\right) = -\frac{1}{c^2}\frac{\partial^2\phi}{\partial t^2}$$

Substituting in eqn (5.42),

$$\nabla^2\phi - \frac{1}{c^2}\frac{\partial^2\phi}{\partial t^2} = -\frac{\rho}{\varepsilon_0} \tag{5.43}$$

Collecting the equations for ϕ and $\mathbf{A}$, we have

$$\nabla^2\phi - \frac{1}{c^2}\frac{\partial^2\phi}{\partial t^2} = -\frac{\rho}{\varepsilon_0} \tag{5.43}$$

$$\nabla^2\mathbf{A} - \frac{1}{c^2}\frac{\partial^2\mathbf{A}}{\partial t^2} = -\frac{\mathbf{J}}{\varepsilon_0 c^2} = -\mu_0\mathbf{J} \tag{5.41}$$

$$\mathbf{V} \cdot \mathbf{A} + \frac{1}{c^2}\frac{\partial\phi}{\partial t} = 0 \tag{5.40}$$

$$\mathbf{E} = -\nabla\phi - \frac{\partial\mathbf{A}}{\partial t} \tag{5.38}$$

$$\mathbf{B} = \mathbf{V} \times \mathbf{A} \tag{5.36}$$

These equations were developed from Maxwell's equations for continuous macroscopic charge and current distributions. It is sometimes more convenient to use these equations for the potentials ϕ and $\mathbf{A}$ to solve problems rather than use Maxwell's equations. In eqns (5.43) and (5.41) ρ and $\mathbf{J}$ are assumed to be continuous functions of position. It is assumed in classical electromagnetism that Maxwell's equations are valid for accelerating charge distributions, so that the above equations for the potentials ϕ and $\mathbf{A}$ should also hold for the fields of accelerating charge distributions.

It can be shown (e.g. Feynman, Leighton and Sands[1], or Ferraro[2]) that, the general solutions of eqns (5.43) and (5.41) are given by the retarded potentials, namely

$$\phi(x, y, z, t) = \frac{1}{4\pi\varepsilon_0}\int\frac{\rho(x', y', z', t-[r]/c)}{[r]}dx'dy'dz'$$

$$= \frac{1}{4\pi\varepsilon_0}\int\frac{[\rho]}{[r]}dV' \tag{5.44}$$

and

$$A(x, y, z, t) = \frac{1}{4\pi\varepsilon_0 c^2} \int \frac{\mathbf{J}(x', y', z', t - [r]/c)}{[r]} \, dx'dy'dz'$$

$$= \frac{1}{4\pi\varepsilon_0 c^2} \int \frac{[\mathbf{J}]}{[r]} \, dV' = \frac{\mu_0}{4\pi} \int \frac{[\mathbf{J}]}{[r]} \, dV' \qquad (5.45)$$

where x, y, z are the co-ordinates of the field point where the potentials are required at a time t. Eqns (5.44) and (5.45) hold for accelerating charge distributions. They were developed via relativity in Section 5.6 for the special case of charges moving with uniform velocities. The element of volume dV' in eqns (5.44) and (5.45) has co-ordinates x', y', and z', and is at a distance $[r]$ from the field point. The charge density $[\rho]$ at the position of the volume element dV' is the value at the retarded time $t - [r]/c$. It is common practice to put quantities, such as ρ, which are measured at the retarded time $t - [r]/c$ inside square brackets. The retarded times are different, for different volume elements of charge in the system, since the various volume elements of charge are at different distances $[r]$ from the field point.

For *purposes of exposition* it is useful to re-introduce the *imaginary* 'information collecting sphere' introduced previously in Section 5.5. Let the imaginary 'information collecting sphere' have its centre at the field point, and let the 'information collecting sphere' collapse with a velocity c such that it arrives at the field point at the time t the potentials are required. The 'information collecting sphere' passes the various volume elements dV' in eqns (5.44) and (5.45) at the correct retarded times. It is useful to imagine this 'information collecting sphere' collecting information about the charge and current densities at the appropriate retarded positions and times. This data on $[\rho]$ and $[\mathbf{J}]$ can be used to calculate ϕ and A from eqns (5.44) and (5.45).

Since Maxwell's equations cannot normally be used conveniently to relate fields back to their sources, the retarded potentials are used extensively in such subjects as antennae theory. When thinking in terms of antennae theory, the dependence of ϕ and A on the charge and current densities at the retarded position and retarded time can be readily accepted. Experiments show that it takes $\sim 2\cdot56$ seconds to send radar signals to the moon and back. The signals received on the earth depend on the electric currents in the transmitter antenna at a time $2\cdot56$ seconds earlier. The 'information collecting sphere' can be illustrated in terms of radio waves. If one had several radio transmitters distributed in empty space, the radio signals

F*   143

reaching a field point would have left the transmitters at different retarded times, and one can imagine the signals from the various transmitters 'collapsing' on a field point in empty space with the imaginary 'information collecting sphere'.

The propagation of electrical signals with a maximum velocity of c in empty space is consistent with the theory of special relativity, since according to that theory the maximum speed for propagating energy and momentum is the speed of light in empty space.

For an account of the applications of eqns (5.44) and (5.45) the reader is referred to the textbooks by Panofsky and Phillips[3], Corson and Lorrain[4], Jordan[5], etc.

5.8. THE LIÉNARD–WIECHERT POTENTIALS

In Section 5.2, the potentials ϕ and $\mathbf{A}$ for a 'point' charge moving with uniform velocity were developed from Coulomb's law for the electric field of a 'point' charge, using the transformations of the theory of special relativity. This was a *microscopic* or *atomistic* approach. In Section 5.7 the equations for the potentials ϕ and $\mathbf{A}$ of continuous macroscopic charge distributions were developed from Maxwell's equations, without the restriction to charges moving with uniform velocities. The formulae developed in Section 5.7 will now be used to develop the potentials of a 'point' charge moving in an arbitrary way. It will be assumed, initially, that the 'point' charge has finite dimensions as shown in *Figure 5.4*. At a later stage, these dimensions can be made very small on the laboratory scale. It will be assumed that electric charge is *continuous* and is distributed through-out the volume of the 'point' charge. It will be assumed initially that the 'point' charge is moving towards P, the field point, as shown in *Figure 5.4*. Let $\mathbf{r}$ be a vector *from* the charge to the field point P.

The 'information collecting sphere' is at a distance $[r]$ from the field point P at the time $t - [r]/c$. Consider the surface element δS of the 'information collecting sphere', which is crossing the continuous charge distribution shown in *Figure 5.4*. Let the surface element δS of the 'information collecting sphere' be at $CBB'C'$ at the time $t - [r]/c$ as shown in *Figure 5.4*. Let it be at the position $EFF'E'$ at a time δt later, where

$$\delta t = \delta r / c \qquad (5.46)$$

and where

$$\delta r = CE$$

Let $\delta V'$ be the volume element swept out by the area δS in the time δt, so that

$$\delta V' = \delta S \delta r \qquad (5.47)$$

In the time the 'information collecting sphere' moves from $CBB'C'$ to $EFF'E'$, charge on the surface $RQQ'R'$ moves to $EFF'E'$ so that the amount of charge actually passed by the element δS of the 'information collecting sphere' inside $\delta V'$ is the charge, which at the

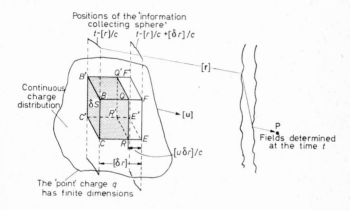

Figure 5.4. *The calculation of the scalar and vector potentials (the Liénard –Wiechert potentials) at the field point P due to an accelerating 'point' charge q moving with non-uniform velocity. In Figure 5.4, $\mathbf{u}$ is parallel to $[\mathbf{r}]$. It is assumed that the 'point' charge q has finite dimensions and is made up from a continuous charge distribution*

time $t - [r]/c$, is inside the volume between $CBB'C'$ and $RQQ'R'$ in *Figure 5.4* where

$$RE = u\delta t = u\delta r/c$$

The volume between $CBB'C'$ and $RQQ'R'$ is $\delta S(\delta r - u\delta r/c) = \delta V'(1 - u/c)$. If $[\rho]$ is the charge density, that is the charge per unit volume measured at a fixed time, say $t - [r]/c$, the amount of charge δQ passed by δS in a time δt is

$$\delta Q = [\rho]\delta V'(1 - u/c) \qquad (5.48)$$

{It is interesting to compare the approach in this section with that in Sections 5.5 and 5.6, where the expressions for the potentials ϕ and $\mathbf{A}$ for a macroscopic charge distribution, made up from a large number of 'point' charges, were developed from the expressions for ϕ and $\mathbf{A}$ for a single 'point' charge moving with uniform velocity. Exactly the same *geometrical* conditions are shown in *Figures 5.4* and *5.3*. In *Figure 5.3*, the charge density $[\rho]$ was determined by counting the atomic charges (dots) per unit volume at a fixed time, say $t - [r]/c$. The number of charges passed by the element δS of the 'information collecting sphere' in a time δt in *Figure 5.3* was equal to the number of charges, which at the time $t - [r]/c$, were between $CBB'C'$ and $RQQ'R'$ in *Figure 5.3*. From eqn (5.31), we had

$$N_0 = n\delta V'(1 - u/c) \tag{5.31}$$

where N_0 was the total number of charges 'counted' by the area δS of the 'information collecting sphere' inside $\delta V'$ and n was the number of charges per unit volume counted at a fixed time. Multiplying both sides of eqn (5.31) by q, the charge of each atomic charge in *Figure 5.3*, since $\delta Q = N_0 q$ and $[\rho] = nq$, we had for the total charge δQ passed by δS in a time δt

$$\delta Q = [\rho]\delta V'(1 - u/c)$$

This is in agreement with eqn (5.48) which was developed for the *continuous* charge distribution shown in *Figure 5.4*.}

In general when $\mathbf{u}$ is not parallel to $[\mathbf{r}]$, in *Figure 5.4*, in eqn (5.48) u must be replaced by the component of $\mathbf{u}$ parallel to $\mathbf{r}$, so that for the general case, eqn (5.48) for the total charge passed by δS in a time δt becomes

$$\delta Q = [\rho]\delta V'[1 - \mathbf{u} \cdot \mathbf{r}/rc] \tag{5.49}$$

Rearranging eqn (5.49), we have

$$[\rho]\delta V' = \frac{\delta Q}{\left[1 - \dfrac{\mathbf{u} \cdot \mathbf{r}}{rc}\right]} \tag{5.50}$$

Substituting in eqn (5.44),

$$\phi = \frac{1}{4\pi\varepsilon_0}\int \frac{[\rho]\mathrm{d}V'}{[r]} = \frac{1}{4\pi\varepsilon_0}\int \frac{\mathrm{d}Q}{\left[r - \dfrac{\mathbf{u} \cdot \mathbf{r}}{c}\right]} \tag{5.51}$$

If it is assumed that the dimensions of the charge are very small, the variation of the denominator inside the integral can be neglected. Also $\int \mathrm{d}Q$ is the total quantity of charge passed by the 'information collecting sphere' which should be equal to q, the total charge of the

146

'point' charge. Hence eqn (5.51) becomes

$$\phi = \frac{1}{4\pi\varepsilon_0}\left[\frac{q}{r-\dfrac{\mathbf{u}\cdot\mathbf{r}}{c}}\right] = \frac{q}{4\pi\varepsilon_0 s} \tag{5.52}$$

where now

$$s = [r - \mathbf{u}\cdot\mathbf{r}/c] \tag{5.53}$$

For the 'continuous charge distribution' shown in Figure 5.4,

$$[\mathbf{J}]\mathrm{d}V' = [\rho\mathbf{u}]\mathrm{d}V'$$

Substituting for $[\rho]\mathrm{d}V'$ from eqn (5.50), we have

$$[\mathbf{J}]\mathrm{d}V' = \left[\frac{\mathbf{u}\mathrm{d}Q}{1-\mathbf{u}\cdot\mathbf{r}/rc}\right]$$

Substituting in eqn (5.45), and proceeding as for ϕ, we obtain

$$\mathbf{A} = \frac{1}{4\pi\varepsilon_0 c^2}\left[\frac{q\mathbf{u}}{r-\dfrac{\mathbf{u}\cdot\mathbf{r}}{c}}\right] = \frac{[q\mathbf{u}]}{4\pi\varepsilon_0 c^2 s} = \frac{\mu_0[q\mathbf{u}]}{4\pi s} \tag{5.54}$$

Eqns (5.52) and (5.54) are known as the Liénard–Wiechert potentials. All the quantities inside the square brackets are measured at the retarded time. Eqns (5.52) and (5.54) are independent of the structure of the 'point' charge. If the velocity [**u**] is constant, eqns (5.52) and (5.54) are the same as eqns (5.22) and (5.23) which were developed in Section 5.4 for a 'point' charge moving with uniform velocity and are the same as eqns (5.7) and (5.8), as they should be. Thus the Liénard–Wiechert potentials are consistent with the theory of special relativity. In classical electromagnetism, it is assumed that eqns (5.52) and (5.54) hold for accelerating charges also. It is interesting to note, that, though the expressions for ϕ and **A** depend on the velocity of the charge, they do not depend on the acceleration of the charge. The expressions for **E** and **B** do however depend on the acceleration of the charge.

The expressions for the electric and magnetic fields of an accelerating 'point' charge can be calculated from the Liénard–Wiechert potentials using the equations

$$\mathbf{E} = -\nabla\phi - \frac{\partial\mathbf{A}}{\partial t} \tag{5.38}$$

and

$$\mathbf{B} = \nabla\times\mathbf{A} \tag{5.36}$$

147

The differentiation must be carried out with respect to the co-ordinates of the field point and the time of observation. It can be shown, e.g. Panofsky and Phillips[3], that the application of eqns (5.38) and (5.36) to eqns (5.52) and (5.54) gives eqns (3.41) and (3.44) for the total electric and magnetic fields of an accelerating 'point' charge. This is the conventional way of developing eqns (3.41) and (3.44). Eqns (5.52) and (5.54) have been developed from Gauss' law using the theory of special relativity by Page and Adams[6]. For a discussion of eqns (3.41) and (3.44) the reader is referred to Appendix 3, and to Panofsky and Phillips[3]. A discussion of how various macroscopic phenomena are related to these equations is given in Appendix 3(d) (cf. eqns (A3.28) and (A3.29)).

REFERENCES

[1] FEYNMAN, R. P., LEIGHTON, R. B. and SANDS, M. *The Feynman Lectures in Physics*, Vol. II p. 21–2. Addison-Wesley, Reading, Mass., 1964

[2] FERRARO, V. C. A., *Electromagnetic Theory*, p. 524. University of London, The Athlone Press, 1954

[3] PANOFSKY, W. K. H. and PHILLIPS, M. *Classical Electricity and Magnetism.* 2nd Ed. Addison-Wesley, Reading, Mass., 1962

[4] CORSON, D. and LORRAIN, P. *Introduction to Electromagnetic Fields and Waves.* W. H. Freeman and Co., San Francisco, 1962

[5] JORDAN, E. C. *Electromagnetic Waves and Radiating Systems.* Prentice-Hall Inc., New York, 1950

[6] PAGE, L. and ADAMS, N. I. *Electrodynamics.* Van Nostrand, New York, 1945

PROBLEMS

Problem 5.1—In the presence of magnetic monopoles, $\nabla \cdot \mathbf{B}$ is not zero. Hence, one cannot use the vector potential $\mathbf{A}$ defined by the relation $\mathbf{B} = \nabla \times \mathbf{A}$. If one did have a system of magnetic monopoles, one could introduce ϕ_m and $\mathbf{A}_m$, which are the magnetic analogues of ϕ and $\mathbf{A}$, defined for a system of magnetic monopoles in empty space by the relations

$$\mathbf{D}_m = \varepsilon_0 \mathbf{E}_m = -\nabla \times \mathbf{A}_m \qquad (5.55)$$

$$\mathbf{H}_m = \frac{\mathbf{B}_m}{\mu_0} = -\nabla \phi_m - \frac{\partial \mathbf{A}_m}{\partial t} \qquad (5.56)$$

The suffix m is used to denote the fact that the fields and potentials are due to magnetic monopoles. Show that, if the potentials for a magnetic monopole moving with uniform velocity $\mathbf{u}$ are

$$\phi_m = \frac{q_m}{4\pi\mu_0 s} = \frac{q_m^*}{4\pi s} \qquad (5.57)$$

and

$$\mathbf{A}_m = \frac{\varepsilon_0 q_m \mathbf{u}}{4\pi s} = \frac{\varepsilon_0 \mu_0 q_m^* \mathbf{u}}{4\pi s} \tag{5.58}$$

where s is given by eqns (5.6) and q_m and q_m^* are defined by eqns (3.48) and (3.49), then the use of eqns (5.55) and (5.56) gives the expressions for the electric and magnetic fields of a magnetic monopole given by eqns (3.56) and (3.57) respectively.

Show that ϕ_m and $\mathbf{A}_m$ given by eqns (5.57) and (5.58) satisfy the relation

$$\mathbf{V} \cdot \mathbf{A}_m + \frac{1}{c^2} \frac{\partial \phi_m}{\partial t} = 0$$

Problem 5.2—Assume that for a system of magnetic monopoles, in empty space, ϕ_m and $\mathbf{A}_m$ are defined by eqns (5.55) and (5.56) of Problem 5.1. Show that Maxwell's equations, in the form developed in Problem 4.2, can be expressed in the form

$$\nabla^2 \mathbf{A}_m - \frac{1}{c^2} \frac{\partial^2 \mathbf{A}_m}{\partial t^2} = -\varepsilon_0 \mathbf{J}_m = -\mu_0 \varepsilon_0 \mathbf{J}_m^*$$

$$\nabla^2 \phi_m - \frac{1}{c^2} \frac{\partial^2 \phi_m}{\partial t^2} = -\frac{\rho_m}{\mu_0} = -\rho_m^*$$

subject to the auxiliary condition

$$\mathbf{V} \cdot \mathbf{A}_m + \frac{1}{c^2} \frac{\partial \phi_m}{\partial t} = 0$$

Show that the solutions of these equations are

$$\mathbf{A}_m = \frac{\varepsilon_0}{4\pi} \int \frac{[\mathbf{J}_m]}{[r]} \, dV' = \frac{\mu_0 \varepsilon_0}{4\pi} \int \frac{[\mathbf{J}_m^*]}{[r]} \, dV'$$

$$\phi_m = \frac{1}{4\pi\mu_0} \int \frac{[\rho_m]}{[r]} \, dV' = \frac{1}{4\pi} \int \frac{[\rho_m^*]}{[r]} \, dV'$$

Problem 5.3—A solenoid of infinite length has a radius of a metres, and has n turns per metre. Show that, if the current in the solenoid is I amperes, the magnetic induction inside the solenoid is equal to $\mu_0 n I$, whereas the magnetic induction outside the solenoid is zero. Show that, if the current in the solenoid is varying with time, the induced electric field lines outside the solenoid are closed circles, concentric with the axis of the solenoid. Show that, if r is the distance of the field point from the axis of the solenoid, then if $r > a$

$$E = -\frac{\mu_0 n a^2 (dI/dt)}{2r}$$

(Hint: Use $\oint \mathbf{E} \cdot d\mathbf{l} = -(\partial/\partial t) \int \mathbf{B} \cdot \mathbf{n} dS$.)

If $\mathbf{B}$ is zero outside the solenoid $\partial \mathbf{B}/\partial t$ is zero, hence $\nabla \times \mathbf{E}$ is zero. How do you account for the electric field outside the solenoid? (Hint: Calculate $\nabla \times \mathbf{E}$ in cylindrical co-ordinates, and show that, even though the lines of $\mathbf{E}$ are circles, $\nabla \times \mathbf{E}$ is zero if $E \propto r^{-1}$.)

Show that the vector potential outside the solenoid is $\mu_0 n a^2 I/2r$. Check your previous value of E using $\mathbf{E} = -\partial \mathbf{A}/\partial t$.

(Hint: Use $\oint \mathbf{A} \cdot d\mathbf{l} = \int (\nabla \times \mathbf{A}) \cdot \mathbf{n} dS = \int \mathbf{B} \cdot \mathbf{n} dS$.)

A beam of electrons is passed close to, but outside such a solenoid, when it is carrying a *steady* current I. In this case $\mathbf{E} = 0$ and $\mathbf{B} = 0$. Would you expect the electrons to be deflected,

    (a)   according to classical electromagnetism?

    (b)   according to quantum mechanics? Discuss the role of the vector potential $\mathbf{A}$ in quantum mechanics. (Reference: *The Feynman Lectures on Physics*, Vols. II and III.)

6

RELATIVISTIC ELECTROMAGNETISM

6.1. INTRODUCTION

In this monograph, Maxwell's equations were developed in Chapter 4 from the formulae for the electric and magnetic fields of a 'point' charge moving with uniform velocity. These formulae had been developed previously in Chapter 3, from Coulomb's law taking the transformations of the theory of special relativity as axiomatic. This approach is the reverse of the historical approach. The theory of special relativity evolved from classical optics and electromagnetism. Maxwell's equations were developed before the theory of special relativity. In 1905, Einstein[1] assumed as his first postulate that all the laws of physics, including the laws of electromagnetism, obeyed the principle of relativity. As his second postulate, Einstein could have assumed that the appropriate laws of electromagnetism were Maxwell's equations. Instead, Einstein chose the principle of the constancy of the speed of light as his second postulate. (It is shown by Rosser[2a, 15], that, if it is assumed that Maxwell's equations are correct and obey the principle of relativity, the principle of the constancy of the speed of light follows.) After developing the Lorentz transformations, Einstein[1] went on to show that the Lorentz transformations did leave Maxwell's equations invariant in mathematical form. This approach is developed in this chapter.

Experiments have confirmed that the *macroscopic* electric and magnetic fields, both in empty space and inside *stationary* material bodies, are adequately described by Maxwell's equations, which in an inertial frame Σ', in which the materials are at rest, take the form

$$\nabla' \times \mathbf{E}' = -\frac{\partial \mathbf{B}'}{\partial t'} \tag{6.1}$$

$$\nabla' \times \mathbf{H}' = \frac{\partial \mathbf{D}'}{\partial t'} + \mathbf{J}' \tag{6.2}$$

$$\nabla' \cdot \mathbf{D}' = \rho' \tag{6.3}$$

$$\nabla' \cdot \mathbf{B}' = 0 \tag{6.4}$$

where $\mathbf{E}'$ is the electric intensity, $\mathbf{D}'$ the electric displacement, $\mathbf{B}'$

151

the magnetic induction, $\mathbf{H}'$ the magnetizing force, ρ' the true macroscopic charge density and $\mathbf{J}'$ the true macroscopic conduction current density. All the quantities appearing in eqns (6.1), (6.2), (6.3) and (6.4) apply to the same point x', y', z' at a time t' in the inertial frame Σ'. All the differential coefficients in the equations are with respect to x', y', z' or t', for example, eqn (6.4) is

$$\frac{\partial B'_x}{\partial x'}+\frac{\partial B'_y}{\partial y'}+\frac{\partial B'_z}{\partial z'} = 0$$

The quantities $\mathbf{D}'$ and $\mathbf{H}'$ are defined for the case of stationary matter in Appendix 2 (cf. eqns (A2.18) and (A2.47)). In order to solve problems these equations have to be supplemented by the constitutive equations relating $\mathbf{D}'$ and $\mathbf{E}'$, $\mathbf{B}'$ and $\mathbf{H}'$, and $\mathbf{E}'$ and the conduction current density $\mathbf{J}'$. These constitutive equations depend on the properties of the materials present in the system.

One of the postulates of the theory of special relativity is that the laws of physics have the same mathematical form in all inertial frames of reference. If Maxwell's equations are correct and obey the principle of relativity, then in the inertial frame Σ moving with velocity v in the negative Ox' direction relative to Σ', one should have

$$\nabla\times\mathbf{E} = -\frac{\partial\mathbf{B}}{\partial t} \tag{6.5}$$

$$\nabla\times\mathbf{H} = \frac{\partial\mathbf{D}}{\partial t}+\mathbf{J} \tag{6.6}$$

$$\nabla.\mathbf{D} = \rho \tag{6.7}$$

$$\nabla.\mathbf{B} = 0 \tag{6.8}$$

These equations should hold for a point x, y, z at a time t in Σ. If all the material bodies are stationary in Σ', then they are moving with uniform velocity v relative to Σ. The constitutive equations may take a different mathematical form for moving matter. This is discussed in detail in Section 6.7.3. The current density $\mathbf{J}$ in Σ includes the convection current density as well as the conduction current density. If Maxwell's equations do obey the principle of relativity, when the co-ordinates and time in eqns (6.5), (6.6), (6.7) and (6.8) are transformed according to the Lorentz transformations, then eqns (6.5)–(6.8) should be changed into eqns (6.1)–(6.4). It will be shown in this chapter that this is so. Equations which have the same mathematical form in both Σ and Σ', when the co-ordinates and time are transformed according to the Lorentz transformations, are said to be *Lorentz covariant*.

152

6.2. THE TRANSFORMATION OF E AND B

Eqns (6.5) and (6.8) only involve the field quantities **E** and **B**. They have the same mathematical form in empty space as in material media. First, the covariance of these equations is discussed. Writing eqns (6.5) and (6.8) out into components one has

$$\frac{\partial E_z}{\partial y} - \frac{\partial E_y}{\partial z} = -\frac{\partial B_x}{\partial t} \tag{6.9}$$

$$\frac{\partial E_x}{\partial z} - \frac{\partial E_z}{\partial x} = -\frac{\partial B_y}{\partial t} \tag{6.10}$$

$$\frac{\partial E_y}{\partial x} - \frac{\partial E_x}{\partial y} = -\frac{\partial B_z}{\partial t} \tag{6.11}$$

$$\frac{\partial B_x}{\partial x} + \frac{\partial B_y}{\partial y} + \frac{\partial B_z}{\partial z} = 0 \tag{6.12}$$

The rule for changing partial differential coefficients with respect to one set of variables into partial differential coefficients with respect to another set of variables, can be obtained as follows. Consider a function F, which is a function of x', y', z' and t' in Σ'. The total differential of F is

$$dF = \frac{\partial F}{\partial x'} dx' + \frac{\partial F}{\partial y'} dy' + \frac{\partial F}{\partial z'} dz' + \frac{\partial F}{\partial t'} dt' \tag{6.13}$$

Now for a given event, x', y', z' and t', are all functions of x, y, z and t. The total differential of x' can be expressed as

$$dx' = \frac{\partial x'}{\partial x} dx + \frac{\partial x'}{\partial y} dy + \frac{\partial x'}{\partial z} dz + \frac{\partial x'}{\partial t} dt \tag{6.14}$$

According to the Lorentz transformations

$$x' = \gamma(x - vt); \quad y' = y; \quad z' = z$$
$$t' = \gamma(t - vx/c^2); \quad \gamma = (1 - v^2/c^2)^{-\frac{1}{2}}$$

It will be assumed throughout this section that Σ' moves with uniform velocity v relative to Σ along their common x axis, so that v and $\gamma = (1 - v^2/c^2)^{-\frac{1}{2}}$ are constants. Hence

$$\frac{\partial x'}{\partial x} = \gamma; \quad \frac{\partial x'}{\partial y} = 0; \quad \frac{\partial x'}{\partial z} = 0; \quad \frac{\partial x'}{\partial t} = -\gamma v$$

Substituting in eqn (6.14)

$$dx' = \gamma dx' - \gamma v dt' \qquad (6.15)$$

Similarly,

$$dy' = dy \qquad (6.16)$$

$$dz' = dz \qquad (6.17)$$

$$dt' = \frac{\partial t'}{\partial x}dx + \frac{\partial t'}{\partial y}dy + \frac{\partial t'}{\partial z}dz + \frac{\partial t'}{\partial t}dt = -\frac{\gamma v}{c^2}dx + \gamma dt \qquad (6.18)$$

Substituting for dx', dy', dz' and dt' from eqns (6.15), (6.16), (6.17) and (6.18) respectively into eqn (6.13), we have

$$dF = \frac{\partial F}{\partial x'}(\gamma dx' - \gamma v dt) + \frac{\partial F}{\partial y'}dy + \frac{\partial F}{\partial z'}dz + \frac{\partial F}{\partial t'}\left(\gamma dt - \frac{\gamma v}{c^2}dx\right)$$

Rearranging,

$$dF = \left(\gamma \frac{\partial F}{\partial x'} - \frac{\gamma v}{c^2}\frac{\partial F}{\partial t'}\right)dx + \frac{\partial F}{\partial y'}dy + \frac{\partial F}{\partial z'}dz + \left(\gamma \frac{\partial F}{\partial t'} - \gamma v \frac{\partial F}{\partial x'}\right)dt$$
$$(6.19)$$

But, if F is a function of x, y, z and t, the total differential of F can be written as

$$dF = \frac{\partial F}{\partial x}dx + \frac{\partial F}{\partial y}dy + \frac{\partial F}{\partial z}dz + \frac{\partial F}{\partial t}dt \qquad (6.20)$$

Comparing the coefficients of dx, dy, dz and dt in eqns (6.19) and (6.20), we obtain

$$\frac{\partial}{\partial x} = \gamma \left(\frac{\partial}{\partial x'} - \frac{v}{c^2}\frac{\partial}{\partial t'}\right) \qquad (6.21)$$

Similarly,

$$\frac{\partial}{\partial y} = \frac{\partial}{\partial y'} \qquad (6.22)$$

and

$$\frac{\partial}{\partial z} = \frac{\partial}{\partial z'} \qquad (6.23)$$

$$\frac{\partial}{\partial t} = \gamma \left(\frac{\partial}{\partial t'} - v \frac{\partial}{\partial x'}\right) \qquad (6.24)$$

Now consider eqn (6.10), namely

$$\frac{\partial E_x}{\partial z} - \frac{\partial E_z}{\partial x} = -\frac{\partial B_y}{\partial t} \tag{6.10}$$

Substituting from eqns (6.23), (6.21) and (6.24),

$$\frac{\partial E_x}{\partial z'} - \gamma\left(\frac{\partial E_z}{\partial x'} - \frac{v}{c^2}\frac{\partial E_z}{\partial t'}\right) = -\gamma\left(\frac{\partial B_y}{\partial t'} - v\frac{\partial B_y}{\partial x'}\right)$$

Rearranging,

$$\frac{\partial E_x}{\partial z'} - \frac{\partial}{\partial x'}\gamma(E_z + vB_y) = -\frac{\partial}{\partial t'}\gamma\left(B_y + \frac{v}{c^2}E_z\right) \tag{6.25}$$

If Maxwell's equations are to be Lorentz covariant, that is invariant in mathematical form in all inertial frames of reference, then in Σ' one must have

$$\frac{\partial E'_x}{\partial z'} - \frac{\partial E'_z}{\partial x'} = -\frac{\partial B'_y}{\partial t'} \tag{6.26}$$

Eqns (6.25) and (6.26) have the same mathematical form showing that the y-component of eqn (6.5) is Lorentz covariant. In fact, if one puts

$$E'_x = E_x \tag{6.27}$$

$$E'_z = \gamma(E_z + vB_y) \tag{6.28}$$

$$B'_y = \gamma\left(B_y + \frac{v}{c^2}E_z\right) \tag{6.29}$$

then eqns (6.25) and (6.26) are exactly the same. Similarly, eqn (6.11) becomes

$$\frac{\partial}{\partial x'}\gamma(E_y - vB_z) - \frac{\partial}{\partial y'}E_x = -\frac{\partial}{\partial t'}\gamma\left(B_z - \frac{v}{c^2}E_y\right)$$

This has the same mathematical form as the equation

$$\frac{\partial E'_y}{\partial x'} - \frac{\partial E'_x}{\partial y'} = -\frac{\partial B'_z}{\partial t'}$$

and the two equations are the same if one puts

$$E'_y = \gamma(E_y - vB_z) \tag{6.30}$$

$$E'_x = E_x$$

$$B'_z = \gamma\left(B_z - \frac{v}{c^2}E_y\right) \tag{6.31}$$

RELATIVISTIC ELECTROMAGNETISM

The only undetermined transformation is that for B_x. To determine this, one must consider both eqns (6.9) and (6.12). Substituting from eqns (6.21), (6.22) and (6.23) into eqn (6.12) one obtains

$$\gamma\left(\frac{\partial B_x}{\partial x'}-\frac{v}{c^2}\frac{\partial B_x}{\partial t'}\right)+\frac{\partial B_y}{\partial y'}+\frac{\partial B_z}{\partial z'}=0 \tag{6.32}$$

The inverses of eqns (6.29) and (6.31) are

$$B_y=\gamma\left(B_y'-\frac{v}{c^2}E_z'\right)$$

and

$$B_z=\gamma\left(B_z'+\frac{v}{c^2}E_y'\right)\quad\text{respectively}$$

Substituting into eqn (6.32),

$$\gamma\left(\frac{\partial B_x}{\partial x'}-\frac{v}{c^2}\frac{\partial B_x}{\partial t'}\right)+\gamma\frac{\partial}{\partial y'}\left(B_y'-\frac{v}{c^2}E_z'\right)+\gamma\frac{\partial}{\partial z'}\left(B_z'+\frac{v}{c^2}E_y'\right)=0$$

Rearranging and cancelling γ,

$$\frac{\partial B_x}{\partial x'}+\frac{\partial B_y'}{\partial y'}+\frac{\partial B_z'}{\partial z'}=\frac{v}{c^2}\left\{\frac{\partial E_z'}{\partial y'}-\frac{\partial E_y'}{\partial z'}+\frac{\partial B_x}{\partial t'}\right\} \tag{6.33}$$

Similarly, after substituting from eqns (6.22), (6.23) and (6.24) into the equation

$$\frac{\partial E_z}{\partial y}-\frac{\partial E_y}{\partial z}=-\frac{\partial B_x}{\partial t}$$

one obtains

$$\frac{\partial E_z}{\partial y'}-\frac{\partial E_y}{\partial z'}=-\gamma\left(\frac{\partial B_x}{\partial t'}-v\frac{\partial B_x}{\partial x'}\right)$$

Substituting for E_z and E_y from the inverses of eqns (6.28) and (6.30),

$$\gamma\left(\frac{\partial E_z'}{\partial y'}-v\frac{\partial B_y'}{\partial y'}\right)-\gamma\left(\frac{\partial E_y'}{\partial z'}+v\frac{\partial B_z'}{\partial z'}\right)=-\gamma\left(\frac{\partial B_x}{\partial t'}-v\frac{\partial B_x}{\partial x'}\right)$$

Rearranging and cancelling γ,

$$-v\left(\frac{\partial B_x}{\partial x'}+\frac{\partial B_y'}{\partial y'}+\frac{\partial B_z'}{\partial z'}\right)+\frac{\partial E_z'}{\partial y'}-\frac{\partial E_y'}{\partial z'}+\frac{\partial B_x}{\partial t'}=0$$

that is

$$\frac{\partial B_x}{\partial x'}+\frac{\partial B'_y}{\partial y'}+\frac{\partial B'_z}{\partial z'} = \frac{1}{v}\left(\frac{\partial E'_z}{\partial y'}-\frac{\partial E'_y}{\partial z'}+\frac{\partial B_x}{\partial t'}\right) \tag{6.34}$$

Comparing eqns (6.33) and (6.34) it is concluded that, since v must be less than c, one must have

$$\frac{\partial B_x}{\partial x'}+\frac{\partial B'_y}{\partial y'}+\frac{\partial B'_z}{\partial z'} = 0 \tag{6.35}$$

and

$$\frac{\partial E'_z}{\partial y'}-\frac{\partial E'_y}{\partial z'} = -\frac{\partial B_x}{\partial t'} \tag{6.36}$$

If Maxwell's equations obey the principle of relativity, then in the inertial frame Σ' corresponding to eqns (6.12) and (6.9) one has

$$\frac{\partial B'_x}{\partial x'}+\frac{\partial B'_y}{\partial y'}+\frac{\partial B'_z}{\partial z'} = 0 \tag{6.37}$$

and

$$\frac{\partial E'_z}{\partial y'}-\frac{\partial E'_y}{\partial z'} = -\frac{\partial B'_x}{\partial t'} \tag{6.38}$$

Eqns (6.35) and (6.37) have the same mathematical form, and if

$$B'_x = B_x \tag{6.39}$$

then they are exactly the same. Eqns (6.36) and (6.38) are then also exactly the same. This completes the field transformations. Collecting the formulae:

$$
\left.
\begin{array}{ll}
E'_x = E_x & E_x = E'_x \\[4pt]
E'_y = \gamma(E_y - vB_z) & E_y = \gamma(E'_y + vB'_z) \\[4pt]
E'_z = \gamma(E_z + vB_y) & E_z = \gamma(E'_z - vB'_y)
\end{array}
\right\} \tag{6.40}
$$

$$
\left.
\begin{array}{ll}
B'_x = B_x & B_x = B'_x \\[8pt]
B'_y = \gamma\left(B_y + \dfrac{v}{c^2} E_z\right) & B_y = \gamma\left(B'_y - \dfrac{v}{c^2} E'_z\right) \\[12pt]
B'_z = \gamma\left(B_z - \dfrac{v}{c^2} E_y\right) & B_z = \gamma\left(B'_z + \dfrac{v}{c^2} E'_y\right)
\end{array}
\right\} \tag{6.41}
$$

These transformations relate the numerical values of the electric and magnetic fields at a point x, y, z at a time t in Σ with the values of the electric and magnetic fields at a point x', y', z' at a time t' in Σ', where x, y, z and t are related to x', y', z' and t' by the Lorentz transformations. If the subscripts $\parallel$ and $\perp$ are used to represent directions parallel and perpendicular to the direction in which Σ' is moving relative to Σ with uniform velocity $\mathbf{v}$, then, remembering that $(\mathbf{v} \times \mathbf{B})_\parallel$ and $(\mathbf{v} \times \mathbf{E}/c^2)_\parallel$ are both zero, the above equations can be rewritten in the vector form

$$\mathbf{E}'_\parallel = (\mathbf{E} + \mathbf{v} \times \mathbf{B})_\parallel, \qquad \mathbf{E}'_\perp = \gamma(\mathbf{E} + \mathbf{v} \times \mathbf{B})_\perp \qquad (6.40\text{a})$$

$$\mathbf{B}'_\parallel = (\mathbf{B} - \mathbf{v} \times \mathbf{E}/c^2)_\parallel, \qquad \mathbf{B}'_\perp = \gamma(\mathbf{B} - \mathbf{v} \times \mathbf{E}/c^2)_\perp \qquad (6.41\text{a})$$

It has been shown that when the co-ordinates and time are transformed according to the Lorentz transformations, then two of Maxwell's equations, namely (6.5) and (6.8), are covariant (that is, invariant in mathematical form), and obey the principle of relativity if the electric and magnetic fields are transformed according to eqns (6.40) and (6.41). It is assumed in classical electromagnetism that eqns (6.5) and (6.8), Maxwell's equations, apply to the radiation fields of accelerating charges. Hence the field transformations, eqns (6.40) and (6.41), should also apply to the radiation fields of accelerating charges.

Problem 6.1—A charge of magnitude q coulombs is moving with uniform velocity $\mathbf{u}$ in the inertial reference frame Σ. In the inertial frame Σ' moving with velocity $\mathbf{u}$, relative to Σ, the charge q is at rest, and the force on it is $q\mathbf{E}'$. Use the force transformations and the transformations for $\mathbf{E}$ and $\mathbf{B}$ to show that the force on the particle in Σ is $q\mathbf{E} + q\mathbf{u} \times \mathbf{B}$.

6.3. ALTERNATIVE DERIVATION OF THE TRANSFORMATIONS FOR E AND B

Some readers may prefer to develop the field transformations from different postulates, using the force transformations of the theory of special relativity. In Section 1.10 of Chapter 1, the total electric intensity $\mathbf{E}$ and the total magnetic induction $\mathbf{B}$ at a point in empty space, were *defined* relative to Σ in terms of the expression for the Lorentz force acting on a non-radiating test charge q moving with velocity $\mathbf{u}$, namely

$$\mathbf{f} = q\mathbf{E} + q\mathbf{u} \times \mathbf{B} \qquad (6.42)$$

In eqn (6.42), the velocity of the test charge q can have any magnitude less than the speed of light in empty space, and can have any direction, without changing the mathematical form of eqn (6.42). In classical electromagnetism, it is assumed that eqn (6.42) is also valid for the motion of the charge q in the radiation fields of accelerating charges.

The total electric intensity $\mathbf{E}'$ and the total magnetic induction $\mathbf{B}'$ relative to Σ' were defined in terms of the equation

$$\mathbf{f}' = q\mathbf{E}' + q\mathbf{u}' \times \mathbf{B}' \qquad (6.43)$$

The same value is used for q in eqns (6.42) and (6.43) so as to be in accord with the principle of constant charge (cf. Section 1.9 of Chapter 1).

It will be assumed that eqns (6.42) and (6.43) refer to the same act of measurement of the fields, so that, if the fields are measured at x, y, z at a time t in Σ and at x', y', z' at a time t' in Σ', these coordinates and time are related by the Lorentz transformations, and $\mathbf{u}$ and $\mathbf{u}'$ are related by the relativistic velocity transformations.

From eqn (1.54) for the force transformations

$$f_y' = \frac{f_y}{\gamma(1 - vu_x/c^2)}$$

Rearranging,

$$f_y = \gamma(1 - vu_x/c^2)f_y' \qquad (6.44)$$

where

$$\gamma = \frac{1}{\sqrt{1 - v^2/c^2}}$$

Substituting the y and y' components of eqns (6.42) and (6.43) respectively into eqn (6.44), we have

$$qE_y + q(u_zB_x - u_xB_z) = \gamma q(1 - vu_x/c^2)(E_y' + u_z'B_x' - u_x'B_z') \qquad (6.45)$$

Now from eqns (1.26) and (1.28)

$$u_x' = \frac{u_x - v}{(1 - vu_x/c^2)} \quad \text{and} \quad u_z' = \frac{u_z}{\gamma(1 - vu_x/c^2)}$$

Substituting in eqn (6.45), cancelling q and rearranging, we obtain

$$E_y + u_zB_x - u_xB_z \equiv \gamma(1 - vu_x/c^2)E_y' + u_zB_x' - \gamma(u_x - v)B_z'$$

That is,

$$E_y + u_zB_x - u_xB_z \equiv \gamma(E_y' + vB_z') + u_zB_x' - u_x\gamma(B_z' + vE_y'/c^2) \qquad (6.46)$$

159

Eqn (6.46) must be valid whatever the value of **u**, the velocity of the test charge relative to Σ. If u_x and u_z can have various values, then the terms containing u_x, u_z and the terms independent of u, the speed of the test charge, on the left-hand side of eqn (6.46) must be equal to the terms containing u_x, u_z and the terms independent of u on the right-hand side of eqn (6.46) respectively. Thus

$$E_y = \gamma(E_y' + vB_z')$$

$$B_x = B_x'$$

$$B_z = \gamma(B_z' + vE_y'/c^2)$$

Similarly, from eqn (1.55)

$$f_z = f_z'\gamma(1 - vu_x/c^2)$$

Substituting from eqns (6.42) and (6.43) for f_z and f_z', using the velocity transformations for u_x' and u_y', and comparing coefficients of the terms containing u_x, u_y and the terms independent of the velocity, we find

$$B_y = \gamma(B_y' - vE_z'/c^2)$$

$$B_x = B_x'$$

$$E_z = \gamma(E_z' - vB_y')$$

Now from eqn (1.56)

$$f_x = f_x' + \frac{vu_y'}{c^2(1 + vu_x'/c^2)}f_y' + \frac{vu_z'}{c^2(1 + vu_x'/c^2)}f_z'$$

$$= f_x' + K_1 u_y' + K_2 u_z' \tag{6.47}$$

Substituting the x components of eqns (6.42) and (6.43) into eqn (6.47), we obtain

$$qE_x + q(u_y B_z - u_z B_y) = qE_x' + q(u_y' B_z' - u_z' B_y') + K_1 u_y' + K_2 u_z' \tag{6.48}$$

Since, from the velocity transformations, eqns (1.27) and (1.28),

$$u_y' = \frac{\sqrt{1 - v^2/c^2}}{(1 - vu_x/c^2)}u_y \quad \text{and} \quad u_z' = \frac{\sqrt{1 - v^2/c^2}}{(1 - vu_x/c^2)}u_z$$

the only terms in eqn (6.48) independent of the velocity of the test charge are qE_x and qE_x'. Hence, we conclude that

$$E_x = E_x'$$

This completes the transformations for **E** and **B**. They agree with the transformations developed in Section 6.2. If the expression for the Lorentz force holds for the motion of a charge in the radiation fields of accelerating charges, the field transformations developed in this section should also hold for the radiation fields of accelerating charges. Thus the field transformations can be developed, either from Maxwell's equations, or from the expression for the Lorentz force. This illustrates the internal consistency of classical electromagnetism, which is based on Maxwell's equations and the Lorentz force equation. Actually the Lorentz force equation was used to define the quantities **E** and **B** appearing in Maxwell's equations, eqns (6.5) and (6.8).

Problem 6.2—A charge of magnitude q is at rest at the origin of an inertial reference frame Σ', which is moving with uniform velocity **v** relative to Σ. In Σ' the electric and magnetic fields at a distance **r'** from the origin are $\mathbf{E}' = q\mathbf{r}'/4\pi\varepsilon_0 r'^3$, $\mathbf{B}' = 0$. Using the transformations for **E** and **B** determine the electric and magnetic fields in Σ. Show that they can be written in the vector form

$$\mathbf{E} = \frac{q\mathbf{r}(1-\beta^2)}{4\pi\varepsilon_0 r^3(1-\beta^2\sin^2\theta)^{3/2}}; \quad \mathbf{B} = \frac{\mathbf{v}\times\mathbf{E}}{c^2}$$

where **v** is the velocity of the charge relative to Σ, **r** a vector from the 'present' position of the charge to the field point, θ is the angle between **v** and **r** and $\beta \doteq v/c$ (Reference: Rosser[2b]. Hint: Some of the geometrical relations developed in Section 3.1 may prove useful.)

Repeat the above calculation for the case of a magnetic monopole q_m (or q_m^*) moving with uniform velocity **v** relative to Σ. Show that its electric and magnetic fields are given by eqns (3.56) and 3.57).

Problem 6.3—By direct substitution of the field transformations show that the following are invariants under a Lorentz transformation: (a) $E^2 - c^2B^2$, (b) **B** . **E**

Hence, or otherwise, show that (i) if $E > cB$ in any inertial frame, then $E > cB$ in any other inertial frame and *vice versa*; (ii) if **E** and **B** are orthogonal in one Lorentz frame, then they are orthogonal in any other Lorentz frame and *vice versa*; (iii) if **E** and **B** are orthogonal, it is possible to choose a set of moving axes to eliminate **E** or c**B** whichever is the smaller; (iv) if the angle between **E** and **B** is acute (or obtuse) in one inertial frame, it is acute (or obtuse) in any other inertial frame.

Problem 6.4—The electric field of an infinite line charge of magnitude λ' C/m at rest along the x' axis of the system Σ' is given by $E_x' = B_x' = B_y' = B_z' = 0$;

$$E_y' = \frac{\lambda'y'}{2\pi\varepsilon_0(y'^2 + z'^2)}; \quad E_z' = \frac{\lambda'z'}{2\pi\varepsilon_0(y'^2 + z'^2)}$$

Use this result to determine the magnetic field of a linear current I along the x axis of the Σ-system. What is the electric field in this instance? Compare the magnitudes of the electric and magnetic forces on a test charge q moving with velocity $\mathbf{u}$ at the point x, y, z.

6.4. MOTIONAL E.M.F.

It is pointed out in Appendix 2(g) that, when an electrical circuit or part of a circuit moves in a magnetic field, a motional e.m.f. is produced in the circuit. This is the principle of the dynamo. Consider a conductor RS which is moving with uniform velocity v parallel to the x axis in the laboratory system Σ, as shown in *Figure 6.1(a)*. In Σ there is a steady uniform magnetic induction $\mathbf{B}$ in the negative z direction, that is away from the reader in *Figure 6.1(a)*. There is no external applied electric field in Σ. Relative to the laboratory system Σ, all the electrons and positive nuclei in the conductor RS move with the conductor with a velocity v, and should experience a magnetic force $q\mathbf{v} \times \mathbf{B}$ relative to the laboratory. The positive nuclei and tightly bound electrons in the moving conductor will be prevented from moving by the cohesive forces in the solid, but the conduction electrons and positive holes are relatively free to move. The positive holes will move in the positive Oy direction and the negative electrons in the opposite direction in *Figure 6.1(a)*. These charges will move until there is a separation of electric charge of such a distribution, that for an isolated moving conductor the electric force on any charge inside and moving with the moving conductor, due to the electric charges displaced in the conductor, is equal and opposite to the magnetic force $q\mathbf{v} \times \mathbf{B}$ on the charge, as illustrated in *Figure 6.1(a)*. The displaced charges also give rise to an electric field extending into the space outside the moving conductor. When the moving conductor RS slides on the curved stationary conducting rail in *Figure 6.1(a)* making electrical contact with the rail, then the external electric field due to the electric charges displaced on the moving conductor RS give rise to a conduction current in the stationary rail. The initial current flow in the stationary rail reduces the magnitudes of the electric charge distributions at the ends of the moving conductor. The magnetic force on the charges inside the moving conductor then exceeds the electric force on the charges inside the moving conductor, and the conduction electrons and positive holes in the moving conductor RS start to move again, and tend to replenish the free charges removed from the ends of the moving conductor by the conduction current in the stationary rail. If the conductor RS moves with uniform velocity and, if the changes

MOTIONAL E.M.F.

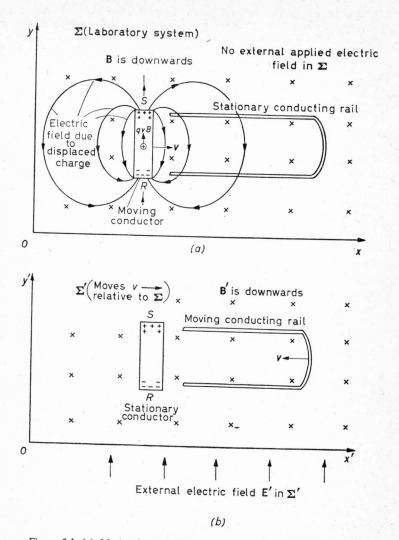

Figure 6.1. (a) Motional e.m.f. There is a magnetic force $q\mathbf{v}\times\mathbf{B}$ on the charges moving with the conductor RS which leads to a displacement of electric charge in the moving conductor RS which in turn gives rise to an electric field. (b) In Σ' the source of the external magnetic field also gives rise to an electric field which gives rise to the displacement of electric charge in the conductor RS, which is 'stationary' in Σ'

163

in the resistance of the complete circuit associated with the movement of *RS* can be neglected, then the current becomes steady, and a state of dynamic equilibrium is reached. The current flow is sustained by the magnetic force $q\mathbf{v} \times \mathbf{B}$ on the charges inside the moving conductor *RS* and which are moving with the conductor *RS*. Before the steady state is reached, there is a redistribution of electric charge on the moving conductor and stationary rail such that the potential differences along all parts of the circuit agree with Kirchhoff's laws (cf. *Figures A5.1* and *A5.2* of Appendix 5(a)).

Following Cullwick[3a], the e.m.f. of a complete circuit can be defined as

$$\varepsilon = \oint \frac{\mathbf{f}}{q} \cdot d\mathbf{l}$$

where **f** is the force, measured relative to Σ, on a charge q which is at rest relative to the section $d\mathbf{l}$ of the circuit. When the section $d\mathbf{l}$ of the circuit moves with velocity **v** relative to Σ, the test charge q must move with velocity **v** with $d\mathbf{l}$, so that

$$\mathbf{f} = q\mathbf{E} + q\mathbf{v} \times \mathbf{B}$$

For the case shown in *Figure 6.1(a)*, there is no applied electric field in Σ, so that the only electric field relative to Σ is associated with surface charge distributions on the conductors. If $v \ll c$, $\nabla \times \mathbf{E} \simeq 0$ and $\oint \mathbf{E} \cdot d\mathbf{l}$ is approximately zero in Σ. Hence,

$$\varepsilon = \oint (\mathbf{E} + \mathbf{v} \times \mathbf{B}) \cdot d\mathbf{l} = \oint (\mathbf{v} \times \mathbf{B}) \cdot d\mathbf{l}$$

$$= - \oint \mathbf{B} \cdot (\mathbf{v} \times d\mathbf{l}) \tag{6.49}$$

For the case shown in *Figure 6.1(a)*, ε is equal to the rate at which Φ the flux of **B** through the circuit formed by the moving conductor *RS*, and the stationary rail, is decreasing with time, so that the equation $\varepsilon = -d\Phi/dt$ is valid in this case. This latter rule is not universally true for motional e.m.f.s. For example, with the Faraday disk generator, an e.m.f. is produced when a circular disk rotates in a steady magnetic field. In this case, there is no change in the total flux of **B** through the circuit. Any radius of the disk is 'cutting' lines of **B** giving a motional e.m.f. All the radii of the disk are in parallel (Reference: Sears[4]). Eqn (6.49) should be interpreted as a 'flux cutting' rule, that is the contribution from any part of a circuit to the total motional e.m.f. is equal to the rate at which that part of the circuit is 'cutting' lines of **B**. In the absence of the moving conductor *RS* in *Figure 6.1(a)* there would be no motional e.m.f. and no external potential difference. The presence of the moving conductor is essential for the current flow to take place.

Now in Σ', the conductor RS is at rest and the conducting rail, shown in *Figure 6.1(b)*, moves with velocity $-v$. The displacement of charge in the isolated conductor RS must be recorded in both Σ and Σ'. Since the conductor RS is at rest in Σ', the displacement of 'stationary' charges relative to Σ' must be interpreted in terms of an applied external electric field relative to Σ'. From eqns (6.40) and (6.41), if $\mathbf{B} = (0, 0, -B)$ and $\mathbf{E} = 0$ in Σ, in Σ' we have

$$E_x' = E_z' = 0; \quad E_y' = \gamma vB; \quad B_x' = B_y' = 0; \quad B_z' = -\gamma B$$

The electric field E_y' in Σ' gives rise to the separation of charges in the 'stationary' conductor. It is left as an exercise for the reader at a later state (Problem 6.13) to interpret the origin of the electric field present in Σ'. The interpretation of a motional e.m.f. in the laboratory system in terms of an electric field in the 'moving' reference frame in which the 'moving' conductor is at rest, is of some historical interest. For example, in a letter to Shankland[5] in 1952, Einstein wrote

What lead me more or less directly to the special theory of relativity was the conviction that the electromotive force acting on a body in motion was nothing less but an electric field.

6.5. TRANSFORMATION OF CHARGE AND CURRENT DENSITIES

Consider a volume distribution of charge, which is moving with velocities $\mathbf{u}$ and $\mathbf{u}'$ relative to Σ and Σ' respectively, as shown in *Figure 6.2*. Let the charge densities in Σ and Σ' be ρ and ρ' respectively. The convection density in Σ is given by

$$\mathbf{J} = \rho\mathbf{u} \qquad (6.50)$$

having components

$$J_x = \rho u_x; \quad J_y = \rho u_y; \quad J_z = \rho u_z \qquad (6.51)$$

Similarly, the convection current density in Σ' is

$$\mathbf{J}' = \rho'\mathbf{u}'$$

having components

$$J_x' = \rho' u_x'; \quad J_y' = \rho' u_y'; \quad J_z' = \rho' u_z' \qquad (6.52)$$

In an inertial reference frame Σ_0 moving with velocity $-\mathbf{u}$ relative to Σ and velocity $-\mathbf{u}'$ relative to Σ', the charge distribution is at rest. Let its proper volume measured in Σ_0 be V_0 and let the total charge, measured in Σ_0, be Q_0. The principle of total electric charge will be taken as axiomatic, so that the total charge in Σ_0, Σ and Σ' is Q_0. [On an atomistic level, if there is a total of N_0 atomic charges of magnitude q each in the system, then the total charge Q_0 is qN_0. Since the *total* number of atomic charges is the same in all reference frames, if q, the charge on each atomic particle, is the same in all inertial reference frames, then Q_0 the *total* charge of the system is equal to qN_0 in every inertial reference frame.]

In the inertial reference frame Σ, the charge distribution of proper volume V_0 moves with a velocity $\mathbf{u}$ so that according to eqn (1.21)

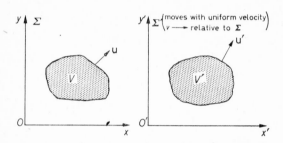

Figure 6.2. Calculation of the transformations for charge and current density. The charge distribution moves with velocities $\mathbf{u}$ and $\mathbf{u}'$ relative to Σ and Σ' respectively

the total charge Q_0 is measured to be in a volume $V_0\sqrt{1-u^2/c^2}$. Hence in Σ, the charge density is given by

$$\rho = \frac{Q_0}{V_0\sqrt{1-u^2/c^2}} \qquad (6.53)$$

In Σ' the total charge Q_0 is measured to be in a volume $V_0\sqrt{1-u'^2/c^2}$. Hence in Σ' the charge density is

$$\rho' = \frac{Q_0}{V_0\sqrt{1-u'^2/c^2}} \qquad (6.54)$$

From eqns (6.53) and (6.54)

$$\rho' = \rho\sqrt{\frac{1-u^2/c^2}{1-u'^2/c^2}} \qquad (6.55)$$

TRANSFORMATION OF CHARGE AND CURRENT DENSITIES

From eqn (1.32)

$$\sqrt{\frac{1-u^2/c^2}{1-u'^2/c^2}} = \frac{1-vu_x/c^2}{\sqrt{1-v^2/c^2}}$$

Substituting in eqn (6.55) since, $J_x = \rho u_x$, we have

$$\rho' = \rho\gamma(1-vu_x/c^2) \qquad (6.56)$$

or,

$$\rho' = \gamma(\rho - vJ_x/c^2) \qquad (6.57)$$

Now in Σ', from eqn (6.52)

$$J'_x = \rho' u'_x$$

Substituting for ρ' from eqn (6.56) and for u'_x using the velocity transformation, eqn (1.26), we obtain

$$J'_x = \rho\gamma(1-vu_x/c^2)\frac{(u_x-v)}{(1-vu_x/c^2)}$$

$$= \rho\gamma(u_x-v) = \gamma(\rho u_x-v\rho) \qquad (6.58)$$

or

$$J'_x = \gamma(J_x-v\rho) \qquad (6.59)$$

Similarly,

$$J'_y = \rho' u'_y = \rho\gamma(1-vu_x/c^2)\frac{u_y}{\gamma(1-vu_x/c^2)} = \rho u_y$$

or

$$J'_y = J_y \qquad (6.60)$$

Similarly,

$$J'_z = J_z \qquad (6.61)$$

This completes the transformation for $\mathbf{J}$ and ρ.

In practice, the most common type of electric current is a conduction current in a metal containing both positive ions and electrons. Consider a simplified model for a conduction current, which consists of the superposition of a positive distribution of charge of charge density ρ_+ moving with velocity $\mathbf{u}_+$ relative to Σ (charge density ρ'_+ moving with velocity $\mathbf{u}'_+$ relative to Σ') and a negative charge distribution of charge density ρ_- moving with velocity $\mathbf{u}_-$ relative to

G   167

Σ (charge density ρ'_- and velocity $\mathbf{u}'_-$ relative to Σ'). From eqn (6.56)

$$\rho'_+ = \gamma[\rho_+ - v\rho_+(u_+)_x/c^2] \tag{6.62}$$

and

$$\rho'_- = \gamma[\rho_- - v\rho_-(u_-)_x/c^2] \tag{6.63}$$

Let

$$\rho_+ + \rho_- = \rho_{\text{total}} = \rho \tag{6.64}$$

and

$$\rho_+(u_+)_x + \rho_-(u_-)_x = (J_{\text{total}})_x = J_x \tag{6.65}$$

with similar expressions in Σ'. Adding eqns (6.62) and (6.63), we have

$$\rho' = \gamma[\rho - vJ_x/c^2]$$

From eqn (6.58)

$$\rho'_+(u'_+)_x = \gamma[\rho_+(u_+)_x - v\rho_+]$$

and

$$\rho'_-(u'_-)_x = \gamma[\rho_-(u_-)_x - v\rho_-]$$

Adding, and using eqns (6.64) and (6.65) and the equivalent expressions in Σ', we obtain

$$J'_x = \gamma(J_x - v\rho)$$

Similarly, it can be shown that for the superposition of the positive and negative charge distributions

$$J'_y = J_y \quad \text{and} \quad J'_z = J_z$$

Collecting the transformation formulae,

$$
\begin{aligned}
J'_x &= \gamma(J_x - v\rho) & J_x &= \gamma(J'_x + v\rho') \\
J'_y &= J_y & J_y &= J'_y \\
J'_z &= J_z & J_z &= J'_z \\
\rho' &= \gamma\left(\rho - \frac{vJ_x}{c^2}\right) & \rho &= \gamma\left(\rho' + \frac{vJ'_x}{c^2}\right)
\end{aligned} \tag{6.66}
$$

It is left as an exercise for the reader to extend the treatment to include the case when the electrons in the negative charge distribution have a distribution of velocities. The current and charge densities $\mathbf{J}$ and ρ are the values at a point x, y, z at a time t in Σ, whilst $\mathbf{J}'$ and ρ'

are the values at the point x', y', z' at a time t' in Σ', where the co-ordinates and time in Σ and Σ' are related by the Lorentz transformations. It is of interest to note that J_x, J_y, J_z and ρ transform in the same way as x, y, z and t, the mathematical form of eqns (6.66) being the same as the mathematical form of the Lorentz transformations.

If in Σ' the charge density ρ' is zero, there is no convection current density in Σ', the current density being a conduction current density only. In this case eqns (6.66) reduce to

$$J_x = \gamma J_x'; \quad J_y = J_y'; \quad J_z = J_z'; \quad \rho = \gamma \frac{vJ_x'}{c^2} \qquad (6.67)$$

According to these equations, there should be a resultant electric charge density ρ equal to $\gamma v J_x'/c^2$ in Σ. Thus a current carrying conductor which is electrically neutral and at rest relative to Σ' should have an electric charge density relative to Σ. This effect will be illustrated by considering a simple example.

Consider a conductor consisting of equal numbers of positive and negative charges of magnitude $\pm q$ each, which are distributed along the $O'x'$ axis of an inertial frame Σ' at a time $t' = 0$, as shown in *Figure 6.3(a)*. Let the negative charges move with uniform velocity $+u_x'$ along the $O'x'$ axis and let the positive charges be at rest in Σ'. In Σ' the charge density ρ' is zero. The total current, which is a conduction current in the negative x' direction, is equal to $-qu_x'/a$, where a is the distance between successive charges. (The number of charges per unit length is equal to $1/a$ in Σ'.) If A' is the area of cross-section of the wire measured in Σ', then

$$J_x' = -\frac{qu_x'}{A'a}; \quad J_y' = J_z' = \rho' = 0 \qquad (6.68)$$

In Σ, which moves along the negative $O'x'$ axis with velocity v relative to Σ', the positions and times of the charges corresponding to the events shown in *Figure 6.3(a)* are:

$$x = 0; \quad \gamma a; \quad 2\gamma a; \quad \ldots$$

$$t = 0; \quad \gamma \frac{va}{c^2}; \quad 2\gamma \frac{va}{c^2}; \quad \ldots$$

These events are shown in *Figure 6.3(b)*; they are not simultaneous in the inertial frame Σ. In order to estimate the current and charge densities in Σ one must count the positive and negative charges at the same time as measured in Σ. This time will be taken to be $t = 0$. In Σ, it follows from the velocity transformations that the positive charges move with a uniform velocity v whilst the negative

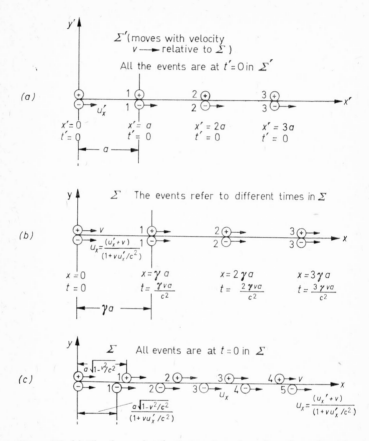

Figure 6.3. (a) The charge distribution consists of equal numbers of positive and negative charges in Σ'. The positive charges are at rest whilst the negative charges move with uniform velocity u'_x relative to Σ'. (b) The events in Σ, corresponding to the positions of the charges in Σ' shown in (a); these events are not simultaneous in Σ. (c) The positions of the same charges at a time $t = 0$ in Σ. Notice that there is a resultant negative electric charge density in Σ

charges move with a uniform velocity $(u'_x + v)/(1 + vu'_x/c^2)$. Since these velocities are uniform, the position of each particle at a time $t = 0$ can be estimated from the positions and times of the events

170

shown in *Figure 6.3(b)*. For example, in a time $\gamma va/c^2$, the charge $+1$ moves a distance $\gamma v^2 a/c^2$, whilst the charge -1 moves a distance

$$\left(\frac{u'_x + v}{1 + vu'_x/c^2}\right) \gamma \frac{va}{c^2}$$

The distance of the charge $+1$ from the origin at the time $t = 0$ is therefore equal to $(\gamma a - \gamma v^2 a/c^2) = (1 - v^2/c^2)^{\frac{1}{2}}a$, whilst the distance of charge -1 from the origin at the time $t = 0$ in Σ is equal to

$$\left[\gamma a - \frac{\gamma av(u'_x + v)}{c^2(1 + vu'_x/c^2)}\right]$$

which is equal to $a(1 - v^2/c^2)^{\frac{1}{2}}/(1 + vu'_x/c^2)$. The distribution of charges at the time $t = 0$ in Σ is therefore as shown in *Figure 6.3(c)*. It can be seen that there is a net negative charge density. The number of positive charges per unit length in Σ is equal to $1/a(1 - v^2/c^2)^{\frac{1}{2}}$, whilst the number of negative charges per unit length is equal to $(1 + u'_x v/c^2)/a(1 - v^2/c^2)^{\frac{1}{2}}$. The resultant electric charge per unit length is equal to $-\gamma qu'_x v/ac^2$. This charge is distributed over the area of cross-section of the wire. Since this area is perpendicular to the direction of the relative motion of Σ and Σ', it is unaffected by a Lorentz transformation. Hence,

$$\rho = -\frac{\gamma qu'_x v}{A'ac^2}$$

But from eqn (6.68) the current density in Σ' is equal to

$$J'_x = -\frac{qu'_x}{A'a}$$

Hence,

$$\rho = +\gamma vJ'_x/c^2 \tag{6.69}$$

This is in agreement with eqn (6.67). It can be seen that this result arises from the difference in the measures of simultaneity in Σ and Σ'. It is a relativistic effect which has no classical analogue.

One consequence of this result is that a current loop, which is electrically neutral when it is at rest in Σ', has an electric dipole moment when observed in the inertial frame Σ in which it is moving. Consider a rectangular loop of wire, of dimensions a' by b', which is at rest in the $x'y'$ plane of an inertial frame Σ'. Let it carry a *conduc-*

171

tion current I' and let the area of cross-section of the wire in Σ' be equal to A', such that the current density inside the wire is equal to I'/A'. The magnetic moment of a stationary plane coil is a vector equal to the product of the area of the coil and the current flowing in the coil, the direction of the vector being normal to the area in such a direction that a right-handed corkscrew rotated in the direction of current flow would advance in the direction of the magnetic moment, that is

$$\mathbf{m}' = \mathbf{n}I'a'b' \tag{6.70}$$

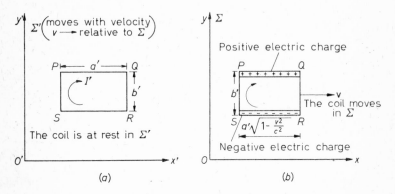

Figure 6.4. (a) The current carrying coil is at rest in the inertial frame Σ'. (b) The coil is moving with uniform velocity v in Σ. According to the theory of special relativity the coil should have an electric dipole moment relative to Σ

where $\mathbf{n}$ is a unit vector normal to the plane of the coil. For the coil shown in Figure 6.4(a), we have in Σ'

$$m'_z = -I'a'b'$$

The dimensions of the loop in Σ can be obtained by applying the Lorentz transformations. Since the sides PQ and RS are in the direction of motion they are 'contracted' to $a'(1-v^2/c^2)^{\frac{1}{2}}$ whilst the area of cross-section is unchanged and equal to A'. The sides PS and QR are unchanged but their areas of cross-section are reduced to $A'(1-v^2/c^2)^{\frac{1}{2}}$. According to eqn (6.67) there is in Σ an electric charge density $\rho = \gamma v J'_x/c^2$ along the arm PQ. The total charge along the arm PQ is equal to $\rho a'(1-v^2/c^2)^{\frac{1}{2}}A'$, which is equal to $\gamma v J'_x A'a'(1-v^2/c^2)^{\frac{1}{2}}/c^2 = I'a'v/c^2$. Similarly, there is a negative charge $-I'a'v/c^2$ along the arm RS as shown in Figure 6.4(b). There is an electric dipole moment $\mathbf{p}$ in the y direction in Σ, which is equal

172

to the magnitude of the charges times the separation of the charges, that is

$$p_y = \frac{I'a'v}{c^2} b' = -\frac{v}{c^2} m_z'$$

In general a current loop, which has a magnetic moment **m′** in the inertial frame in which it is at rest, has an electric dipole moment equal to

$$\mathbf{p} = \frac{\mathbf{v} \times \mathbf{m}'}{c^2} \tag{6.71}$$

when it is moving with uniform velocity **v** relative to the laboratory. The Amperian approach is often adopted for interpreting the magnetic properties of matter. In this approach the magnetic properties of materials are attributed to atomic current loops. If eqn (6.71) were applicable to these atomic current loops then a magnetized body which is moving relative to the laboratory should have a finite electric polarization relative to the laboratory. This contribution to the polarization of a moving body is discussed in Section 6.7.

Problem 6.5—A toroid lies flat in the $x'\,y'$ plane of an inertial reference frame Σ', which is moving with uniform velocity **v** relative to Σ along their common x axis. Relative to Σ', there is a steady conduction current in the toroid. According to the transformations for charge, the toroid has a charge density relative to Σ. Sketch the charge distribution relative to Σ (Reference: Becker and Sauter[6]).

6.6. THE TRANSFORMATION OF **D** AND **H**

Consider the y component of the equation

$$\nabla \times \mathbf{H} = \mathbf{J} + \frac{\partial \mathbf{D}}{\partial t} \tag{6.6}$$

that is

$$\frac{\partial H_x}{\partial z} - \frac{\partial H_z}{\partial x} = J_y + \frac{\partial D_y}{\partial t}$$

Substituting for $\partial/\partial z$, $\partial/\partial x$ and $\partial/\partial t$ from eqns (6.23), (6.21) and (6.24) and for J_y from eqn (6.66), we obtain

$$\frac{\partial H_x}{\partial z'} - \gamma \left(\frac{\partial}{\partial x'} - \frac{v}{c^2} \frac{\partial}{\partial t'} \right) H_z = J_y' + \gamma \left(\frac{\partial}{\partial t'} - v \frac{\partial}{\partial x'} \right) D_y$$

or

$$\frac{\partial H_x}{\partial z'} - \frac{\partial}{\partial x'} \gamma (H_z - vD_y) = J_y' + \frac{\partial}{\partial t'} \gamma \left(D_y - \frac{v}{c^2} H_z \right) \tag{6.72}$$

If Maxwell's equations were valid in Σ', one would have

$$\frac{\partial H'_x}{\partial z'} - \frac{\partial H'_z}{\partial x'} = J'_y + \frac{\partial D'_y}{\partial t'} \tag{6.73}$$

Eqns (6.72) and (6.73) have the same mathematical form, and the two are exactly the same, if

$$H'_x = H_x; \quad H'_z = \gamma(H_z - vD_y); \quad D'_y = \gamma\left(D_y - \frac{v}{c^2}H_z\right)$$

It is left as an exercise for the reader to show by similar methods that the x and z components of eqn (6.6) transform into the x' and z' components of eqn (6.2) and that the equation $\mathbf{V} \cdot \mathbf{D} = \rho$ transforms into the equation $\mathbf{V}' \cdot \mathbf{D}' = \rho'$, if $\mathbf{J}$ and ρ are transformed using eqns (6.66) and if $\mathbf{H}$ and $\mathbf{D}$ satisfy the transformations

$$
\left.
\begin{array}{ll}
D'_x = D_x & D_x = D'_x \\[2mm]
D'_y = \gamma\left(D_y - \dfrac{v}{c^2}H_z\right) & D_y = \gamma\left(D'_y + \dfrac{v}{c^2}H'_z\right) \\[2mm]
D'_z = \gamma\left(D_z + \dfrac{v}{c^2}H_y\right) & D_z = \gamma\left(D'_z - \dfrac{v}{c^2}H'_y\right)
\end{array}
\right\} \tag{6.74}
$$

$$
\left.
\begin{array}{ll}
H'_x = H_x & H_x = H'_x \\[2mm]
H'_y = \gamma(H_y + vD_z) & H_y = \gamma(H'_y - vD'_z) \\[2mm]
H'_z = \gamma(H_z - vD_y) & H_z = \gamma(H'_z + vD'_y)
\end{array}
\right\} \tag{6.75}
$$

Or in vector form

$$\mathbf{D}'_\| = \left(\mathbf{D} + \frac{1}{c^2}\mathbf{v} \times \mathbf{H}\right)_\| \quad \text{and} \quad \mathbf{D}'_\perp = \gamma\left(\mathbf{D} + \frac{1}{c^2}\mathbf{v} \times \mathbf{H}\right)_\perp \tag{6.74a}$$

and

$$\mathbf{H}'_\| = (\mathbf{H} - \mathbf{v} \times \mathbf{D})_\| \quad \text{and} \quad \mathbf{H}'_\perp = \gamma(\mathbf{H} - \mathbf{v} \times \mathbf{D})_\perp \tag{6.75a}$$

where

$$\gamma = 1/\sqrt{(1 - v^2/c^2)}$$

The quantities $\mathbf{H}$ and $\mathbf{D}$ refer to a point x, y, z at a time t in Σ and the quantities $\mathbf{H}'$ and $\mathbf{D}'$ refer to the point x', y', z' at a time t' in Σ', where x', y', z' and t' are related to x, y, z and t by the Lorentz transformations.

It has been shown that if Maxwell's equations are valid in Σ, then if the co-ordinates and time are transformed using the Lorentz transformations, and if one takes the principle of constant charge as axiomatic, then the transformed equations have the same mathematical form as Maxwell's equations would have if they were valid in Σ'. This is true whether there are material bodies present at the point or not. Thus Maxwell's equations satisfy the principle of relativity when the co-ordinates and time are transformed according to the Lorentz transformations.

In our approach to electromagnetism via special relativity, $\mathbf{E}$ and $\mathbf{B}$ are the basic field vectors. The properties of dielectrics and magnetic materials are introduced in terms of the polarization $\mathbf{P}$ and the magnetization $\mathbf{M}$. The vectors $\mathbf{D}$ and $\mathbf{H}$ are merely abbreviations for $\varepsilon_0\mathbf{E}+\mathbf{P}$ and $(\mathbf{B}/\mu_0)-\mathbf{M}$ respectively. The *assumption* that eqns (6.5), (6.6), (6.7) and (6.8) are Lorentz invariant, which leads to the transformations for $\mathbf{D}$ and $\mathbf{H}$, implies some interesting properties for moving dielectrics and moving magnetic media.

Problem 6.6—Show that the following quantities are invariants under a Lorentz transformation:

(a) $H^2-c^2D^2$;   (b) $\mathbf{H} \cdot \mathbf{D}$;   (c) $\mathbf{B} \cdot \mathbf{H}-\mathbf{E} \cdot \mathbf{D}$;

(d) $c\mathbf{B} \cdot \mathbf{D}+(1/c)\,\mathbf{E} \cdot \mathbf{H}$

6.7. THE ELECTRODYNAMICS OF MOVING MEDIA

To a large extent the theory of special relativity developed initially from investigations on the electrodynamics of moving media. For example, the title of Einstein's paper of 1905 was 'On the electrodynamics of moving bodies'.

The electrodynamics of moving media will now be developed using the transformations of the theory of special relativity. The approach to be used was developed, initially, by Minkowski in 1908.

In the presence of a material medium the values of the macroscopic fields depend on the properties of the medium. To solve problems, Maxwell's equations have to be supplemented by the constitutive equations. For a large class of substances, when the substances are at rest, the constitutive relations take the form

$$\mathbf{B} = \mu_r\mu_0\mathbf{H}; \quad \mathbf{D} = \varepsilon_r\varepsilon_0\mathbf{E} \quad \text{and} \quad \mathbf{J}_{\text{cond.}} = \sigma\mathbf{E} \qquad (6.76)$$

where μ_r is the relative permeability, ε_r is the dielectric constant and σ is the electrical conductivity. For single crystals $\mathbf{B}$ and $\mathbf{H}$ are not necessarily parallel, neither are $\mathbf{D}$ and $\mathbf{E}$ nor $\mathbf{J}$ and $\mathbf{E}$. In these

G*

cases μ_r, ε_r and σ are second rank tensors. For some materials, such as ferromagnetic materials, μ_r is not a constant but depends on **H**. There are, however, a wide range of materials for which μ_r, ε_r and σ can be considered as constants when the media are at rest.

If the material medium is at rest in the inertial frame Σ', then it is moving with uniform velocity v in the inertial frame Σ. If Maxwell's equations obey the principle of relativity, then they are also valid in the inertial frame Σ in which the material medium is moving. The approach initiated by Minkowski was to assume that the properties of materials and the constitutive equations are known in the inertial frame Σ' in which the material medium is at rest. The transformations of the theory of special relativity are then applied in order to obtain equations which are valid in the inertial frame Σ, in which the material medium is moving with uniform velocity. This approach will now be applied to various electromagnetic quantities.

6.7.1. *The Transformation of the Polarization Vector* **P**

The polarization vector **P** is defined for stationary matter as the dipole moment induced per unit volume of a dielectric due to the influence of an applied electric field. The polarization of a dielectric arises partly from the alignment of polar molecules in the direction of the applied electric field and partly from the separation of the positive and negative charges in the atoms and molecules of the dielectric due to the applied electric field. In interpreting polarization a simplified classical model is introduced in Appendix 2(b). In this model it is assumed that the atomic electric dipoles consist of equal and opposite electric charges which are separated in the presence of an electric field. For purposes of calculation it is convenient to introduce the displacement vector **D** defined by the relation

$$\mathbf{D} = \varepsilon_0 \mathbf{E} + \mathbf{P} \qquad (6.77)$$

It is the vector **D** which generally appears in Maxwell's equations and not the vector **P**. The transformations for **P** can be obtained from the relation

$$\mathbf{P} = \mathbf{D} - \varepsilon_0 \mathbf{E}$$

and the corresponding relation in Σ'

$$\mathbf{P}' = \mathbf{D}' - \varepsilon_0 \mathbf{E}'$$

using the transformations for **D** and **E** given by eqn (6.74) and (6.40) respectively. The same value is used for the electric space constant ε_0 in Σ and Σ' (Reference: Rosser[2a]). For the x component of **P**,

$$P_x = D_x - \varepsilon_0 E_x$$

Substituting the transformations for D_x and E_x from eqns (6.74) and (6.40) respectively, one obtains

$$P_x = D'_x - \varepsilon_0 E'_x$$

Hence,

$$P_x = P'_x$$

Substituting the transformations for D_y and E_y into the equation $P_y = D_y - \varepsilon_0 E_y$, one obtains

$$P_y = \gamma\left(D'_y + \frac{v}{c^2} H'_z\right) - \varepsilon_0\gamma(E'_y + vB'_z)$$

$$= \gamma\left\{(D'_y - \varepsilon_0 E'_y) - v\left(\varepsilon_0 B'_z - \frac{1}{c^2} H'_z\right)\right\}$$

$$= \gamma\left\{P'_y - \frac{v}{c^2}(\varepsilon_0 c^2 B'_z - H'_z)\right\}$$

But $c^2 = 1/\varepsilon_0\mu_0$, hence

$$P_y = \gamma\left\{P'_y - \frac{v}{c^2}\left(\frac{B'_z}{\mu_0} - H'_z\right)\right\}$$

But by definition $\mathbf{B}' = \mu_0\mathbf{H}' + \mu_0\mathbf{M}'$ (cf. Appendix 2(f)). Hence

$$P_y = \gamma\left(P'_y - \frac{v}{c^2} M'_z\right)$$

Similarly,

$$P_z = D_z - \varepsilon_0 E_z = \gamma\left(D'_z - \frac{v}{c^2} H'_y\right) - \varepsilon_0\gamma(E'_z - vB'_y)$$

$$= \gamma\left\{(D'_z - \varepsilon_0 E'_z) + \frac{v}{c^2}\left(\frac{B'_y}{\mu_0} - H'_y\right)\right\}$$

or

$$P_z = \gamma\left(P'_z + \frac{v}{c^2} M'_y\right)$$

This completes the transformation for **P**. Collecting the transformations,

$$
\left.
\begin{array}{ll}
P_x = P'_x & P'_x = P_x \\[2mm]
P_y = \gamma\left(P'_y - \dfrac{v}{c^2}\,M'_z\right) & P'_y = \gamma\left(P_y + \dfrac{v}{c^2}\,M_z\right) \\[2mm]
P_z = \gamma\left(P'_z + \dfrac{v}{c^2}\,M'_y\right) & P'_z = \gamma\left(P_z - \dfrac{v}{c^2}\,M_y\right)
\end{array}
\right\}
\qquad (6.78)
$$

If $\mathbf{P}_{\parallel}$ and $\mathbf{P}_{\perp}$ are the components of **P** parallel and perpendicular to **v**, eqns (6.78) can be rewritten as

$$
\mathbf{P}_{\parallel} = \mathbf{P}'_{\parallel}; \quad \mathbf{P}_{\perp} = \gamma\left(\mathbf{P}' + \frac{\mathbf{v}\times\mathbf{M}'}{c^2}\right)_{\perp}
\qquad (6.78a)
$$

The transformations for **P** are now interpreted in terms of simplified classical models. It will be assumed that the dielectric is at rest in Σ', and that in Σ' it consists of n' stationary dipoles/m³, which, for purposes of calculating **P**′, can each be considered as equal positive and negative charges q a distance l_0 apart, such that **p**′, the dipole moment of each atomic dipole in Σ', is equal to ql_0. For purposes of calculating **M**′ the magnetic dipoles will be considered as little current loops. If the electric dipoles are parallel to **v**, then due to length 'contraction' the distance between the two charges in each atomic dipole is reduced from l_0 to $l_0(1-v^2/c^2)^{\frac{1}{2}}$ in Σ, whilst, according to the principle of constant charge, q is a constant, so that the dipole moment of each atomic electric dipole is reduced to $ql_0(1-v^2/c^2)^{\frac{1}{2}} = p'(1-v^2/c^2)^{\frac{1}{2}}$ in Σ. If the volume of the dielectric is V_0 when it is at rest in Σ', then, when it is moving relative to Σ, its volume is $V_0(1-v^2/c^2)^{\frac{1}{2}}$ so that in Σ the number of dipoles/m³ goes up from n' to $n = n'/(1-v^2/c^2)^{\frac{1}{2}}$. In Σ, $\mathbf{P} = n\mathbf{p}$, and if **P**′ is parallel to **v** one has

$$
P = np = \frac{n'}{\sqrt{(1-v^2/c^2)}}\, p'\sqrt{(1-v^2/c^2)} = n'p'
$$

Hence

$$
P_x = P'_x
$$

If the atomic electric dipoles are perpendicular to the direction of the relative velocity between Σ and Σ', the separation between the charges is unchanged so that $p = p'$. The number of dipoles/m³ in Σ is then greater in Σ than Σ', so that the atomic electric dipoles give

178

rise to a polarization $\gamma P'_\perp$ in Σ. There is an extra term equal to $\gamma(\mathbf{v} \times \mathbf{M}')_\perp / c^2$ in eqn (6.78). This term has no non-relativistic counterpart, unless one accepts the concept of magnetic 'poles' (cf. Cullwick[3b]). The simple case of a rectangular current loop of magnetic moment $\mathbf{m}'$ at rest in Σ' was considered in Section 6.5, using the current and charge transformations. It was shown that relative to Σ, the moving coil had an electric dipole moment given by

$$\mathbf{p} = \frac{\mathbf{v} \times \mathbf{m}'}{c^2} \tag{6.71}$$

If this formula holds for each atomic magnetic dipole in the moving dielectric, since $n = n'/(1 - v^2/c^2)^{\frac{1}{2}}$, one has for the electric dipole moment per unit volume in Σ

$$\frac{n'}{\sqrt{1 - v^2/c^2}} \frac{\mathbf{v} \times \mathbf{m}'}{c^2} = \gamma \frac{\mathbf{v} \times n'\mathbf{m}'}{c^2} = \gamma \frac{\mathbf{v} \times \mathbf{M}'}{c^2} = \gamma \frac{(\mathbf{v} \times \mathbf{M}')_\perp}{c^2}$$

Adding this term to the term $\gamma P'_\perp$,

$$\mathbf{P}_\perp = \gamma \left(\mathbf{P}' + \frac{\mathbf{v} \times \mathbf{M}'}{c^2} \right)_\perp$$

in agreement with eqn (6.78).

Problem 6.7—If atomic magnetic dipoles did consist of spatially separated North-seeking and South-seeking magnetic monopoles, how would you interpret the transformation

$$\mathbf{P} = \gamma \mathbf{v} \times \mathbf{M}'/c^2 ?$$

(Hint: Think in terms of the magnetic analogue of *Figure 6.5*.) What is the curl of the electric field associated with this polarization?

6.7.2. *The Transformation of the Magnetization Vector* $\mathbf{M}$

The magnetization vector $\mathbf{M}$ is defined in Appendix 2(f), eqn (A2.40), for stationary matter as the magnetic dipole moment induced per unit volume by an applied magnetic field. From the definition of the magnetizing force $\mathbf{H}$, eqn (A2.47),

$$\mathbf{M} = \frac{\mathbf{B}}{\mu_0} - \mathbf{H} \quad \text{in } \Sigma \tag{6.79}$$

and

$$\mathbf{M}' = \frac{\mathbf{B}'}{\mu_0} - \mathbf{H}' \quad \text{in } \Sigma' \tag{6.80}$$

179

The same value is used for the magnetic space constant in Σ and Σ' since it is equal to $4\pi \times 10^{-7}$ by definition. Substituting the transformations for B_x and H_x from eqns (6.41) and (6.75) into the equation $M_x = [(B_x/\mu_0) - H_x]$

$$M_x = \frac{B'_x}{\mu_0} - H'_x$$

therefore,

$$M_x = M'_x$$

Similarly,

$$M_y = \frac{B_y}{\mu_0} - H_y = \frac{\gamma}{\mu_0}\left(B'_y - \frac{v}{c^2} E'_z\right) - \gamma(H'_y - vD'_z)$$

$$= \gamma\left(\frac{B'_y}{\mu_0} - H'_y\right) + \gamma v\left(D'_z - \frac{1}{c^2\mu_0} E'_z\right)$$

But

$$\frac{1}{c^2\mu_0} = \varepsilon_0$$

Hence,

$$M_y = \gamma\left(\frac{B'_y}{\mu_0} - H'_y\right) + \gamma v(D'_z - \varepsilon_0 E'_z)$$

or

$$M_y = \gamma(M'_y + vP'_z)$$

Similarly,

$$M_z = \frac{B_z}{\mu_0} - H_z = \frac{\gamma}{\mu_0}\left(B'_z + \frac{v}{c^2} E'_y\right) - \gamma(H'_z + vD'_y)$$

$$= \gamma\left(\frac{B'_z}{\mu_0} - H'_z\right) - v\gamma\left(D'_y - \frac{1}{\mu_0 c^2} E'_y\right) = \gamma(M'_z - vP'_y)$$

Collecting the transformations and the inverse transformations,

$$
\left.
\begin{aligned}
M'_x &= M_x & M_x &= M'_x \\
M'_y &= \gamma(M_y - vP_z) & M_y &= \gamma(M'_y + vP'_z) \\
M'_z &= \gamma(M_z + vP_y) & M_z &= \gamma(M'_z - vP'_y)
\end{aligned}
\right\} \quad (6.81)
$$

or, in vector form

$$\mathbf{M}_\parallel = \mathbf{M}'_\parallel; \quad \mathbf{M}_\perp = \gamma(\mathbf{M}' - \mathbf{v} \times \mathbf{P}')_\perp = \gamma(\mathbf{M}' + \mathbf{P}' \times \mathbf{v})_\perp \quad (6.81a)$$

It will be assumed that the moving dielectric material is at rest in Σ'. According to the term $-\gamma \mathbf{v} \times \mathbf{P}'$ there is a contribution to the magnetization in Σ due to the motion of a polarized dielectric relative to Σ. It is called a current of dielectric convection or a Röntgen current. For purposes of field calculation a uniformly polarized dielectric at rest in Σ' can be replaced by surface charge distributions equal to P_n' C/m². These charge distributions are moving relative to the inertial frame Σ and give rise to a magnetic field, as illustrated in *Figure 6.5*.

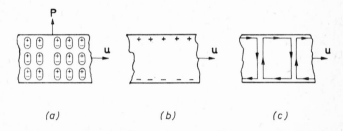

(a)   *(b)*   *(c)*

Figure 6.5. *The illustration of the magnetic field produced by a moving dielectric. When calculating the macroscopic electric field, the uniformly polarized dielectric shown in (a) can be replaced by the surface charge distributions shown in (b). If the dielectric is moving, when calculating the macroscopic magnetic field, the moving dielectric can be replaced by a series of current loops as shown in (c). Such an array of current loops would give rise to a magnetic field perpendicular to the paper in the direction away from the reader*

In the experiment carried out by Röntgen in 1888, a non-magnetic dielectric plate was placed between the plates of a charged parallel plate capacitor. The dielectric was rotated at high speed, whilst the plates of the capacitor remained stationary. The dielectric was polarized, and, according to eqn. (6.81a) whilst it was rotating it should have had a magnetic moment and given rise to a magnetic field which extended into the space outside the dielectric. Röntgen measured the magnetic field just outside the periphery of the rotating dielectric disk and showed that, to first order, the magnetic field was consistent with the equation $\mathbf{M} = \mathbf{P}' \times \mathbf{v}$.

The other terms in eqns (6.81) arise from the different measures of the value of atomic magnetic dipole moments in Σ and Σ' and the different measures of the number of atomic magnetic dipoles per unit volume in Σ and Σ'. It is left as an exercise for the reader to consider a simplified classical model for each atomic magnetic dipole, such as a small square current carrying coil, and show that the different measures of its magnetic dipole moment in Σ and Σ' are consistent with eqns (6.81a).

6.7.3. *The Constitutive Equations*

It will be assumed that when the materials are at rest in Σ' the constitutive equations take the form

$$\mathbf{D}' = \varepsilon_r \varepsilon_0 \mathbf{E}'; \quad \mathbf{B}' = \mu_r \mu_0 \mathbf{H}'; \quad \mathbf{J}' = \sigma \mathbf{E}' \tag{6.82}$$

From eqn (6.40a)

$$\mathbf{E}'_\| = (\mathbf{E} + \mathbf{v} \times \mathbf{B})_\|; \quad \mathbf{E}'_\perp = \gamma(\mathbf{E} + \mathbf{v} \times \mathbf{B})_\perp$$

and from eqn (6.74a),

$$\mathbf{D}'_\| = \left(\mathbf{D} + \frac{1}{c^2}\mathbf{v} \times \mathbf{H}\right)_\|; \quad \mathbf{D}'_\perp = \gamma\left(\mathbf{D} + \frac{1}{c^2}\mathbf{v} \times \mathbf{H}\right)_\perp$$

Substituting in the equation $\mathbf{D}' = \varepsilon_r \varepsilon_0 \mathbf{E}'$ for the parallel components of $\mathbf{D}'$ and $\mathbf{E}'$,

$$\left(\mathbf{D} + \frac{1}{c^2}\mathbf{v} \times \mathbf{H}\right)_\| = \varepsilon_r \varepsilon_0 (\mathbf{E} + \mathbf{v} \times \mathbf{B})_\| \tag{6.83}$$

Substituting for the perpendicular components of $\mathbf{D}'$ and $\mathbf{E}'$,

$$\gamma\left(\mathbf{D} + \frac{1}{c^2}\mathbf{v} \times \mathbf{H}\right)_\perp = \varepsilon_r \varepsilon_0 \gamma(\mathbf{E} + \mathbf{v} \times \mathbf{B})_\perp$$

Cancelling γ,

$$\left(\mathbf{D} + \frac{1}{c^2}\mathbf{v} \times \mathbf{H}\right)_\perp = \varepsilon_r \varepsilon_0 (\mathbf{E} + \mathbf{v} \times \mathbf{B})_\perp \tag{6.84}$$

Eqns (6.83) and (6.84) can be combined into the single equation

$$\left(\mathbf{D} + \frac{1}{c^2}\mathbf{v} \times \mathbf{H}\right) = \varepsilon_r \varepsilon_0 (\mathbf{E} + \mathbf{v} \times \mathbf{B}) \tag{6.85}$$

This is a constitutive equation valid for a medium moving with uniform velocity $\mathbf{v}$ relative to an inertial frame Σ; when $\mathbf{v} = 0$ it reduces to the equation $\mathbf{D} = \varepsilon_r \varepsilon_0 \mathbf{E}$.

Similarly substituting for $\mathbf{B}'$ from eqn (6.41a) and for $\mathbf{H}'$ from eqn (6.75a) into the equation $B' = \mu_r\mu_0 H'$, one obtains

$$\left(B - \frac{1}{c^2}\mathbf{v}\times\mathbf{E}\right) = \mu_r\mu_0(\mathbf{H} - \mathbf{v}\times\mathbf{D}) \tag{6.86}$$

This is a constitutive equation valid in Σ for a medium moving with uniform velocity $\mathbf{v}$ relative to the inertial frame Σ.

In Σ' we have

$$J'_x = \sigma E'_x$$

From eqn (6.66), $J'_x = \gamma(J_x - v\rho)$ and from eqn (6.40) $E'_x = E_x$, hence

$$\gamma(J_x - v\rho) = \sigma E_x \tag{6.87}$$

Similarly substituting in the eqns $J'_y = \sigma E'_y$ and $J'_z = \sigma E'_z$,

$$J_y = \sigma\gamma(E_y - vB_z) \tag{6.88}$$

$$J_z = \sigma\gamma(E_z + vB_y) \tag{6.89}$$

Eqns (6.87), (6.88) and (6.89) can be rewritten in the form

$$\gamma(\mathbf{J} - \mathbf{v}\rho)_{\parallel} = \sigma(\mathbf{E} + \mathbf{v}\times\mathbf{B})_{\parallel} \tag{6.90}$$

$$(\mathbf{J} - \mathbf{v}\rho)_{\perp} = \sigma\gamma(\mathbf{E} + \mathbf{v}\times\mathbf{B})_{\perp} \tag{6.91}$$

where $\mathbf{J}$, the total current density, includes the convection current density $\mathbf{v}\rho$. Hence, $\mathbf{J} - \mathbf{v}\rho$ is equal to the conduction current density in Σ. When $\mathbf{v} = 0$, eqns (6.90) and (6.91) reduce to the equation $\mathbf{J} = \sigma\mathbf{E}$. If $v \ll c$, and ρ is zero,

$$\mathbf{J} = \sigma(\mathbf{E} + \mathbf{v}\times\mathbf{B}) \tag{6.92}$$

This equation can be applied to an electrically neutral conductor moving in an electric and a magnetic field. For example, it can be applied to plasmas moving relative to the laboratory. Eqn (6.92) can also be applied to the moving conductor, in the example of a motional e.m.f. shown in *Figure 6.1(a)*.

For accounts of the complete development of the equations for the electrodynamics of media moving with uniform velocity and for a discussion of boundary conditions the reader is referred to the text-books by Cullwick[3], Møller[7], Pauli[8] and Sommerfeld[9].

Problem 6.8—Show that for a non-magnetic ($\mu_r = 1$) non-conducting ($\sigma = 0$) dielectric, of dielectric constant ε_r moving with uniform velocity $\mathbf{v}$, Maxwell's equations and the constitutive equations can, to first order v/c, be written in the form:

$$\nabla . \mathbf{D} = \rho$$

$$\nabla . \mathbf{B} = 0$$

$$\nabla \times \mathbf{E} = -\frac{\partial \mathbf{B}}{\partial t}$$

$$\nabla \times \mathbf{B} = \mu_0 \left[\mathbf{J} + \varepsilon_0 \frac{\partial \mathbf{E}}{\partial t} + \frac{\partial \mathbf{P}}{\partial t} + \nabla \times (\mathbf{P} \times \mathbf{v}) \right]$$

$$\mathbf{P} = \varepsilon_0(\varepsilon_r - 1)(\mathbf{E} + \mathbf{v} \times \mathbf{B})$$

(Hint: The term $\nabla \times (\mathbf{P} \times \mathbf{v})$ arises by putting $\mathbf{M} \simeq \mathbf{P} \times \mathbf{v}$ in eqn (4.114). Reference: Rosser[2c].)

Problem 6.9—The velocity of light, measured relative to Σ', in a liquid at rest in Σ' is c/n' where n' is the refractive index. Use the velocity transformations to show that the velocity of light parallel to the x axis, measured relative to Σ, is to first order of v/c,

$$u_x = c/n' + v[1 - 1/n'^2]$$

Assume plane wave solutions of the equations developed in Problem (6.8) and show that the velocity of light in a moving dielectric is in agreement with the above value (Reference: Rosser[2d]).

6.7.4. *The Wilson–Wilson Experiment*

The main difference between Minkowski's electrodynamics of moving media and an earlier theory due to Lorentz, is that, according to Minkowski's theory, a moving magnetized body should have an electric dipole moment given by eqns (6.78). In 1908, Einstein and Laub[10] suggested an experiment to test the validity of Minkowski's theory of the electrodynamics of moving bodies. The experiment was carried out in a slightly modified form by Wilson and Wilson[11] in 1913.

Consider a large parallel plate capacitor lying with its plates parallel to the xy plane as shown in *Figure 6.6*. Let the distance between the plates be very small compared with the area of the plates. Let the dielectric between the plates have a dielectric constant ε_r and a relative permeability μ_r, which are both greater than unity in the inertial reference frame in which the magnetic dielectric is at rest. It is assumed that the electrical conductivity of the dielectric is zero. Let a uniform external magnetic field be applied in the positive y direction of the laboratory system Σ, as shown in *Figure 6.6*. Let the applied magnetizing force in the laboratory system be equal to H_y. Let the capacitor and dielectric both move with uniform velocity v in

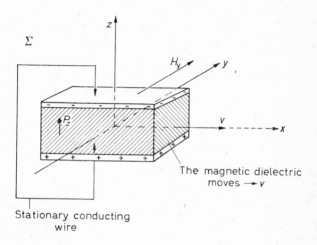

Figure 6.6. The theory of the Wilson–Wilson experiment. The 'magnetic' dielectric, which is between the plates of a parallel plate capacitor, moves in the $+x$ direction. A magnetic field is applied in the $+y$ direction. The moving 'magnetic' dielectric is polarized. If the plates of the moving capacitor are short circuited by a stationary connecting wire, then charge distributions appear on the plates of the capacitor such that the electric intensity $\mathbf{E}$ inside the moving dielectric is zero. If the direction of the magnetic field is reversed, the signs of the charges on the plates of the capacitor must change and a transient current will flow in the connecting wire

the positive x direction relative to the laboratory system Σ, as shown in *Figure 6.6*. The moving dielectric is magnetized in the external magnetic field H_y. According to eqn (6.78), the moving magnetized medium should give a contribution $\gamma \mathbf{v} \times \mathbf{M}'/c^2$ to the polarization of the moving dielectric relative to Σ. This polarization is in the $+z$ direction in Σ. Let the plates of the moving capacitor be short circuited by a wire, which is stationary in Σ, and which makes sliding contact with the outsides of the plates of the moving capacitor, as shown in *Figure 6.6*. Now in Σ, Maxwell's equations should be valid, so that

$$\nabla \times \mathbf{E} = -\frac{\partial \mathbf{B}}{\partial t} \tag{6.93}$$

If the capacitor were of infinite dimensions in the xy plane then $\partial \mathbf{B}/\partial t$ would be zero, once the velocity of the capacitor was uniform. Integrating eqn (6.93) over the area of a loop formed by the stationary wire and completed by a path in the z direction inside the moving 'magnetic' dielectric, one has

$$\int \nabla \times \mathbf{E} \cdot \mathbf{n}\, dS = \oint \mathbf{E} \cdot d\mathbf{l} = -\int \frac{\partial \mathbf{B}}{\partial t} \cdot \mathbf{n}\, dS = 0$$

When the steady state is reached, since the dielectric between the plates is a non-conductor, no steady current can flow in the connecting wire, and the electric intensity $\mathbf{E}$ inside the conducting connecting wire must be zero. If $\oint \mathbf{E} \cdot d\mathbf{l} = 0$, the component E_z of the electric field inside the moving dielectric must also be zero. This is the electric field measured by an observer at rest in Σ, and relative to whom the dielectric is moving with uniform velocity v. During the transient state, when the capacitor starts to move with uniform velocity, current will flow in the connecting wire until steady true (or free) charge distributions build up on the plates of the capacitor, as shown in *Figure 6.6*, such that inside the moving dielectric their electric field compensates the electric field associated with the polarization of the moving dielectric.

The moving capacitor and magnetic dielectric are at rest in the inertial frame Σ′, which moves with uniform velocity $+v$ along the $+x$ axis relative to the laboratory system Σ.

There are three main contributions to the polarization of the moving dielectric: (1) The apparent polarization associated with a moving magnetized body. (2) In the reference frame Σ′, in which the dielectric is at rest, according to eqn (6.40) there is an electric field associated with the external applied magnetic field; this contributes to the polarization of the dielectric in Σ′. (3) The electric field associated with the true charges on the metal plates of the capacitor also affects the total polarization of the dielectric.

According to eqn (6.75), the magnitude of the *applied* magnetic field relative to the inertial frame Σ′ in which the dielectric is at rest is

$$H_y' = \gamma H_y + \gamma v D_z$$

The reader can repeat the following calculation to check that the second term on the right-hand side can be neglected, giving

$$H'_y \simeq \gamma H_y \qquad (6.94)$$

In general, a magnetized body of finite dimensions contributes to the total magnetizing force $\mathbf{H}'$. In practice, this contribution is generally known as the demagnetizing force. According to eqn (A2.42) of Appendix 2(f) this contribution to the total magnetizing force $\mathbf{H}'$ in Σ' can be calculated, using a fictitious volume magnetic pole distribution $-\nabla' \cdot \mathbf{M}'$ and a fictitious surface magnetic pole distribution $\mathbf{M}' \cdot \mathbf{n}$ (cf. Section A2(f)). In the present case these fictitious pole distributions are confined to the vicinity of the thin ends of the 'magnetic' dielectric. If the thickness of the 'magnetic' dielectric is negligible compared with its area in the $x'y'$ plane, the contributions of the fictitious magnetic poles to the total magnetizing force $\mathbf{H}'$ in the y' direction is negligible over most of the magnetic dielectric. Hence, away from the ends of the moving dielectric, the total magnetizing force is

$$H' = H'_y = \gamma H_y \qquad (6.95)$$

In Σ', the 'magnetic' dielectric is at rest, so that the normal constitutive equations should be applicable. Hence in Σ',

$$\mathbf{B}' = \mu_r \mu_0 \mathbf{H}'$$

or

$$B'_y = \mu_r \mu_0 H'_y \qquad (6.96)$$

From the definition of $\mathbf{H}'$, we have

$$\mathbf{B}' = \mu_0 (\mathbf{H}' + \mathbf{M}') \qquad (A2.47)$$

Hence, using eqn (6.95)

$$M'_y = (\mu_r - 1)H'_y = \gamma(\mu_r - 1)H_y \qquad (6.97)$$

Let the surface density of true electric charge on the plates of the capacitor be σ, measured relative to the laboratory system Σ. Since $\nabla \cdot \mathbf{D} = \rho$, relative to Σ, application of Gauss' law shows that the electric displacement $\mathbf{D}$ inside the capacitor is

$$D_z = \sigma \qquad (6.98)$$

Using eqn (6.74), the electric displacement in Σ' is

$$D'_z = \gamma(D_z + vH_y/c^2) \qquad (6.99)$$

187

In Σ', the 'stationary' dielectric is inside this uniform electric displacement D'_z. This is the same problem as the calculation of the polarization of a dielectric, placed between the plates of a stationary parallel plate capacitor. Since the dielectric is stationary,

$$D'_z = \varepsilon_r \varepsilon_0 E'_z$$

From the definition of $\mathbf{D}'$,

$$D'_z = \varepsilon_0 E'_z + P'_z$$

Hence,

$$P'_z = D'_z - \varepsilon_0 E'_z = D'_z(1 - 1/\varepsilon_r)$$

Substituting for D'_z, from eqn (6.99),

$$P'_z = \gamma(D_z + vH_y/c^2)(1 - 1/\varepsilon_r) \tag{6.100}$$

The two contributions to P'_z are associated with γD_z, where D_z arises from the free charge on the plates of the capacitors, and with $\gamma v H_y/c^2$ which arises from the electric field in Σ' associated with the source of the applied external magnetic field.

Relative to the laboratory system Σ, from eqn (6.78)

$$P_z = \gamma P'_z + \gamma v M'_y/c^2 \tag{6.101}$$

Substituting from eqns (6.97) and (6.100)

$$P_z = \gamma^2 \left(D_z + \frac{vH_y}{c^2}\right)\left(1 - \frac{1}{\varepsilon_r}\right) + \frac{\gamma^2 v}{c^2}(\mu_r - 1)H_y \tag{6.102}$$

In the laboratory system Σ, due to the connecting wire in *Figure 6.6*, the electric field is zero inside the moving dielectric, once the steady state is reached. Hence, since E_z is zero,

$$D_z = \varepsilon_0 E_z + P_z = P_z$$

Substituting for P_z in eqn (6.102) neglecting terms of order v^2/c^2, rearranging, and using eqn (6.98), we have

$$\sigma = vH_y(\mu_r \varepsilon_r - 1)/c^2 \tag{6.103}$$

If the polarization associated with $\mathbf{v} \times \mathbf{M}'/c^2$ were ignored, that is omitting the second term on the right-hand side of eqns (6.101) and (6.102), we would have

$$\sigma = vH_y(\varepsilon_r - 1)/c^2 \tag{6.104}$$

If the direction of the magnetic field is reversed, then the direction of the polarization of the dielectric in Σ reverses and if E_z is to be zero inside the dielectric, the charges on the plates of the capacitor must change sign, and a transient current will flow through the stationary short circuiting wire. The total charge flowing could be measured, using a ballistic galvanometer. According to eqn (6.103) the charge flowing should be proportional to $(\mu_r \varepsilon_r - 1)$. It was shown in Section 6.5 that the apparent polarization due to a moving magnetized body was a relativistic effect arising from the different measures of time in Σ and Σ'. This effect was not included in Lorentz's original electron theory, so that according to that theory the charge flowing should be proportional to $(\varepsilon_r - 1)$. To quote Lorentz:

> Now, according to Hertz's theory the effect should be proportional to ε_r. On the other hand, according to the older electron theory, the aether does not share in the motion of matter and is thus not exposed to the action of the induced electric force. The latter produces a polarization only in matter, so that the effect should be proportional to $(\varepsilon_r - 1)$. According to the theory of relativity, the effect should be proportional to $(\mu_r \varepsilon_r - 1)$.

For a more rigorous discussion of the theory of the experiment the reader is referred to Cullwick [3b].

In the experiment performed by Wilson and Wilson in 1913 a cylindrical capacitor rotating at high speed was used, since it was not practical to move a large parallel plate capacitor with sufficient uniform linear velocity to produce an observable effect. Wilson and Wilson constructed an artificial 'magnetic' dielectric by embedding steel balls ~ 3 mm in diameter in wax. The values of μ_r and ε_r for this 'magnetic' dielectric were 3 and 6 respectively. The dielectric was in the form of a hollow cylinder of length 9·5 cm and inner and outer diameters of 2·0 and 3·73 cm respectively. The inside and outside surfaces of the hollow cylinder were coated with metal. The cylindrical capacitor was rotated about its axis and the magnetic field was applied along the axis of the cylinder. Contact was made with the inside and outside metal surfaces with brushes which were connected to a stationary quadrant electrometer. Wilson and Wilson observed the charge flowing when the direction of the magnetic field was reversed. If effects due only to rotational motion could be neglected, the experiment was the same as that proposed by Einstein and Laub. After applying a correction for the e.m.f. induced when the magnetic field was reversed, Wilson and Wilson concluded that the experimental value for $1 - (1/\mu_r \varepsilon_r)$ was 0·96 whereas the value predicted by the relativistic theory was $1 - (\frac{1}{6} \times \frac{1}{3}) = 0.944$ and the value predicted by Lorentz's electron theory was $(1 - 1/\varepsilon_r) = (1 - \frac{1}{6}) = 0.83$. The experimental value was in better agreement with the relativistic theory than with Lorentz's electron theory.

Problem 6.10—Use the constitutive equations for moving media to determine eqn (6.103). (Hint: From eqn (6.85), since E_z is zero and **v** has an x component only

$$D_z + vH_y/c^2 = \varepsilon_r\varepsilon_0 vB_y$$

From eqn (6.86)

$$B_y = \mu_r\mu_0(H_y + vD_z)$$

Eliminate B_y, put $\mu_0\varepsilon_0 = 1/c^2$ and ignore terms of order $> v^2/c^2$.)

6.8. UNIPOLAR INDUCTION

In 1851, Faraday showed that a current is induced in a *stationary* wire which makes sliding contact near the pole and equator of a *rotating* cylindrical permanent magnet. The phenomenon is generally called *unipolar* induction. This method of current generation has been used for large scale electric generators. Unipolar induction can be interpreted in the laboratory system in terms of Minkowski's Electrodynamics of Moving Media. Before proceeding to consider a rotating magnet, we shall consider the linear motion of a permanent magnet.

(a) Moving Permanent Magnet

In Section 6.4 the motion of an isolated conductor in an *externally* applied magnetic field was considered. In this case there was no electric field in the laboratory system Σ. The magnetic force $q\mathbf{v} \times \mathbf{B}$ on charges moving with the conductor gave rise to a displacement of true (or free) charge inside the isolated conductor, as illustrated in *Figure 6.1(a)*. Now a permanent magnet generally gives rise to a magnetic field both inside and outside the magnet. Hence, atomic electric charges inside an isolated moving permanent magnet, such as conduction electrons, should also experience a magnetic force $q\mathbf{v} \times \mathbf{B}$, where q is the magnitude of the charge, **v** the velocity of the magnet and **B** the magnetic induction inside the magnet due to the magnet itself. Does this force lead to a displacement of true (or free) electric charge?

Consider a very large isolated rectangular conducting permanent magnet, which is moving with uniform linear velocity v in the positive x direction relative to the laboratory system, as shown in *Figure 6.7(a)*. It will be assumed, initially, that there is *no* stationary wire connecting the points A and C in *Figure 6.7(a)*. The magnet is at rest in the inertial frame Σ', which is moving with uniform velocity v relative to Σ, as shown in *Figure 6.7(b)*. Let the uniform magnetization of the permanent magnet in Σ' be $\mathbf{M}'$, and let it be in the $+z'$

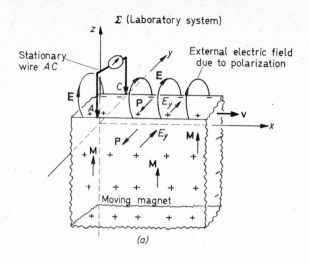

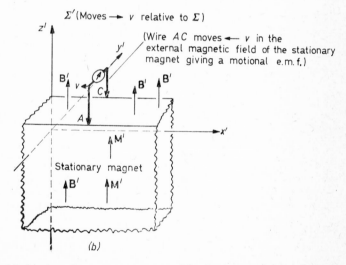

Figure 6.7. (a) *A moving permanent magnet. The dimensions of the magnet are much greater in the xz plane than in the y direction. The moving magnet has an electric polarization $\gamma \mathbf{v} \times \mathbf{M}'/c^2$ in Σ. This polarization gives an electric field outside the moving magnet which starts a conduction current flowing in the stationary wire AC from A to C. (b) The magnet is at rest in Σ' and the moving wire AC gives rise to a motional e.m.f., due to its motion in the external magnetic field of the 'stationary' permanent magnet*

191

direction in *Figure 6.7(b)*. It is being assumed that the thickness of the magnet in the y' direction is negligible compared with its dimensions in the $x'z'$ plane. We shall start by considering a field point inside the magnet well away from the edges of the permanent magnet, where fringing effects are negligible. At such a field point according to eqn (A2.42) of Appendix 2(f) the magnetizing force $\mathbf{H}'$ due to the fictitious magnetic pole distributions $-\nabla'.\mathbf{M}'$ and $\mathbf{M}'.\mathbf{n}$ is negligible, so that the magnetic induction is given by

$$B'_x = B'_y = 0; \quad B'_z = \mu_0 M'_z \qquad (6.105)$$

In the inertial frame Σ', in which the isolated permanent magnet is at rest, there is only a magnetic field, since a stationary, isolated, permanent magnet gives no electric field. Hence *stationary* charges inside the magnet, which is at rest relative to Σ', are not acted upon by either an electric or magnetic force relative to Σ', and so no conduction electrons should be displaced relative to Σ'. From the equation $\rho = \gamma(\rho' + vJ'_x/c^2)$, since ρ' and J'_x are zero in Σ', then ρ is zero in Σ. Hence, there cannot be a displacement of true electric charge in an *isolated* permanent magnet moving with uniform velocity relative to the laboratory system Σ, even though there is a magnetic force $q\mathbf{v}\times\mathbf{B}$ on the charges inside the magnet.

In the laboratory system Σ, using eqns (6.41), we have, if $\mathbf{E} = 0$,

$$B_z = \gamma B'_z = \mu_0 \gamma M'_z \qquad (6.106)$$

Since the charges in the magnet are moving relative to Σ, they are acted upon by a magnetic force $q\mathbf{v}\times\mathbf{B}$, having components relative to Σ of

$$f_x = f_z = 0; \quad f_y = -qvB_z = -qv\mu_0\gamma M'_z \qquad (6.107)$$

However, in Σ, using eqns (6.40) we expect an electric field given by

$$E_x = E_z = 0; \quad E_y = \gamma vB'_z = vB_z \qquad (6.108)$$

This electric field gives rise to an electric force on a charge inside the magnet, having components

$$f_x = f_z = 0; \quad f_y = qE_y = +qvB_z \qquad (6.109)$$

It can be seen that the electric force, given by eqn (6.109), exactly balances the magnetic force given by eqn (6.107). Thus there should be no displacement of electric charge inside an *isolated* permanent magnet moving with uniform linear velocity. The origin of the electric field inside the moving magnet will now be interpreted.

It was illustrated in Section 6.5, *Figure 6.4(b)*, that a moving magnetic dipole has an electric dipole moment given by eqn (6.71). This in turn leads to the transformation

$$\mathbf{P} = \frac{\gamma \mathbf{v} \times \mathbf{M}'}{c^2} \tag{6.110}$$

Since in Σ', $\mathbf{P}' = 0$ and $\mathbf{M}'$ has the component M'_z, we have in the laboratory system Σ,

$$P_x = P_z = 0; \quad P_y = -\frac{\gamma v M'_z}{c^2} \tag{6.111}$$

The direction of the polarization is shown in *Figure 6.7(a)*. For purposes of calculating the macroscopic electric field, according to eqn (A2.14) of Appendix 2(b), a uniformly polarized dielectric can be replaced by a fictitious surface electric charge distribution $\mathbf{P} \cdot \mathbf{n}$ coulombs/metre^2. These fictitious charges are shown in *Figure 6.7(a)*. Away from the ends of the magnet, application of Gauss' law shows that the macroscopic electric field associated with such a charge distribution is in the $+y$ direction and equal to $-P_y/\varepsilon_0 = +\gamma v M'_z/\varepsilon_0 c^2$. Using $\mu_0 = 1/\varepsilon_0 c^2$, it follows that the magnitude of this electric field is

$$E_p = -\frac{P_y}{\varepsilon_0} = +\gamma v \mu_0 M'_z \tag{6.112}$$

This gives rise to an electric force qE_p in the $+y$ direction, which balances the magnetic force $-qvB_z = -qv\gamma\mu_0 M'_z$ given by eqn (6.107). Hence no conduction electrons should be displaced giving rise to true (or free) charge distributions in an isolated, conducting, permanent magnet moving with uniform linear velocity. The polarization of a moving magnetized medium, given by eqn (6.110) plays an important role in the above interpretation.

Near the ends of the moving magnet, the electric field associated with the electric polarization of the moving magnet extends into the space outside the magnet, as shown in *Figure 6.7(a)*.

It will now be assumed that contact is made with the moving conducting magnet in *Figure 6.7(a)* at two points A and C by a conducting wire, which is stationary relative to the laboratory system Σ, as shown in *Figure 6.7(a)*. When the magnet is moving, current flows from A to C in the stationary wire. It is easiest to interpret the effect in the inertial reference frame Σ' in which the magnet is at rest, and in which the wire AC moves in the negative x' direction as shown in *Figure 6.7(b)*. The magnetic field of the magnet extends into the space outside the magnet. In Σ' the wire is moving in this external magnetic field, and this motion gives rise to a motional e.m.f. of the type discussed in Section 6.4. The electric current starts to flow in the wire AC, the direction of current flow being from A to C. Positive charge starts accumulating near C and negative charge near A, giving rise to an electric field relative to Σ' across the magnet in the direction C to A, which in turn gives rise to a return current flow from C to A inside the magnet.

In the laboratory system Σ, the wire AC is at rest as shown in *Figure 6.7(a)*. In Σ the moving magnet has the electric polarization given by eqn (6.111). This polarization gives rise to an electric field both inside and outside the moving magnet, as shown in *Figure 6.7(a)*. In the absence of the connecting wire AC, the magnetic force $q\mathbf{v} \times \mathbf{B}$ on a charge moving with the magnet is balanced by the electric force $q\mathbf{E}_p$ on the same charge, due to the electric field $\mathbf{E}_p$ associated with the electric polarization. When the connecting wire AC is present, the electric field outside the magnet, associated with the electric polarization, gives a conduction current flow in the direction A to C in the connecting wire. The current in the wire joining the points A and C tries to build up true charge distributions, so as to equalize the electric potentials at A and C. It is unable to do so completely. As soon as positive charge starts accumulating at C and negative charge at A, these true charge distributions give rise to an electric field which, inside the magnet, is opposite in direction to $\mathbf{E}_p$. This upsets the balance that would exist between $q\mathbf{v} \times \mathbf{B}$ and $q\mathbf{E}_p$ in an isolated moving magnet. The magnetic force on a charge moving with the magnet now exceeds the total electric force, associated with both the electric polarization, and the true charge distributions built up due to the current flow in the wire AC. Hence, electric current flows from C to A inside the moving conducting magnet. The current flow in the wire does its best to equalize the electric potentials of A and C, but due to the return current flow inside the moving magnet it is unable to do so completely. Only a compromise is reached. For an infinitely long magnet, a state of dynamic equilibrium is reached, when continuous current flows in the stationary wire AC. The magnitude of this current and the electric potential difference between A and C depend on how quickly the wire AC can carry charge. This depends on the electrical resistance of AC. For ohmic conductors the current in and potential drop across AC are consistent with Ohm's law.

Before proceeding to consider a rotating permanent magnet, we must consider one other effect. Assume that instead of the stationary wire AC in *Figure 6.7(a)* there is an extra permanent magnet, just above the original permanent magnet shown in *Figure 6.7(a)*. Let the extra permanent magnet move with a uniform velocity different to **v** relative to the laboratory system Σ. This extra permanent magnet moves in the fringing electric field associated with the electric polarization of the original moving magnet, and also in its magnetic field. There are magnetic and electric forces on the charges moving with the extra magnet leading to a displacement of conduction electrons giving a true electric charge distribution on the extra moving magnet. Similarly, the electric and magnetic fields of the extra magnet lead to a displacement of conduction electrons giving rise to true charge distributions on the original magnet. It is impossible to find an inertial reference frame in which both permanent magnets are at rest. Hence, it is concluded that, when two permanent *conducting* magnets are in uniform *relative* motion, true electric charge distributions build up on both magnets. This effect is important in rotating magnets, since different parts of a rotating magnet have different linear speeds relative to the laboratory.

(b) A Rotating Spherical Magnet

Consider an *isolated* sphere of radius R rotating with uniform angular velocity $\boldsymbol{\omega}$ relative to the laboratory system Σ, as shown in *Figure 6.8*. [It will be assumed, initially, that the wire AC in *Figure 6.8* is *not* present.] Let the rotating sphere be magnetized uniformly in a direction parallel to $\boldsymbol{\omega}$. Let the magnetization vector relative to Σ be **M**. The linear velocity of any element of the sphere is $\mathbf{v} = \boldsymbol{\omega} \times \mathbf{r}$. According to eqn (6.78), associated with this linear velocity relative to Σ, there is an electric polarization relative to Σ given by

$$\mathbf{P} = \gamma \mathbf{v} \times \mathbf{M}'/c^2 = \gamma(\boldsymbol{\omega} \times \mathbf{r}) \times \mathbf{M}'/c^2 \qquad (6.113)$$

From eqn (6.81), if $\mathbf{P}' = 0$ and if $v \ll c$ and $\gamma \simeq 1$, then $\mathbf{M}' \simeq \mathbf{M}$, and eqn (6.113) becomes

$$\mathbf{P} = \mathbf{v} \times \mathbf{M}/c^2 = (\boldsymbol{\omega} \times \mathbf{r}) \times \mathbf{M}/c^2 \qquad (6.114)$$

For a point at a distance (r, θ) from the centre of the sphere

$$P = \omega r M \sin \theta / c^2 \qquad (6.115)$$

The direction of this polarization is outwards from the axis of rotation, as shown in *Figure 6.8*. For purposes of calculating the macroscopic electric field, the isolated polarized rotating sphere can be replaced by a fictitious volume charge distribution $-\mathbf{V} \cdot \mathbf{P}$ and a fictitious surface charge distribution $\mathbf{P} \cdot \mathbf{n}$, where $\mathbf{n}$ is a unit vector normal to the surface of the sphere (cf. Appendix 2(b)). According to eqn (6.115), $\mathbf{P}$ varies with θ and r so that $\mathbf{V} \cdot \mathbf{P}$ is finite. Now

$$\rho_p = -\mathbf{V} \cdot \mathbf{P} = -\mathbf{V} \cdot (\mathbf{v} \times \mathbf{M}/c^2) \qquad (6.116)$$

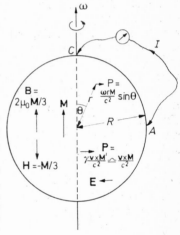

Figure 6.8. A uniformly magnetized sphere rotating with uniform angular velocity ω. If the magnetization is $\mathbf{M}$, the internal magnetic fields are $\mathbf{B} = 2\mu_0\mathbf{M}/3$ and $\mathbf{H} = -\mathbf{M}/3$. Associated with the motion of the magnet there is a polarization $\mathbf{P} = \gamma\mathbf{v} \times \mathbf{M}'/c^2$, which is outwards from the axis of rotation. When the magnet rotates, current flows in the wire AC from A to C. In the first part of the discussion in the text it is assumed that the wire AC is absent

According to eqn (A1.22) of Appendix 1,

$$\mathbf{V} \cdot (\mathbf{v} \times \mathbf{M}) = \mathbf{M} \cdot (\mathbf{V} \times \mathbf{v}) - \mathbf{v} \cdot (\mathbf{V} \times \mathbf{M})$$

Inside the rotating sphere $\mathbf{M}$ is a constant, so that $\mathbf{V} \times \mathbf{M}$ is zero. It is straightforward to show that $\mathbf{V} \times \mathbf{v}$ is equal to 2ω. Hence eqn (6.116) becomes

$$\rho_p = -2\omega M/c^2 \qquad (6.117)$$

Thus the fictitious volume polarization charge ρ_p is negative and of constant magnitude throughout the sphere, as shown in *Figure 6.9(a)*.

The apparent surface polarization charge density is given by

$$\sigma_p = \mathbf{P} \cdot \mathbf{n} = (\mathbf{v} \times \mathbf{M}) \cdot \mathbf{n}/c^2$$

Now for a point on the surface of the sphere, $r = R$, $v = \omega R \sin \theta$

and $|\mathbf{v} \times \mathbf{M}| = \omega M R \sin \theta$. The angle between $\mathbf{v} \times \mathbf{M}$ and $\mathbf{n}$ is $(\pi/2 - \theta)$. Hence,

$$\sigma_p = \frac{\omega M R}{c^2} \sin \theta \cos (\pi/2 - \theta) = \frac{\omega M R}{c^2} \sin^2 \theta \qquad (6.118)$$

The fictitious surface charge distribution σ_p, associated with the polarization of the rotating sphere is positive. It is a maximum at the

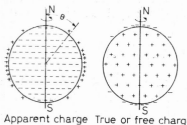

Apparent charge   True or free charge
(a)   (b)

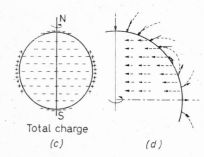

Total charge
(c)   (d)

Figure 6.9. Isolated, rotating, conducting, permanent spherical magnet. (a) Apparent charge distribution due to the electric polarization of the rotating magnet. (b) True or free charge distribution, due to displacement of conduction electrons. (c) Total charge distribution. (d) The resultant electric field. The field intensity is proportional to the length of the arrows. The external field at the surface is shown by solid arrows, and the internal field, which is proportional to the radial distance of the point from the axis, by dotted arrows. The lines of force lie in planes passing through the axis of the sphere. (Reproduced from Electromagnetism and Relativity by E. G. Cullwick, Longmans Green & Co. Ltd., with the permission of the author and of the publisher)

equator where $\theta = \pi/2$ and zero when θ is zero, as shown in Figure 6.9(a).

So far only effects associated with the electric polarization of the rotating magnetized sphere have been considered. Different parts of the rotating sphere have different linear speeds $\boldsymbol{\omega} \times \mathbf{r}$ relative to the laboratory system Σ. Each volume element of the rotating magnet moves in the magnetic field and the electric field (associated with the electric polarization) of all the other volume elements of the rotating magnet. If the rotating magnet is a conductor, this leads to a displacement of conduction electrons, which builds up true charge distributions such that, when the steady state is reached, the total

197

force on a charge moving with the isolated rotating spherical magnet is zero. That is

$$\mathbf{f} = q\mathbf{E} + q\mathbf{v} \times \mathbf{B} = 0 \qquad (6.119)$$

or,

$$\mathbf{E} = -\mathbf{v} \times \mathbf{B} \qquad (6.120)$$

where $\mathbf{E}$ is the total electric intensity due to both the electric polarization of the rotating sphere and the true charge distributions on the rotating sphere. Maxwell's equations can be applied in the laboratory system Σ. Hence,

$$\rho_{\text{true}} = \mathbf{V} \cdot \mathbf{D} = \varepsilon_0 \mathbf{V} \cdot \mathbf{E} + \mathbf{V} \cdot \mathbf{P}$$

When the true charge distributions on the rotating magnet have been built up, eqn (6.120) is valid. Also $\mathbf{P}$ is given by eqn (6.114). Hence,

$$\rho_{\text{true}} = -\varepsilon_0 \mathbf{V} \cdot (\mathbf{v} \times \mathbf{B}) + \mathbf{V} \cdot (\mathbf{v} \times \mathbf{M}/c^2)$$

Using

$$\mu_0 \varepsilon_0 = 1/c^2, \quad \text{and} \quad \mathbf{B}/\mu_0 - \mathbf{M} = \mathbf{H},$$

we have

$$\rho_{\text{true}} = -\mathbf{V} \cdot (\mathbf{v} \times \mathbf{H}/c^2) \qquad (6.121)$$

Inside a stationary uniformly magnetized sphere, $\mathbf{H} = -\mathbf{M}/3$. (Reference: Slater and Frank[12].) The magnetic effects associated with the rotation of the charge distributions in *Figure 6.9* are negligible if $v \ll c$, cf. Problem 6.11, so that for the rotating magnet, $\mathbf{H} \simeq -\mathbf{M}/3$. Using eqn (A1.22) of Appendix 1, remembering $\mathbf{V} \times \mathbf{M} = 0$ and $\mathbf{V} \times \mathbf{v} = 2\boldsymbol{\omega}$, we find

$$\rho_{\text{true}} = 2\omega M/3c^2 \qquad (6.122)$$

The volume distribution of true (or free) charge is positive and uniform throughout the volume of the rotating sphere, as shown in *Figure 6.9(b)*. The surface density of true (or free) charge can be calculated from the total electric field of the rotating magnet. It will be shown later in this section that

$$\sigma_{\text{true}} = -\frac{2\omega M R \cos^2 \theta}{3c^2} \qquad (6.123)$$

This charge distribution is negative. It is a maximum on the axis of

198

rotation, where θ is zero, as shown in *Figure 6.9(b)*. The total charge distributions are

$$\sigma = \sigma_p + \sigma_{\text{true}} = \frac{\omega M R}{c^2} (1 - \tfrac{5}{3} \cos^2 \theta) \qquad (6.124)$$

$$\rho = \rho_p + \rho_{\text{true}} = -\frac{4\omega M}{3c^2} \qquad (6.125)$$

These total charge distributions are shown in *Figure 6.9(c)*.

The calculation of the potential and electric field of the rotating sphere is a boundary value problem (Reference: Schlomka and Schenkel[13]). An outline will now be given. From eqn (6.120) for a field point inside the rotating magnet

$$\mathbf{E} = -\mathbf{v} \times \mathbf{B}$$

where $\mathbf{E}$ is the *total* electric intensity. Inside the sphere, provided $v \ll c$, the magnetic induction is parallel to $\mathbf{M}$ and to $\boldsymbol{\omega}$. Actually

$$\mathbf{B} = \frac{2\mu_0 \mathbf{M}}{3} = \frac{2\mathbf{M}}{3\varepsilon_0 c^2} \qquad (6.126)$$

(Reference: Slater and Frank[12]). The magnetic induction $\mathbf{B}$ is perpendicular to $\mathbf{v} = \boldsymbol{\omega} \times \mathbf{r}$. Hence inside the sphere

$$E = Bv = Br\omega \sin \theta = \frac{2M\omega r \sin \theta}{3\varepsilon_0 c^2} \qquad (6.127)$$

The direction of the total electric field $\mathbf{E} = -\mathbf{v} \times \mathbf{B}$ is towards the axis of rotation of the sphere. Consider $\int \mathbf{E} \cdot d\mathbf{l}$ evaluated from a point (R, θ) on the surface of the sphere to the axis of rotation in a direction parallel to $\mathbf{E}$, that is, perpendicular to the axis of rotation. Now,

$$dl = -d(r \sin \theta)$$

Hence,

$$\int \mathbf{E} \cdot d\mathbf{l} = - \int_{R \sin \theta}^{0} \omega B r \sin \theta \, d(r \sin \theta) = \tfrac{1}{2}\omega B R^2 \sin^2 \theta$$

Now $\int \mathbf{E} \cdot d\mathbf{l}$ is zero for a path along the axis of rotation of the spherical magnet, since there is no component of $\mathbf{E}$ in this direction. Hence $\int \mathbf{E} \cdot d\mathbf{l}$ between a point (R, θ) on the surface of the sphere and any point on the axis of rotation is $\tfrac{1}{2}\omega B R^2 \sin^2 \theta$. Hence the electric potential at the point (R, θ) is higher than at a point on the

H

axis of rotation. The potential difference between the surface of the sphere at the equator and the axis of rotation is $\frac{1}{2}\omega BR^2$. In general, the potential of a point on the surface of the sphere is

$$\phi = K_1 + \tfrac{1}{2}\omega BR^2 \sin^2\theta = K_2 - \tfrac{1}{2}\omega BR^2 \cos^2\theta \qquad (6.128)$$

where K_1 and K_2 are constants.

Outside the rotating sphere, ρ is zero and $\partial\phi/\partial t$ is zero when conditions are steady. Hence eqn (5.43) of Chapter 5 reduces to Laplace's equation, $\nabla^2\phi = 0$. Hence, the potential ϕ can be expanded in terms of Legendre polynomials. The general solution is

$$\phi = \sum_{n=0}^{\infty} \left(\frac{A_n}{r^{n+1}} + B_n r^n \right) P_n(\cos\theta) \qquad (6.129)$$

Now if the potential is to be zero at $r = \infty$, all the B_n must be zero. Eqn (6.129) must reduce to eqn (6.128) on the surface of the sphere. Hence all the coefficients A_n for $n > 2$ must be zero, since Legendre polynomials of degree greater than two contain powers of $\cos\theta$ greater than $\cos^2\theta$. Hence outside the sphere, since

$$P_0 = 1, \quad P_1 = \cos\theta, \quad P_2 = \tfrac{1}{2}(3\cos^2\theta - 1),$$

$$\phi = \frac{A_0}{r} + \frac{A_1}{r^2}\cos\theta + \frac{A_2}{2r^3}(3\cos^2\theta - 1) \qquad (6.130)$$

Putting $r = R$ in eqn (6.130), and comparing the coefficients of the various powers of $\cos\theta$ in eqns (6.128) and (6.130), we find

$$A_1 = 0$$

$$A_2 = -\omega BR^5/3$$

If A_0 is finite, at large values of r, the term A_0/r gives an electric field consistent with Coulomb's law. If the total electric charge on the sphere is zero, then A_0 must be zero. Hence, the electric potential outside the sphere is given by

$$\phi = \frac{A_2}{r^3} P_2(\cos\theta) = -\frac{\omega BR^5}{6r^3}(3\cos^2\theta - 1)$$

Outside the sphere, $\mathbf{E} = -\nabla\phi$. Using spherical polars,

$$E_r = -\frac{\partial\phi}{\partial r}; \quad E_\theta = -\frac{1}{r}\frac{\partial\phi}{\partial\theta}$$

Hence,

$$\mathbf{E} = \frac{\omega B R^5}{2r^4} \left[-\mathbf{a}_r (3 \cos^2 \theta - 1) - \mathbf{a}_\theta \sin 2\theta \right]$$

where $\mathbf{a}_r$ and $\mathbf{a}_\theta$ are unit vectors in the direction of increasing r and θ respectively. Using eqn (6.126)

$$\mathbf{E} = \frac{\omega M R^5}{3\varepsilon_0 c^2 r^4} \left[-\mathbf{a}_r (3 \cos^2 \theta - 1) - \mathbf{a}_\theta \sin 2\theta \right] \tag{6.131}$$

The electric field inside and outside the sphere is sketched in *Figure 6.9(d)*. The electric field extends into the space outside the sphere. For an isolated sphere, $\mathbf{E} = -\mathbf{v} \times \mathbf{B}$ inside the sphere. The directions of the electric field lines outside the sphere are from the equator towards the points where the axis of rotation cuts the surface of the sphere. There are potential differences between different points on the surface of the sphere.

Maxwell's equations can be applied in the laboratory system. From the equation $\nabla \cdot \mathbf{D} = \rho_{\text{true}}$, we have in the usual way (cf. Problem 4.13 of Chapter 4),

$$(D_n)_1 - (D_n)_2 = \sigma_{\text{true}} \tag{6.132}$$

where $(D_n)_1$ and $(D_n)_2$ are the normal components of the electric displacement just outside and just inside the sphere in *Figure 6.8*. Just outside the sphere, from eqn (6.131) for $r = R$

$$(D_n)_1 = \varepsilon_0 E_r = -\frac{\omega M R}{3c^2} (3 \cos^2 \theta - 1)$$

Inside the sphere

$$(D_n)_2 = \varepsilon_0 E_n + P_n$$

where, from eqn (6.127), for $r = R$,

$$E_n = -E \sin \theta = -\frac{2\omega R M}{3\varepsilon_0 c^2} \sin^2 \theta$$

and from eqn (6.118)

$$P_n = \sigma_p = \omega M R \sin^2 \theta / c^2$$

Substituting in eqn (6.132), gives

$$\sigma_{\text{true}} = -\frac{2\omega M R}{3c^2} \cos^2 \theta \tag{6.123}$$

201

Actually, one would have expected the same *total* electric field and the same *total* charge distributions on the older theories in which the electric polarization of a moving magnetic medium was ignored. On this theory, it was assumed that, in a conducting magnet, conduction electrons would be displaced giving rise to true (or free) charge distributions such that their electric field inside the rotating magnet satisfied the relation $\mathbf{E} = -\mathbf{v} \times \mathbf{B}$. The appropriate true charge distributions are given by eqns (6.124) and (6.125) and are shown in *Figure 6.9(c)*. Such distributions of true charge would give the electric fields given by eqns (6.127) and (6.131). Thus, one expects the same *total* electric field and *total* charge distributions in both cases. However, the old theory, in which the electric polarization of a moving magnetic medium was ignored, cannot account for the absence of true (or free) charge distributions on an isolated permanent magnet moving with uniform linear velocity (cf. Section 6.8(a)).

The solution of the problem of a rotating spherical magnet illustrates several differences between moving conductors and stationary conductors. In electrostatics, one expects all of a stationary conductor to be at the same electric potential. This is not true for an isolated rotating permanent magnet. In electrostatics $\mathbf{E}$ is zero inside a stationary conductor. For an isolated rotating magnet, $\mathbf{E}$ is not zero, but equal to $-\mathbf{v} \times \mathbf{B}$. In electrostatics, $\mathbf{E}$ is perpendicular to the surface of a stationary conductor. This is not true in *Figure 6.9(d)*. In electrostatics, all of the true (free) charge is on the surface of the conductor. This is not true in *Figure 6.9(b)*. These differences arise from the fact that inside a moving conductor, it is not $\mathbf{E}$ but $\mathbf{E} + \mathbf{v} \times \mathbf{B}$ which is zero, when $\mathbf{J}$ is zero.

So far only an isolated rotating magnetized sphere has been considered. It will now be assumed that the surface of the sphere at the equator, and the point where the axis of rotation cuts the surface are joined by the wire AC, which is stationary in the laboratory system, as shown in *Figure 6.8*. Due to the electric field outside the rotating magnetized sphere, shown in *Figure 6.9(d)*, electric conduction current will flow in the stationary wire from A to C and will try to build up true (or free) charge distributions to try and equalize the electric potentials at A and C. However, once these extra true charge distributions begin to build up near the points A and C, the balance between $q\mathbf{E}$ and $q\mathbf{v} \times \mathbf{B}$ is upset and current will flow inside the conducting rotating magnet from C to A to try and sustain the charge distribution shown in *Figure 6.9(c)*. A compromise is reached, when a steady current flows in the stationary wire AC, and the potential difference across AC is consistent with Ohm's law.

The examples discussed in Sections 6.8(a) and 6.8(b) illustrate how, in the laboratory system, unipolar induction can be interpreted in terms of Minkowski's electrodynamics of moving media. There is no need to introduce the concept of rotating magnetic field lines. Maxwell's equations can be applied in the laboratory system, which in this instance is a satisfactory approximation to an inertial reference frame, since effects associated with the rotation of the earth are negligible. Maxwell's equations cannot be applied in the rotating non-inertial reference frame in which the magnetized sphere is at rest. Just as the laws of mechanics are different in a rotating reference frame (one must introduce the centrifugal and Coriolis forces) so are the laws of electromagnetism different in a rotating reference frame (Reference: Trocheris[14]).

Problem 6.11—The charge distribution shown in *Figure 6.9(c)* is rotating relative to the laboratory system Σ. Show that the magnetic induction due to the rotating charge distribution is negligible compared with the internal magnetic induction of the magnet. (Hint: From eqn (3.27) the magnetic field of a charge moving with uniform velocity is $\mathbf{v} \times \mathbf{E}/c^2$, where $\mathbf{E}$ is its electric field. Inside the magnet, the electric field due to the charges is related to the magnetic induction due to the magnet by eqn (6.120).)

Problem 6.12—Sketch the electric polarization, the fictitious polarization charge distribution, the true charge distribution, the total charge distribution and the electric field due to a long isolated cylindrical bar magnet rotating about its axis. The magnet is magnetized uniformly parallel to its axis.

Problem 6.13—Interpret the origin of an electric field present in Σ' in *Figure 6.1(b)* if the external magnetic field in Σ is due (*a*) to an air-cored electromagnet stationary in Σ, (*b*) an electromagnet with a ferromagnetic core stationary in Σ, (*c*) a permanent magnet at rest in Σ. Why is there no displacement of charge in the connecting rail moving with velocity $-v$ relative to Σ', when the rail is far away from the 'stationary' conductor in *Figure 6.1(b)*?

6.9. THE TRANSFORMATION OF THE POTENTIALS

The equations for the potentials were developed in Section 5.7 from Maxwell's equations. From eqns (5.43) and (5.41), one has in Σ,

$$\nabla^2 \phi - \frac{1}{c^2} \frac{\partial^2 \phi}{\partial t^2} = -\rho/\varepsilon_0 \qquad (6.133)$$

$$\nabla^2 \mathbf{A} - \frac{1}{c^2} \frac{\partial^2 \mathbf{A}}{\partial t^2} = -\mu_0 \mathbf{J} \qquad (6.134)$$

203

subject to the condition

$$\mathbf{\nabla} \cdot \mathbf{A} + \frac{1}{c^2} \frac{\partial \phi}{\partial t} = 0 \qquad (6.135)$$

It was shown in Section 6.2, that

$$\frac{\partial}{\partial x} = \gamma \left(\frac{\partial}{\partial x'} - \frac{v}{c^2} \frac{\partial}{\partial t'} \right); \; \frac{\partial}{\partial y} = \frac{\partial}{\partial y'}; \; \frac{\partial}{\partial z} = \frac{\partial}{\partial z'}; \; \frac{\partial}{\partial t} = \gamma \left(\frac{\partial}{\partial t'} - v \frac{\partial}{\partial x'} \right)$$

$$(6.136)$$

Now

$$\frac{\partial^2}{\partial x^2} + \frac{\partial^2}{\partial y^2} + \frac{\partial^2}{\partial z^2} - \frac{1}{c^2} \frac{\partial^2}{\partial t^2} = \frac{\partial}{\partial x} \frac{\partial}{\partial x} + \frac{\partial}{\partial y} \frac{\partial}{\partial y} + \frac{\partial}{\partial z} \frac{\partial}{\partial z} - \frac{1}{c^2} \frac{\partial}{\partial t} \frac{\partial}{\partial t}$$

$$= \gamma^2 \left(\frac{\partial}{\partial x'} - \frac{v}{c^2} \frac{\partial}{\partial t'} \right)^2 + \frac{\partial^2}{\partial y'^2} + \frac{\partial^2}{\partial z'^2} - \frac{1}{c^2} \gamma^2 \left(\frac{\partial}{\partial t'} - v \frac{\partial}{\partial x'} \right)^2$$

Remembering that

$$\frac{\partial^2}{\partial t' \partial x'} = \frac{\partial^2}{\partial x' \partial t'},$$

$$\frac{\partial^2}{\partial x^2} + \frac{\partial^2}{\partial y^2} + \frac{\partial^2}{\partial z^2} - \frac{1}{c^2} \frac{\partial^2}{\partial t^2} = \gamma^2 \frac{\partial^2}{\partial x'^2} - \frac{2v}{c^2} \frac{\gamma^2 \partial^2}{\partial x' \partial t'} + \gamma^2 \frac{v^2}{c^4} \frac{\partial^2}{\partial t'^2}$$

$$+ \frac{\partial^2}{\partial y'^2} + \frac{\partial^2}{\partial z'^2} - \frac{\gamma^2}{c^2} \frac{\partial^2}{\partial t'^2} + 2v \frac{\gamma^2}{c^2} \frac{\partial^2}{\partial t' \partial x'} - \gamma^2 \frac{v^2}{c^2} \frac{\partial^2}{\partial x'^2}$$

$$= \gamma^2 (1 - v^2/c^2) \frac{\partial^2}{\partial x'^2} + \frac{\partial^2}{\partial y'^2} + \frac{\partial^2}{\partial z'^2} - \frac{\gamma^2}{c^2} (1 - v^2/c^2) \frac{\partial^2}{\partial t'^2}$$

$$= \frac{\partial^2}{\partial x'^2} + \frac{\partial^2}{\partial y'^2} + \frac{\partial^2}{\partial z'^2} - \frac{1}{c^2} \frac{\partial^2}{\partial t'^2}$$

That is

$$\nabla^2 - \frac{1}{c^2} \frac{\partial^2}{\partial t^2} = \nabla'^2 - \frac{1}{c^2} \frac{\partial^2}{\partial t'^2} \qquad (6.137)$$

or

$$\square^2 = \square'^2 \qquad (6.138)$$

where

$$\square^2 = \nabla^2 - \frac{1}{c^2} \frac{\partial^2}{\partial t^2} \quad \text{etc.}$$

THE TRANSFORMATION OF THE POTENTIALS

From eqn (6.66)

$$\rho = \gamma(\rho' + v J_x'/c^2) \qquad (6.139)$$

Substituting for $\Box^2$ and for ρ in eqn (6.133), one obtains

$$\Box'^2\phi = -\frac{\gamma}{\varepsilon_0}(\rho' + v J_x'/c^2) \qquad (6.140)$$

From eqn (6.66)

$$J_x = \gamma(J_x' + v\rho'); \quad J_y = J_y'; \quad J_z = J_z'$$

Substituting in eqn (6.134), one has for the components of $\mathbf{A}$:

$$\Box'^2 A_x = -\mu_0\gamma(J_x' + v\rho') \qquad (6.141)$$

$$\Box'^2 A_y = -\mu_0 J_y' \qquad (6.142)$$

$$\Box'^2 A_z = -\mu_0 J_z' \qquad (6.143)$$

Multiplying eqn (6.141) by v and subtracting it from eqn (6.140), one obtains

$$\Box'^2(\phi - vA_x) = -\frac{\gamma}{\varepsilon_0}\left(\rho' + v\frac{J_x'}{c^2}\right) + v\mu_0\gamma(J_x' + v\rho')$$

But

$$\frac{1}{\mu_0\varepsilon_0} = c^2 \quad \text{or} \quad \mu_0 = \frac{1}{\varepsilon_0 c^2}$$

Hence,

$$\Box'^2(\phi - vA_x) = -\frac{\gamma}{\varepsilon_0}\left(\rho' + v\frac{J_x'}{c^2}\right) + \frac{v\gamma}{\varepsilon_0 c^2}(J_x' + v\rho')$$

$$= -\frac{\gamma}{\varepsilon_0}\left[\rho'\left(1 - \frac{v^2}{c^2}\right)\right] = -\frac{\rho'}{\gamma\varepsilon_0}$$

Since v and γ are assumed to be constant, this can be rewritten as

$$\Box'^2\gamma(\phi - vA_x) = -\rho'/\varepsilon_0 \qquad (6.144)$$

Multiplying eqn (6.140) by v/c^2 and subtracting it from eqn (6.141)

$$\Box'^2\left(A_x - \frac{v}{c^2}\phi\right) = -\mu_0\gamma(J_x' + v\rho') + \frac{v}{c^2}\frac{\gamma}{\varepsilon_0}(\rho' + v J_x'/c^2)$$

205

But

$$\frac{1}{\varepsilon_0 c^2} = \mu_0$$

Hence,

$$\Box'^2 \left(A_x - \frac{v}{c^2} \phi \right) = -\mu_0 \gamma \left[J_x' + v\rho' - v\rho' - \frac{v^2}{c^2} J_x' \right]$$

Therefore

$$\Box'^2 \gamma \left(A_x - \frac{v}{c^2} \phi \right) = -\mu_0 J_x' \qquad (6.145)$$

Collecting the transformed equations and comparing them with the equations that would hold for $\mathbf{A}'$ and ϕ' in Σ', if the equations of electromagnetism obeyed the principle of relativity, we have

$$\Box'^2 \gamma \left(A_x - \frac{v}{c^2} \phi \right) = -\mu_0 J_x' \qquad \Box'^2 A_x' = -\mu_0 J_x'$$

$$\Box'^2 A_y = -\mu_0 J_y' \qquad \Box'^2 A_y' = -\mu_0 J_y'$$

$$\Box'^2 A_z = -\mu_0 J_z' \qquad \Box'^2 A_z' = -\mu_0 J_z'$$

$$\Box'^2 \gamma(\phi - vA_x) = -\rho'/\varepsilon_0 \qquad \Box'^2 \phi' = -\rho'/\varepsilon_0$$

The equations on the left-hand side were derived from those in Σ using the Lorentz transformations. They have the same mathematical form as those on the right-hand side, which would be valid in Σ' if the laws of electromagnetism obeyed the principle of relativity. The equations for the potentials are therefore covariant under a Lorentz transformation, and if we put

$$A_x' = \gamma \left(A_x - \frac{v}{c^2} \phi \right) \qquad (6.146)$$

$$A_y' = A_y \qquad (6.147)$$

$$A_z' = A_z \qquad (6.148)$$

$$\phi' = \gamma(\phi - vA_x) \qquad (6.149)$$

the two sets of equations become identical.

Now consider the Lorentz condition [eqn (6.135)]. Substituting for $\partial/\partial x$, $\partial/\partial y$, $\partial/\partial z$, $\partial/\partial t$ from eqn (6.136) one obtains

$$0 = \frac{\partial A_x}{\partial x} + \frac{\partial A_y}{\partial y} + \frac{\partial A_z}{\partial z} + \frac{1}{c^2}\frac{\partial \phi}{\partial t}$$

$$= \gamma\left(\frac{\partial}{\partial x'} - \frac{v}{c^2}\frac{\partial}{\partial t'}\right)A_x + \frac{\partial A_y}{\partial y'} + \frac{\partial A_z}{\partial z'}$$

$$+ \frac{\gamma}{c^2}\left(\frac{\partial}{\partial t'} - v\frac{\partial}{\partial x'}\right)\phi$$

$$= \frac{\partial}{\partial x'}\gamma\left(A_x - \frac{v}{c^2}\phi\right) + \frac{\partial A_y}{\partial y'} + \frac{\partial A_z}{\partial z'}$$

$$+ \frac{1}{c^2}\frac{\partial}{\partial t'}\gamma(\phi - vA_x)$$

This is the same as the equation

$$\frac{\partial A_x'}{\partial x'} + \frac{\partial A_y'}{\partial y'} + \frac{\partial A_z'}{\partial z'} + \frac{1}{c^2}\frac{\partial \phi'}{\partial t'} = 0$$

if $\mathbf{A}$ and ϕ transform according to eqns (6.146), (6.147), (6.148) and (6.149). Thus with the choice for $\mathbf{V}.\mathbf{A}$ given by the Lorentz condition [eqn (6.135)], the differential equations for ϕ and $\mathbf{A}$ and the Lorentz condition itself, are Lorentz covariant. Notice A_x, A_y, A_z, ϕ/c^2 transform in the same way as x, y, z and t.

Problem 6.14—Develop the transformations for $\mathbf{E}$ and $\mathbf{B}$ from the transformations for ϕ and $\mathbf{A}$ using the relations $\mathbf{B} = \mathbf{V}\times\mathbf{A}$ and $\mathbf{E} = -\mathbf{V}\phi - \partial\mathbf{A}/\partial t$. For example

$$B_x' = \frac{\partial A_z'}{\partial y'} - \frac{\partial A_y'}{\partial z'}$$

Using eqns (6.22), (6.148), (6.23) and (6.147)

$$B_x' = \frac{\partial A_z}{\partial y} - \frac{\partial A_y}{\partial z} = B_x$$

(Reference: Rosser[2e]).

Problem 6.15—In Σ' a 'point' charge of magnitude q is at rest at the origin. The potentials at a distance r' from the origin are

$$\mathbf{A}' = 0; \quad \phi' = q/4\pi\varepsilon_0 r'$$

Use the transformations of ϕ' and $\mathbf{A}'$ to show that in Σ

$$\phi = \frac{q}{4\pi\varepsilon_0 s}; \quad \mathbf{A} = \frac{q\mathbf{v}}{4\pi\varepsilon_0 c^2 s}$$

where $s = r[1-(v^2/c^2)\sin^2\theta]^{\frac{1}{2}}$, $\mathbf{r}$ is the distance from the 'present' position of the charge to the field point, $\mathbf{v}$ is the velocity of the charge and θ the angle between $\mathbf{r}$ and $\mathbf{v}$.

(Hint: Use eqns (3.6) and (3.8).)

Problem 6.16—Show that the following are invariants:

$$(a)\ A^2 - \phi^2/c^2; \quad (b)\ J^2 - c^2\rho^2; \quad (c)\ \mathbf{A}.\mathbf{J} - \rho\phi$$

6.10. DISCUSSION

It was shown in Sections 6.2 and 6.6 that Maxwell's equations for the macroscopic fields obey the principle of relativity when the co-ordinates and time are transformed according to the Lorentz transformations, that is Maxwell's equations are Lorentz covariant. Since the magnetic forces between moving charges are of second order compared with the electric forces between the charges, it is important to include all second order effects when interpreting magnetic phenomena. For example, in Section 2.3 of Chapter 2 (cf. *Figure 2.3*) we had to allow for the Lorentz length contraction. In Section 6.5, when interpreting the electric dipole moment of a moving magnetic dipole in terms of the Amperian theory of magnetism, we had to allow for the different measures of time in Σ and Σ'. Thus classical electromagnetism is a fully relativistic theory. The relativistic invariance of Maxwell's equations and the equations for the potentials ϕ and $\mathbf{A}$ is more apparent when the equations are expressed in terms of four vectors and tensors. The interested reader is referred to Rosser[2f].

REFERENCES

[1] EINSTEN, A. *Ann. Phys. Lpz.* **17** (1905) 891
[2] ROSSER, W. G. V. *An Introduction to the Theory of Relativity.* (a) Section 3.3, (b) p. 311, (c) p. 337, (d) pp. 147 and 339, (e) p. 356, (f) Ch. 6 and 10. Butterworths, London, 1964
[3] CULLWICK, E. G. *Electromagnetism and Relativity*, 2nd Ed. (a) p. 123, (b) p. 161. Longmans, London, 1959
[4] SEARS, F. W. *Principles of Physics II, Electricity and Magnetism*, Ch. 12. Addison-Wesley, Reading, Mass., 1946
[5] SHANKLAND, R. S. *Scientific American*, November 1964, p. 114

REFERENCES

[6] BECKER, R. and SAUTER, F. *Electromagnetic Fields and Interactions, Volume 1, Electromagnetic Theory and Relativity*, p. 372. Blackie & Son Ltd., Glasgow and London, 1964

[7] MØLLER, C. *The Theory of Relativity.* Oxford University Press, 1952

[8] PAULI, W. *Theory of Relativity.* Pergamon Press, 1958

[9] SOMMERFELD, A. *Electrodynamics; Lectures on Theoretical Physics.* Vol. 3. Academic Press, New York, 1952

[10] EINSTEN, A. and LAUB, J. *Ann. Phys. Lpz.* **26** (1908) 536

[11] WILSON, M. and WILSON, A. A. *Proc. Roy. Soc.* **A89** (1913) 99

[12] SLATER, J. C. and FRANK, N. H. *Electromagnetism.* McGraw-Hill, New York and London, 1947

[13] SCHLOMKA, T. and SCHENKEL, G. *Ann. Phys.* (Folge 6) **5** (1949) 51

[14] TROCHERIS, M. G. *Phil Mag.* 7th Series, **40** (1949) 1143

[15] ROSSER, W. G. V. *Introductory Relativity.* Butterworths, London, 1967, Appendix 2

APPENDIX 1

A SUMMARY OF THE FORMULAE OF VECTOR ANALYSIS

The symbols $\mathbf{A}$, $\mathbf{B}$ or $\mathbf{C}$ will be used to denote vectors and the symbols ϕ and ψ will be used to denote scalars. The magnitude of the vector $\mathbf{A}$ will be denoted by A, or by $|\mathbf{A}|$.

The *scalar product* of two vectors $\mathbf{A}$ and $\mathbf{B}$ will be written as $\mathbf{A} \cdot \mathbf{B}$. The magnitude of the scalar product is equal to the product of the magnitudes of $\mathbf{A}$ and $\mathbf{B}$ and the cosine of the angle between them, i.e.

$$\mathbf{A} \cdot \mathbf{B} = AB \cos (\mathbf{A}, \mathbf{B}) = A_x B_x + A_y B_y + A_z B_z \qquad (A1.1)$$

The scalar product obeys the distributive law of addition, e.g.

$$\mathbf{A} \cdot (\mathbf{B} + \mathbf{C}) = \mathbf{A} \cdot \mathbf{B} + \mathbf{A} \cdot \mathbf{C} \qquad (A1.2)$$

The *vector product* of two vectors $\mathbf{A}$ and $\mathbf{B}$ will be denoted by $\mathbf{C} = \mathbf{A} \times \mathbf{B}$. It is a vector of magnitude $AB \sin (\mathbf{A}, \mathbf{B})$ and it is in a direction perpendicular to $\mathbf{A}$ and $\mathbf{B}$ such that $\mathbf{A}$, $\mathbf{B}$ and $\mathbf{C}$ form a right-handed set. If $\mathbf{i}$, $\mathbf{j}$ and $\mathbf{k}$ are unit vectors in the positive x, y and z directions of a Cartesian co-ordinate system, then

$$\mathbf{A} \times \mathbf{B} = \mathbf{i}(A_y B_z - A_z B_y) + \mathbf{j}(A_z B_x - A_x B_z) + \mathbf{k}(A_x B_y - A_y B_x) \quad (A1.3)$$

The order of the vectors is important, since

$$\mathbf{A} \times \mathbf{B} = -\mathbf{B} \times \mathbf{A}$$

The vector product obeys the distributive law of addition, e.g.

$$\mathbf{A} \times (\mathbf{B} + \mathbf{C}) = \mathbf{A} \times \mathbf{B} + \mathbf{A} \times \mathbf{C} \qquad (A1.4)$$

The triple scalar product is written as $(\mathbf{A} \times \mathbf{B}) \cdot \mathbf{C}$. We have,

$$(\mathbf{A} \times \mathbf{B}) \cdot \mathbf{C} = (\mathbf{B} \times \mathbf{C}) \cdot \mathbf{A} = (\mathbf{C} \times \mathbf{A}) \cdot \mathbf{B} \qquad (A1.5)$$

The triple vector product is written as $\mathbf{A} \times (\mathbf{B} \times \mathbf{C})$. We have

$$\mathbf{A} \times (\mathbf{B} \times \mathbf{C}) = \mathbf{B}(\mathbf{A} \cdot \mathbf{C}) - \mathbf{C}(\mathbf{A} \cdot \mathbf{B}) \qquad (A1.6)$$

The formulae of vector analysis are often expressed in terms of the operator ∇, called del or nabla. In Cartesian co-ordinates

$$\nabla = \mathbf{i} \frac{\partial}{\partial x} + \mathbf{j} \frac{\partial}{\partial y} + \mathbf{k} \frac{\partial}{\partial z} \qquad (A1.7)$$

A SUMMARY OF THE FORMULAE OF VECTOR ANALYSIS

The *gradient* of a scalar ϕ, $\nabla\phi$ or grad ϕ, can be defined as that vector operation on ϕ that gives the total differential $\mathrm{d}\phi$ when it forms a scalar product with $\mathrm{d}\mathbf{l}$, that is

$$\mathrm{d}\phi = \nabla\phi \cdot \mathrm{d}\mathbf{l} \tag{A1.8}$$

In a Cartesian co-ordinate system,

$$\nabla\phi = \mathbf{i}\frac{\partial\phi}{\partial x}+\mathbf{j}\frac{\partial\phi}{\partial y}+\mathbf{k}\frac{\partial\phi}{\partial z} \tag{A1.9}$$

The gradient of a scalar is a vector.

The *divergence* of a vector $\mathbf{A}$, div $\mathbf{A}$ or $\nabla\cdot\mathbf{A}$ can be defined by the relation

$$\nabla\cdot\mathbf{A} = \underset{\Delta V\to 0}{\mathrm{Limit}}\frac{\int\mathbf{A}\cdot\mathbf{n}\mathrm{d}S}{\Delta V} \tag{A1.10}$$

where ΔV is a small volume element and $\mathbf{n}$ is a unit vector normal to the surface $\mathrm{d}S$. The surface integral in eqn (A1.10) is over the surface of ΔV. Thus the divergence of a vector $\mathbf{A}$ is the ratio of the flux of the vector $\mathbf{A}$ from the surface of a small volume element ΔV (that is $\int\mathbf{A}\cdot\mathbf{n}\mathrm{d}S$) to the volume of the element. In Cartesian co-ordinates

$$\nabla\cdot\mathbf{A} = \frac{\partial A_x}{\partial x}+\frac{\partial A_y}{\partial y}+\frac{\partial A_z}{\partial z} \tag{A1.11}$$

The divergence of a vector is a scalar.

The *curl* of a vector $\mathbf{A}$, curl $\mathbf{A}$ or $\nabla\times\mathbf{A}$, can be defined by the equation

$$(\nabla\times\mathbf{A})\cdot\mathbf{n} = \underset{\Delta S\to 0}{\mathrm{Limit}}\frac{\oint\mathbf{A}\cdot\mathrm{d}\mathbf{l}}{\Delta S} \tag{A1.12}$$

In eqn (A1.12), $\mathbf{n}$ is a unit vector normal to a small element of surface ΔS. The direction of $\mathbf{n}$ is the direction a right-handed cork-screw would advance, if it were rotated in the direction of $\mathrm{d}\mathbf{l}$, as shown in *Figure A1.1*. In eqn (A1.12), $\oint\mathbf{A}\cdot\mathrm{d}\mathbf{l}$ is evaluated around the boundary of ΔS. According to eqn (A1.12), the component of $(\nabla\times\mathbf{A})$ parallel to $\mathbf{n}$ is the limit as $\Delta S\to 0$ of the ratio of the line integral of $\mathbf{A}$ around the boundary of ΔS to the area of ΔS. In Cartesian co-ordinates

$$\nabla\times\mathbf{A} = \mathbf{i}\left(\frac{\partial A_z}{\partial y}-\frac{\partial A_y}{\partial z}\right)+\mathbf{j}\left(\frac{\partial A_x}{\partial z}-\frac{\partial A_z}{\partial x}\right)+\mathbf{k}\left(\frac{\partial A_y}{\partial x}-\frac{\partial A_x}{\partial y}\right) \tag{A1.13}$$

The curl of a vector is a vector.

APPENDIX 1

If ϕ and ψ are scalar fields and $\mathbf{A}$, $\mathbf{B}$ and $\mathbf{C}$ are vector fields, then the following relations are valid

$$\nabla(\phi+\psi) = \nabla\phi + \nabla\psi \qquad (A1.14)$$

$$\nabla\phi\psi = \phi\nabla\psi + \psi\nabla\phi \qquad (A1.15)$$

$$\nabla \cdot (\mathbf{A}+\mathbf{B}) = \nabla \cdot \mathbf{A} + \nabla \cdot \mathbf{B} \qquad (A1.16)$$

$$\nabla \times (\mathbf{A}+\mathbf{B}) = \nabla \times \mathbf{A} + \nabla \times \mathbf{B} \qquad (A1.17)$$

$$\nabla \times (\nabla\phi) = 0 \qquad (A1.18)$$

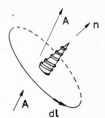

Figure A1.1. In the definition of $\nabla \times \mathbf{A}$ in terms of eqn (A1.12) the positive direction for the unit vector $\mathbf{n}$ is chosen to be the direction in which a right-handed corkscrew would advance, if rotated in the direction of $d\mathbf{l}$

$$\nabla \cdot (\nabla \times \mathbf{A}) = 0 \qquad (A1.19)$$

$$\nabla \cdot (\phi\mathbf{A}) = \phi(\nabla \cdot \mathbf{A}) + \mathbf{A} \cdot (\nabla\phi) \qquad (A1.20)$$

$$\nabla \times (\phi\mathbf{A}) = \phi(\nabla \times \mathbf{A}) + (\nabla\phi) \times \mathbf{A} \qquad (A1.21)$$

$$\nabla \cdot (\mathbf{A} \times \mathbf{B}) = \mathbf{B} \cdot (\nabla \times \mathbf{A}) - \mathbf{A} \cdot (\nabla \times \mathbf{B}) \qquad (A1.22)$$

$$\nabla \times (\mathbf{A} \times \mathbf{B}) = \mathbf{A}(\nabla \cdot \mathbf{B}) - \mathbf{B}(\nabla \cdot \mathbf{A})$$
$$+ (\mathbf{B} \cdot \nabla)\mathbf{A} - (\mathbf{A} \cdot \nabla)\mathbf{B} \qquad (A1.23)$$

In Cartesian co-ordinates

$$\nabla \times (\nabla \times \mathbf{A}) = \nabla(\nabla \cdot \mathbf{A}) - \nabla^2\mathbf{A} \qquad (A1.24)$$

where ∇^2 is the Laplacian operator, which in Cartesian co-ordinates is given by

$$\nabla^2 = \frac{\partial^2}{\partial x^2} + \frac{\partial^2}{\partial y^2} + \frac{\partial^2}{\partial z^2}$$

Stokes' Theorem

According to Stokes' theorem, which can be developed from eqn (A1.12), the surface integral of the curl of a vector field $\mathbf{A}$, taken over any finite surface, is equal to the line integral of **A** around the boundary of that surface, that is

$$\int (\nabla \times A) \cdot \mathbf{n} dS = \oint A \cdot d\mathbf{l} \qquad (A1.25)$$

Gauss' Mathematical Theorem

According to Gauss' mathematical theorem, which can be developed from eqn (A1.10), the volume integral of the divergence of a vector field **A**, taken over any finite volume, is equal to the surface integral of **A** over the surface of that volume, that is

$$\int (\nabla \cdot A) \, dV = \int A \cdot \mathbf{n} dS \qquad (A1.26)$$

Spherical Polar Co-ordinates

In spherical polar co-ordinates r, θ and ϕ, if $\mathbf{a}_r$, $\mathbf{a}_\theta$ and $\mathbf{a}_\phi$ are unit vectors in the directions of increasing r, θ and ϕ respectively

$$\nabla = \mathbf{a}_r \frac{\partial}{\partial r} + \mathbf{a}_\theta \frac{1}{r} \frac{\partial}{\partial \theta} + \mathbf{a}_\phi \frac{1}{r \sin \theta} \frac{\partial}{\partial \phi} \qquad (A1.27)$$

$$\nabla \cdot A = \frac{1}{r^2} \frac{\partial}{\partial r} (r^2 A_r) + \frac{1}{r \sin \theta} \frac{\partial}{\partial \theta} (A_\theta \sin \theta) + \frac{1}{r \sin \theta} \frac{\partial A_\phi}{\partial \phi} \qquad (A1.28)$$

$$\nabla \times A = \mathbf{a}_r \frac{1}{r \sin \theta} \left[\frac{\partial}{\partial \theta} (A_\phi \sin \theta) - \frac{\partial A_\theta}{\partial \phi} \right]$$

$$+ \mathbf{a}_\theta \left[\frac{1}{r \sin \theta} \frac{\partial A_r}{\partial \phi} - \frac{1}{r} \frac{\partial}{\partial r} (r A_\phi) \right]$$

$$+ \mathbf{a}_\phi \left[\frac{1}{r} \frac{\partial}{\partial r} (r A_\theta) - \frac{1}{r} \frac{\partial A_r}{\partial \theta} \right] \qquad (A1.29)$$

GENERAL REFERENCES

PUGH, E. M. and PUGH, E. W. *Principles of Electricity*. Addison-Wesley, Reading, Mass., 1960

SPIEGEL, M. R. *Theory and Problems of Vector Analysis*. Shaum Publishing Co., New York, 1959

APPENDIX 2

A CRITIQUE OF CONVENTIONAL APPROACHES
TO MAXWELL'S EQUATIONS

In this Appendix a brief review is given of typical contemporary approaches to Maxwell's equations, in order to point out the assumptions made in such approaches, and to discuss the experimental evidence, on which they are based. A reader interested in the historical background to the development of classical electromagnetism is referred to Whittaker[1] and Tricker[2].

(a) The equation $\nabla \cdot \mathbf{E} = \rho/\varepsilon_0$

This equation is normally developed from Coulomb's law. The existence of static electrification by friction has been known since early times, but quantitative electrostatics did not develop until the eighteenth century. Priestley (1767) was probably the first to develop the inverse square law from the observation that there is no electric field inside a hollow charged conductor. This approach was extended, amongst others, by Maxwell and by Plimpton and Lawton (1936)[3]. The latter showed that, if the force between two 'point' charges is proportional to $1/r^n$, where r is the separation of the charges, then experimentally $n = 2$ to within 1 part in 10^9. The inverse square law was checked directly by Robison (1769) and by Coulomb (1785). An account of Coulomb's experiment is given by Magie[4]. On the basis of these experiments the force between two 'point' charges q and q_1 a distance r apart in empty space is assumed to be

$$\mathbf{f} = \frac{qq_1\mathbf{r}}{4\pi\varepsilon_0 r^3} \tag{A2.1}$$

This equation is generally known as Coulomb's law. By 'point' charges, one means charges whose dimensions are very much smaller than their spatial separation. In the M.K.S.A. system of units, the unit of charge, the coulomb, is the charge passing any point per second, when the current is one ampere. On page 226, the ampere is defined in terms of the forces between electric currents. In eqn (A2.1), ε_0 the electric space constant (or permittivity of vacuum as it is sometimes called) is a constant that must be determined by experiment. If f is in Newtons, q and q_1 in coulombs and r in metres, then

214

the *experimental* value of ε_0 is $8 \cdot 854 \times 10^{-12}$ farads/metre. Coulomb's law, eqn (A2.1), has been used extensively in atomic physics. It is assumed to be the basic force in Bohr's theory of the hydrogen atom, Rutherford scattering, etc., and it is now universally accepted as the expression for the electric force between stationary electric charges. In Chapter 3, Coulomb's law is used to develop the electric and magnetic fields of a 'point' charge moving with uniform velocity.

Eqn (A2.1) is generally separated into

$$= q \left(\frac{q_1 \mathbf{r}}{4\pi\varepsilon_0 r^3} \right) = q\mathbf{E}_1$$

It is then said that the charge q_1 gives rise to an electric field $\mathbf{E}_1$ at the position of the test charge q. In general the electric field intensity at a point is defined in terms of eqn (1.61), as described in Section 1.10. The electric field of a system of 'point' charges and continuous charge distributions can be obtained by applying the *principle of superposition*. It is sometimes easier to determine the electric fields of electrostatic charge distributions, using the electrostatic potential ϕ defined such that

$$\mathbf{E} = -\nabla\phi \tag{A2.2}$$

The potential at a distance r from a stationary 'point' charge q is

$$\phi = \frac{q}{4\pi\varepsilon_0 r} \tag{A2.3}$$

By considering the electric flux from an isolated 'point' charge, Gauss' law can be developed from Coulomb's law. According to Gauss' law the flux of $\mathbf{E}$ from any surface is equal to $(1/\varepsilon_0)$ times the total charge inside the surface. For a continuous stationary charge distribution, of charge density ρ,

$$\int \mathbf{E} \cdot \mathbf{n} dS = \frac{1}{\varepsilon_0} \int \rho dV \tag{A2.4}$$

In eqn (A2.4), $\mathbf{E}$ is the resultant electric field due to all the charges in the system, some of which may be outside the Gaussian surface used to evaluate $\int \mathbf{E} \cdot \mathbf{n} dS$. Using eqn (A1.26)

$$\int \nabla \cdot \mathbf{E} dV = \frac{1}{\varepsilon_0} \int \rho dV \tag{A2.5}$$

According to eqn (A1.10), $\nabla \cdot \mathbf{E}$ can be defined by the relation

$$\nabla \cdot \mathbf{E} = \operatorname*{Limit}_{\Delta V \to 0} \frac{\int \mathbf{E} \cdot \mathbf{n} dS}{\Delta V} \qquad (A1.10)$$

Hence, by making the volume, over which the integration is carried out in eqn (A2.5), very small on the laboratory scale, we have

$$\nabla \cdot \mathbf{E} = \rho/\varepsilon_0 \qquad (A2.6)$$

This equation should be true inside a continuous charge distribution. It is now known that 'continuous' charge distributions are made up of discrete positive and negative ions and electrons. In this case ρ is defined as in Section 4.1, by eqn (4.5), namely

$$\rho = \Sigma q_i/\Delta V \qquad (A2.7)$$

where in eqn (A2.7) ΔV is small on the laboratory scale, but large on the atomic scale. Similarly, in eqns (A1.10) and (A2.6) the volume ΔV used to define $\nabla \cdot \mathbf{E}$ must be large on the atomic scale, but kept small on the laboratory scale (cf. Section 4.2). Eqn (A2.6) then holds for the *macroscopic* electric field, which is the local space and time average of the microscopic field. In the above approach, the physical content of eqn (A2.6) is precisely the same as Coulomb's law. When it is developed in this way, eqn (A2.6) is strictly valid only for *stationary* charges. Yet it is *assumed* in classical electromagnetism that eqn (A2.6) holds for the electric fields of moving and accelerating charges. It is clear that the use of eqn (A2.6) for the electric fields of accelerating charges is an example of the application of Maxwell's equations in a context beyond the experimental evidence on which they are normally developed. Effectively, eqn (A2.6) becomes one of the axioms of the theory, and the validity of the use of eqn (A2.6) in the case of moving and accelerating charges is that predictions based on the use of eqn (A2.6) are in agreement with the experimental results.

(b) The Equation $\nabla \cdot \mathbf{D} = \rho_{\text{true}}$

Consider a *stationary* dielectric which is polarized under the influence of an applied electric field. The applied electric field tends to separate the positive and negative charges in the molecules inside the dielectric, such that each molecule has a resultant dipole moment. It is beyond the limits imposed by the Uncertainty Principle to give a precise classical model for atomic electric dipoles. Naively one can say that each charge can be described by a wave function which can be used to calculate the probability of finding

A CRITIQUE OF MAXWELL'S EQUATIONS

the charge at any particular point in space. For non-polar molecules, in the absence of an electric field, the positive 'centre of charge' coincides with the negative 'centre of charge' in the atom. Under the influence of the electric field the two 'centres of charge' are displaced giving rise to a dipole moment which is denoted by $\mathbf{p}_i$. Any polar molecules will tend to align in the direction of the applied electric field, and any ions in crystals will tend to be displaced in the direction of the applied field. Thus due to the influence of the applied electric field a large number of atomic dipole moments are induced or aligned in the direction of the applied field (for isotropic dielectrics). Consider a small element of volume ΔV. Let

$$\sum_{i=1}^{n} \mathbf{p}_i = \mathbf{P}\Delta V \quad \text{or} \quad \mathbf{P} = \frac{1}{\Delta V} \sum_{i=1}^{n} \mathbf{p}_i \qquad (A2.8)$$

where the vector summation is carried out over all the atomic electric dipoles in the volume element ΔV. Eqn (A2.8) is the definition of the polarization vector $\mathbf{P}$, in terms of the microscopic atomic dipole moments. The volume element ΔV can be made small on the laboratory scale (say a 10^{-5} cm cube) and contain many molecules ($\sim 10^9$ for a 10^{-5} cm cube). Thus in a macroscopic theory $\mathbf{P}$ can be treated as a continuous function of position, and defined as the dipole moment per unit volume.

It can be shown that the electrostatic potential at a distance $\mathbf{r}$ from an electric dipole consisting of 'point' charges $+q$ and $-q$ a distance l apart (such that $p = ql$) is

$$\phi = \frac{\mathbf{p} \cdot \mathbf{r}}{4\pi\varepsilon_0 r^3} \qquad (A2.9)$$

provided $r \gg l$. This is the dipole approximation. The direction of $\mathbf{p}$ is from the negative to the positive charge (Reference: Rosser[5]). The electric field due to the polarized dielectric shown in *Figure A2.1(a)* will be calculated at the point T. Consider a volume element ΔV inside the dielectric, at a point where the polarization vector is $\mathbf{P}$. Let the distance from ΔV to T be $\mathbf{r}$ as shown in *Figure A2.1(a)*. Consider one atomic electric dipole of dipole moment $\mathbf{p}_i$ inside the volume element ΔV. Its contribution to the electrostatic potential at the point T is equal to $\mathbf{p}_i \cdot \mathbf{r}/4\pi\varepsilon_0 r^3$. The electrostatic potential at T, due to all the atomic dipoles in ΔV, which are all at a distance r from T, is equal to

$$\Delta\phi = \sum_{i=1}^{n} \frac{\mathbf{p}_i \cdot \mathbf{r}}{4\pi\varepsilon_0 r^3}$$

the summation being over all the atomic dipoles in ΔV. For any three vectors one has

$$\mathbf{A} \cdot \mathbf{C} + \mathbf{B} \cdot \mathbf{C} = (\mathbf{A} + \mathbf{B}) \cdot \mathbf{C} \tag{A1.2}$$

so that

$$(\mathbf{p}_1 \cdot \mathbf{r} + \mathbf{p}_2 \cdot \mathbf{r} + \ldots + \mathbf{p}_n \cdot \mathbf{r}) = (\mathbf{p}_1 + \mathbf{p}_2 + \ldots + \mathbf{p}_n) \cdot \mathbf{r}$$

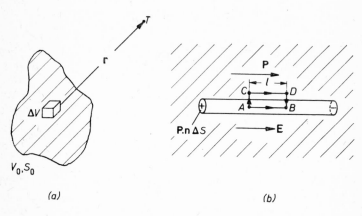

(a)   (b)

Figure A2.1. (a) A polarized dielectric of volume V_0 and surface area S_0. The electric field is calculated at the field point T. (b) A needle-shaped cavity is cut in the dielectric in a direction parallel to the electric intensity E. The line integral of the electric intensity is evaluated along the path AB direct, and along the path ACDB. It is shown that E is equal to the local space average of the microscopic electric field inside the dielectric

But from eqn (A2.8)

$$\sum_{i=1}^{n} \mathbf{p}_i = \mathbf{P} \Delta V$$

Hence,

$$\Delta \phi = \frac{\mathbf{P} \cdot \mathbf{r} \, \Delta V}{4 \pi \varepsilon_0 r^3}$$

so that

$$\phi = \frac{1}{4 \pi \varepsilon_0} \int_{V_0} \frac{\mathbf{P} \cdot \mathbf{r}}{r^3} \, dV \tag{A2.10}$$

But, if the position of ΔV is changed when the field point T is fixed,

$$\frac{\mathbf{r}}{r^3} = \nabla\left(\frac{1}{r}\right)$$

so that eqn (A2.10) becomes

$$\phi = \frac{1}{4\pi\varepsilon_0}\int_{V_0}\mathbf{P}\cdot\nabla\left(\frac{1}{r}\right)dV \qquad (A2.11)$$

For any scalar ϕ and vector $\mathbf{A}$,

$$\nabla\cdot\phi\mathbf{A} = \phi\nabla\cdot\mathbf{A} + \mathbf{A}\cdot\nabla\phi \qquad (A1.20)$$

Substituting $\mathbf{A} = \mathbf{P}$ and $\phi = 1/r$, one obtains

$$\mathbf{P}\cdot\nabla\left(\frac{1}{r}\right) = \nabla\cdot\left(\frac{\mathbf{P}}{r}\right) - \frac{1}{r}\nabla\cdot\mathbf{P}$$

Substituting in eqn (A2.11),

$$\phi = \frac{1}{4\pi\varepsilon_0}\int_{V_0}\nabla\cdot\left(\frac{\mathbf{P}}{r}\right)dV + \frac{1}{4\pi\varepsilon_0}\int_{V_0}\frac{(-\nabla\cdot\mathbf{P})}{r}dV \qquad (A2.12)$$

From Gauss' mathematical theorem, for any vector $\mathbf{A}$,

$$\int_{V_0}\nabla\cdot\mathbf{A}\,dV = \int_{S_0}\mathbf{A}\cdot\mathbf{n}\,dS \qquad (A1.26)$$

Hence, eqn (A2.12) can be rewritten as

$$\phi = \frac{1}{4\pi\varepsilon_0}\int_{S_0}\frac{\mathbf{P}\cdot\mathbf{n}}{r}dS + \frac{1}{4\pi\varepsilon_0}\int_{V_0}\frac{(-\nabla\cdot\mathbf{P})}{r}dV \qquad (A2.13)$$

Comparing with eqn (A2.3), it can be seen that, if one had a surface charge distribution $\mathbf{P}\cdot\mathbf{n}$ and a volume charge distribution $(-\nabla\cdot\mathbf{P})$ *in vacuo*, the geometrical configuration of the surface and volume charge distributions being the same as that of the dielectric, then the potential at the point T would also be given by eqn (A2.13). Thus for purposes of electric field calculation the dielectric can be replaced by a fictitious surface charge distribution

$$\sigma' = \mathbf{P}\cdot\mathbf{n} \qquad (A2.14)$$

and a volume charge distribution

$$\rho' = -\nabla\cdot\mathbf{P} \qquad (A2.15)$$

219

Thus the problem of the polarized dielectric can be replaced, for purposes of field calculation, by a distribution of fictitious charges, given by eqns (A2.14) and (A2.15), placed *in vacuo*. Coulomb's law, and equations derived from it, can be applied to this fictitious charge distribution to calculate the electric field and potential. This method is valid provided the dipole approximation is valid for *all* the dipoles in the dielectric. It is therefore valid for a point outside a dielectric, or for a point in the middle of a cavity, large on the atomic scale, cut inside the dielectric. If one wishes to calculate the microscopic field near or inside a molecule in the dielectric, in a region of space where the dipole approximation breaks down, one must use a microscopic theory based on a particular atomic model.

The microscopic electric field varies enormously on the atomic scale. One can have electric fields exceeding 10^8 volts/metre just outside atoms, and very much stronger fields inside atoms. It can be shown that inside a dielectric, the electric field calculated using eqn (A2.13) is equal to the local space average of the microscopic field, that is eqn (A2.13) gives the macroscopic electric field inside the dielectric. A rigorous proof is given by Corson and Lorrain[6a]. We shall merely give a short plausibility argument. Consider a needle-shaped cavity cut inside an isotropic linear dielectric in a direction parallel to the polarization vector $\mathbf{P}$ and the electric intensity $\mathbf{E}$, calculated using eqn (A2.13), as shown in *Figure A2.1(b)*. The length of the cavity is made small on the laboratory scale, so that the variations of $\mathbf{E}$ and $\mathbf{P}$ are negligible over the length of the cavity. The length of the cavity is, however, made very large on the atomic scale. The length of the cavity is also much greater than the diameter of the cavity, though the diameter is kept equal to many atomic diameters. At points A and B inside the cavity at distances equal to many atomic diameters away from all the surfaces of the cavity, the dipole approximation for the electric field of a dipole should be valid for *all* the atomic dipoles in the dielectric, so that eqn (A2.13) can be used to calculate the electric field at A and B. In addition to the fictitious volume charge distribution $-\nabla \cdot \mathbf{P}$ and the surface charge distribution $\mathbf{P} \cdot \mathbf{n}$ on the outer surface of the dielectric in *Figure A2.1(b)*, we must include in eqn (A2.13) the fictitious surface charge distributions $\mathbf{P} \cdot \mathbf{n} \Delta S$ at each end of the needle-shaped cavity, where ΔS is the area of cross-section of the cavity. If the diameter of the cavity is very much less than its length, it follows from Coulomb's law that the relative contributions of these latter surface charge distributions to the total electric field intensities at A and B is negligible. Hence the calculated field $\mathbf{E}$ at A and B is the same as that calculated by eqn (A2.13) for the case when there was no needle-shaped cavity present. This field is also equal to the value calculated using eqn (A2.13) for points inside the dielectric just outside the cavity in *Figure A2.1(b)*.

Consider the line integral of the electric intensity between *A* and *B*, that is the potential difference between the points *A* and *B* which are a distance *l* apart. For a direct path from *A* to *B* inside the cavity, since the microscopic electric intensity **e** is equal to **E** inside the cavity,

$$\int_A^B \mathbf{e} \cdot d\mathbf{l} = \int_A^B \mathbf{E} \cdot d\mathbf{l} = El$$

Now consider the line integral of the microscopic electric field **e** along the path *ACDB* in *Figure A2.1(b)*. Inside the cavity, *AC* and *DB* are perpendicular to *E*. Hence if *C* and *D* are only a distance $\ll l$ just inside the cavity,

$$\int_A^C \mathbf{e} \cdot d\mathbf{l} \quad \text{and} \quad \int_D^B \mathbf{e} \cdot d\mathbf{l}$$

are negligible compared with

$$\int_C^D \mathbf{e} \cdot d\mathbf{l}$$

so that for the path *ACDB* in *Figure A2.1(b)*,

$$\int_A^B \mathbf{e} \cdot d\mathbf{l} \simeq \int_C^D \mathbf{e} \cdot d\mathbf{l}$$

Now the electric field of the stationary electric dipoles in a dielectric is conservative, even at positions where the dipole approximation breaks down. Hence

$$\int_A^B \mathbf{e} \cdot d\mathbf{l}$$

should be independent of the path chosen. There will be large fluctuations at various points along various paths inside the dielectric but these will average out in the line integral over a total path length large on the atomic scale.

Hence,

$$\int_C^D \mathbf{e} \cdot d\mathbf{l} \simeq \int_A^B \mathbf{E} \cdot d\mathbf{l} = El$$

or,

$$E = \frac{1}{l} \int_C^D \mathbf{e} \cdot d\mathbf{l}$$

Since the electric intensity **E**, calculated using eqn (A2.13) has the same value in both the cavity and just inside the dielectric, the above equation suggests that **E** is equal to the space average between C and D of the microscopic electric intensity **e** inside the dielectric. Hence, the electric intensity **E** at field points inside a dielectric, calculated using eqn (A2.13) should be equal to the macroscopic electric intensity in the dielectric.

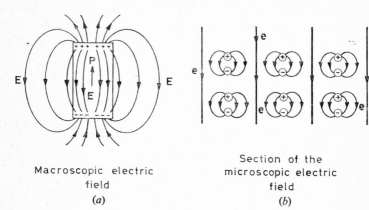

Macroscopic electric
field

(a)

Section of the
microscopic electric
field

(b)

Figure A2.2. (a) The macroscopic electric intensity **E** due to a polarized cylindrical dielectric. The uniform polarization **P** is parallel to the axis of the cylinder. The macroscopic electric intensity **E** can be calculated by replacing the dielectric by fictitious charge distributions **P** . **n** at each end of the dielectric. (b) A magnified section of the dielectric showing a rough sketch of the electric intensity on the atomic scale. If these microscopic electric fields **e** are averaged over volumes large on the atomic scale, but small on the laboratory scale, one obtains the smoothly varying macroscopic (or local space average) electric intensity **E** shown in (a)

As an example consider the uniformly polarized cylindrical dielectric shown in *Figure A2.2(a)*. The polarization vector **P** is parallel to the axis of the cylinder. The macroscopic electric intensity **E** can be calculated by replacing the dielectric by a fictitious surface charge distribution **P** . **n** per unit area at each end. (If **P** is uniform ∇ . **P** is zero, so that there are no fictitious volume charge distributions.) The macroscopic electric intensity is shown in *Figure*

A2.2(a). If one were able to look at individual atomic dipoles, the microscopic electric intensity **e** would be roughly as shown in *Figure A2.2(b)*. The local microscopic electric intensity **e** varies enormously in magnitude and direction on the atomic scale. However, if the microscopic electric intensity **e** shown in *Figure A2.2(b)* is averaged over regions of space large on the atomic scale, but small on the laboratory scale, one obtains the smoothly varying macroscopic electric intensity **E** shown in *Figure A2.2(b)*.

An ion in a crystal lattice does not experience the local space average of the microscopic electric field, since it is restrained from moving far from its equilibrium position by the restoring forces in the crystal. Hence, if one wanted to calculate the electric force on an ion in a crystal lattice, one would have to use a microscopic theory taking into account the crystal structure, and eqn (A2.13) would not be applicable. For a field point inside a stationary dielectric we have for the divergence of the *macroscopic*, or local space average electric field,

$$\mathbf{V} \cdot \mathbf{E} = \frac{\rho_{\text{total}}}{\varepsilon_0} = \frac{1}{\varepsilon_0}\left(\rho_{\text{true}} + \rho'\right) \qquad \text{(A2.16)}$$

where ρ_{true} is the true charge density due to charges one puts on the dielectric, for example, by actually rubbing it; ρ' is the fictitious polarization charge density given by eqn (A2.15). Substituting $\rho' = -\mathbf{V} \cdot \mathbf{P}$ into eqn (A2.16),

$$\mathbf{V} \cdot \mathbf{E} = \frac{1}{\varepsilon_0}\left(\rho_{\text{true}} - \mathbf{V} \cdot \mathbf{P}\right)$$

or

$$\mathbf{V} \cdot \left(\varepsilon_0 \mathbf{E} + \mathbf{P}\right) = \rho_{\text{true}} \qquad \text{(A2.17)}$$

Now the quantity $\varepsilon_0 \mathbf{E} + \mathbf{P}$ appears so frequently in electromagnetic theory, that it is convenient, in the interests of simplicity, to give it a special name, namely, the electric displacement, and to give it a special symbol, namely **D**, so that *by definition* one has

$$\mathbf{D} = \varepsilon_0 \mathbf{E} + \mathbf{P} \qquad \text{(A2.18)}$$

Eqn (A2.17) can be rewritten in the form

$$\mathbf{V} \cdot \mathbf{D} = \rho_{\text{true}} \qquad \text{(A2.19)}$$

This is one of Maxwell's equations. It is developed for *stationary* dielectrics. It relates two macroscopic quantities **D** and ρ_{true}. One could just as well use eqn (A2.17), and never introduce **D** into the theory.

In the above theory of dielectrics, it was not necessary to use a precise model for atomic electric dipoles. It was only necessary to assume that the atoms of a dielectric have a resultant electric dipole moment in the presence of an applied electric field. In elementary treatments a simplified classical model is often used. One often assumes that the atomic dipoles consist of classical 'point' charges a small distance apart as illustrated for example in *Figure A2.2(b)*.

The theory developed so far is related to possible experiments by applying the theory to calculate the force on stationary test charges placed inside cavities of various geometrical configurations cut in the polarized dielectric. Consider the needle-shaped cavity parallel to the electric field in an isotropic dielectric shown in *Figure A2.1(b)*. The diameter of the cavity can be made small on the macroscopic scale, and yet be equal to many atomic diameters. The force on a stationary test charge q' at the centre of the cavity is equal to $q'\mathbf{E}$, where $\mathbf{E}$ is equal to the theoretical macroscopic electric field intensity [calculated using eqn (A2.13)] present in the dielectric before the cavity was cut. It can be shown that the force on a test charge q' at rest at the centre of a disk-shaped cavity, whose axis is parallel to the field is equal to $q'\mathbf{D}/\varepsilon_0$, where $\mathbf{D}$ is the theoretical electric displacement in the isotropic dielectric before the cavity was cut.

In principle, one could measure the macroscopic electric field intensity $\mathbf{E}$ at a 'point' in a liquid dielectric in terms of the total force of *electrical* origin on a stationary test charge, large on the atomic scale but small on the laboratory scale. On the atomic scale, there would be enormous differences in the force on different portions of the surface of the test charge, but the total electrical force should be a measure of the macroscopic field $\mathbf{E}$ as spatial fluctuations in the microscopic electric field should average out over a surface large on atomic dimensions.

Eqn (A2.18) is often rewritten in the form

$$\mathbf{D} = \varepsilon_r \varepsilon_0 \mathbf{E} \qquad (A2.20)$$

where ε_r is the dielectric constant or relative permittivity. For a large class of materials ε_r is a constant. For some dielectrics ε_r may be a function of $\mathbf{E}$. Some dielectrics exhibit remanence, in which cases ε_r depends on the past history of the dielectric. For some crystals, ε_r is a tensor. Eqn (A2.20) is one of the constitutive equations, ε_r depending on the properties of the material at the field point.

(c) The Biot–Savart Law

It was illustrated in Chapter 3 how the magnetic forces between moving charges are of second order, compared with the electric forces between the charges. One cannot carry out experiments with a moving charge to determine, experimentally, the precise expression for its magnetic field. One can only carry out accurate experiments with complete electric circuits or magnets. Experiments have confirmed that the magnetic force on a charge q moving with velocity $\mathbf{u}$ in a magnetic field $\mathbf{B}$ is given by the Lorentz force, namely

$$\mathbf{f}_{mag} = q\mathbf{u} \times \mathbf{B} \qquad (A2.21)$$

The magnetic field of a *complete* circuit is generally calculated using the Biot–Savart law, according to which the magnetic induction at the field point P in *Figure A2.3*, having co-ordinates x, y, z is

$$\mathbf{B} = \frac{\mu_0 I}{4\pi} \oint \frac{d\mathbf{l} \times \mathbf{r}}{r^3} \qquad (A2.22)$$

where $d\mathbf{l}$ is an element of the circuit at x', y', z' and $\mathbf{r}$ is a vector from $d\mathbf{l}$ to P. The forces between two complete circuits are generally calculated using Grassmann's formula (cf. *Figure A4.1* of Appendix 4),

$$\mathbf{f}_2 = \frac{\mu_0 I_1 I_2}{4\pi} \oint_2 \oint_1 \frac{d\mathbf{l}_2 \times (d\mathbf{l}_1 \times \mathbf{r}_{12})}{r_{12}^3} \qquad (A2.23)$$

It is illustrated in Appendix 4 that eqn (A2.23) is a combination of eqns (A2.21) and (A2.22). Since one cannot carry out experiments with isolated current elements, eqns (A2.22) and (A2.23) have only been verified experimentally for steady currents in *complete* circuits.

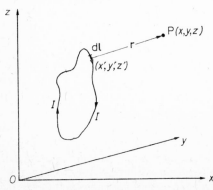

Figure A2.3. The calculation of the magnetic induction $\mathbf{B}$ due to a steady current in an electric circuit, using the Biot–Savart law

The frequency of the normal commercial electricity supply is about 50 cycles per second. For all experiments on the laboratory scale, the changes in current are generally slow enough at this frequency for retardation effects to be neglected when calculating the magnetic field. At such frequencies it is assumed that the current has the same value in all parts of the circuit, and the Biot–Savart law is applied to calculate the associated magnetic fields. These conditions are called quasi-stationary conditions.

Since it is not possible to carry out experiments with steady currents in isolated current elements, and there have as yet been no accurate direct experimental determinations of the magnetic field of a moving charge, eqns (A2.22) and (A2.23) have not been confirmed directly for isolated current elements in the forms

$$d\mathbf{B} = \frac{\mu_0 I}{4\pi} \frac{d\mathbf{l} \times \mathbf{r}}{r^3} \tag{A2.24}$$

and

$$d\mathbf{f}_2 = \frac{\mu_0 I_1 I_2}{4\pi} \frac{d\mathbf{l}_2 \times (d\mathbf{l}_1 \times \mathbf{r}_{12})}{r_{12}^3} \tag{A2.25}$$

Various alternative expressions for the forces between current elements have been suggested. For example, Ampère (cf. Tricker[2]) suggested that

$$d\mathbf{f}_2 = -\mathbf{r}_{12} \frac{\mu_0 I_1 I_2}{4\pi r_{12}^3} \left\{ 2d\mathbf{l}_1 \cdot d\mathbf{l}_2 - \frac{3}{r_{12}^2} (d\mathbf{l}_1 \cdot \mathbf{r}_{12})(d\mathbf{l}_2 \cdot \mathbf{r}_{12}) \right\}$$

where $\mathbf{r}_{12}$ is a vector from $d\mathbf{l}_1$ to $d\mathbf{l}_2$. According to Ampère's equation, the forces between the current elements $I_1 d\mathbf{l}_1$ and $I_2 d\mathbf{l}_2$ are equal and opposite. This is not true of Grassmann's equation (A2.25) (cf. Appendix 4). Ampère's equation gives the same result as eqn (A2.25) when integrated around complete circuits. Only with hind sight, after developing Maxwell's equations or the theory of special relativity, can it be shown that eqns (A2.24) and (A2.25) are very good approximations in the quasi-stationary approximation. A proof using the theory of special relativity is given in Appendix 4. A derivation of the magnetic field of a moving charge from Maxwell's equations is given in Section 4.7. It is shown in Appendix 4 that eqn (A2.22) arises from the velocity dependent term $\mathbf{B}_V$ in the expression for the magnetic field of an accelerating charge, namely eqn (3.44). Eqn (A2.22) is not applicable to the radiation fields of accelerating charges. If the Biot–Savart law in the integral form given by eqn (A2.22) is developed from experiments with *steady* currents in *complete* circuits, then equations developed from eqn (A2.22) need only apply to steady currents in complete circuits. Eqn (A2.23) has been verified for steady currents in complete circuits to better than 1 part in 10^5.

Using eqn (A2.23), it can be shown that the forces between two thin, long, straight, parallel wires a distance r apart in empty space and carrying steady currents of I_1 and I_2 amperes are $\mu_0 I_1 I_2 / 2\pi r$ newtons per metre length. Now the ampere is defined as 'that un-varying current which, if present in each of two infinitely thin parallel conductors of infinite length and one metre apart in empty space, causes each conductor to experience a force of exactly 2×10^{-7} newtons per metre of length'. Hence, it follows from the *definition* of the ampere that μ_0 is equal to $4\pi . 10^{-7}$.

(d) The equation $\mathbf{V} . \mathbf{B} = 0$

Consider the electric circuit shown in *Figure A2.3*. Let the current in the circuit be I. The magnetic induction at a field point P having co-ordinates x, y, z is given by eqn (A2.22), namely

$$\mathbf{B} = \frac{\mu_0 I}{4\pi} \oint \frac{d\mathbf{l} \times \mathbf{r}}{r^3} \qquad (A2.22)$$

where $d\mathbf{l}$ is an element of the circuit at x', y', z' and $\mathbf{r}$ is a vector from $d\mathbf{l}$ to the field point P. Now

$$r = [(x-x')^2 + (y-y')^2 + (z-z')^2]^{1/2} \qquad (A2.26)$$

so that r is a function of x, y, z and of x', y', z'. Now,

$$\mathbf{V} . \mathbf{B} = \frac{\mu_0 I}{4\pi} \mathbf{V} . \oint \frac{d\mathbf{l} \times \mathbf{r}}{r^3} = \frac{\mu_0 I}{4\pi} \oint \mathbf{V} . \left(\frac{d\mathbf{l} \times \mathbf{r}}{r^3} \right)$$

where $\mathbf{V} . \mathbf{B}$ depends on how $\mathbf{B}$ varies in the vicinity of the field point P, when the circuit is fixed, that is the differential operators in $\mathbf{V}$ are with respect to x, y and z, the co-ordinates of the field point. Using eqn (A2.26)

$$\frac{\partial}{\partial x} \left(\frac{1}{r} \right) = -\frac{(x-x')}{r^3}$$

Hence,

$$\frac{\mathbf{r}}{r^3} = -\mathbf{V} \left(\frac{1}{r} \right) \qquad (A2.27)$$

Hence,

$$\mathbf{V} . \mathbf{B} = -\frac{\mu_0 I}{4\pi} \oint \mathbf{V} . \left[d\mathbf{l} \times \mathbf{V} \left(\frac{1}{r} \right) \right] \qquad (A2.28)$$

Now for any two vectors **A** and **C**

$$\mathbf{V} \cdot (\mathbf{A} \times \mathbf{C}) = \mathbf{C} \cdot (\mathbf{V} \times \mathbf{A}) - \mathbf{A} \cdot (\mathbf{V} \times \mathbf{C}) \qquad (A1.22)$$

Put $\mathbf{A} = d\mathbf{l}$ and $\mathbf{C} = \mathbf{V}(1/r)$ in eqn (A1.22) and substitute in eqn (A2.28), which then becomes

$$\mathbf{V} \cdot \mathbf{B} = -\frac{\mu_0 I}{4\pi} \left\{ \oint \mathbf{V}\left(\frac{1}{r}\right) \cdot (\mathbf{V} \times d\mathbf{l}) - \oint d\mathbf{l} \cdot \mathbf{V} \times \left[\mathbf{V}\left(\frac{1}{r}\right) \right] \right\}$$

Since $d\mathbf{l}$ is a function of x', y' and z' only, $\mathbf{V} \times d\mathbf{l}$ is zero. The curl of the gradient of any scalar function is zero, so that

$$\mathbf{V} \times \left[\mathbf{V}\left(\frac{1}{r}\right) \right]$$

is zero. Hence,

$$\mathbf{V} \cdot \mathbf{B} = 0 \qquad (A2.29)$$

Since it is developed from the Biot–Savart law, eqn (A2.29) is developed for the magnetic fields of steady currents in complete circuits. In classical electromagnetism it is *assumed* that eqn (A2.29) is also valid for the magnetic fields of varying currents in incomplete circuits, and for the radiation fields of accelerating charges. This is another example of the use of one of Maxwell's equations in wider experimental conditions than contained in the experimental evidence, on which they are developed, eqn (A2.22) in this instance. As explained in Section 4.5. the assumption $\mathbf{V} \cdot \mathbf{B} = 0$ implies the absence of magnetic monopoles.

If $\mathbf{V} \cdot \mathbf{B}$ is zero, then, since the divergence of the curl of any vector is zero, we can write

$$\mathbf{B} = \mathbf{V} \times \mathbf{A} \qquad (A2.30)$$

where **A** is the vector potential.

From eqns (A2.22) and (A2.27)

$$\mathbf{B} = \frac{\mu_0 I}{4\pi} \oint \mathbf{V}\left(\frac{1}{r}\right) \times d\mathbf{l} \qquad (A2.31)$$

Now for any vector **C** and scalar ϕ

$$\mathbf{V} \times (\phi \mathbf{C}) = \phi(\mathbf{V} \times \mathbf{C}) + \mathbf{V}\phi \times \mathbf{C} \qquad (A1.21)$$

or,

$$\mathbf{V}\phi \times \mathbf{C} = \mathbf{V} \times (\phi \mathbf{C}) - \phi(\mathbf{V} \times \mathbf{C})$$

Putting $\phi = 1/r$ and $\mathbf{C} = \mathrm{d}\mathbf{l}$ and substituting in eqn (A2.31), we obtain

$$\mathbf{B} = \frac{\mu_0 I}{4\pi}\left[\oint \nabla \times \left(\frac{\mathrm{d}\mathbf{l}}{r}\right) - \oint \frac{1}{r}\nabla \times \mathrm{d}\mathbf{l}\right]$$

Now since $\mathrm{d}\mathbf{l}$ is a function of x', y', z' only, $\nabla \times \mathrm{d}\mathbf{l}$ is zero.

Hence,

$$\mathbf{B} = \frac{\mu_0 I}{4\pi}\oint \nabla \times \left(\frac{\mathrm{d}\mathbf{l}}{r}\right) = \nabla \times \frac{\mu_0}{4\pi}\oint \frac{I\,\mathrm{d}\mathbf{l}}{r} \qquad (A2.32)$$

Comparing eqns (A2.30) and (A2.32), we conclude that the vector potential $\mathbf{A}$ is given by

$$\mathbf{A} = \frac{\mu_0}{4\pi}\oint \frac{I\,\mathrm{d}\mathbf{l}}{r} \qquad (A2.33)$$

For a volume distribution of current, if $\mathbf{J}\,(x',\,y',\,z')$ is the current density at the point x', y', z', the vector potential at a field point x, y, z is

$$\mathbf{A} = \frac{\mu_0}{4\pi}\int \frac{\mathbf{J}\,\mathrm{d}x'\,\mathrm{d}y'\,\mathrm{d}z'}{r} \qquad (A2.34)$$

where r is the distance from x', y', z' to x, y, z.

(e) *The equation* $\nabla \times \mathbf{B} = \mu_0 \mathbf{J}$

In this section, the discussion will be limited to *steady* currents in *complete* circuits. For the conditions shown in *Figure A2.3*, we conclude *from experiments* with current balances that

$$\mathbf{B} = \frac{\mu_0 I}{4\pi}\oint \frac{\mathrm{d}\mathbf{l} \times \mathbf{r}}{r^3} \qquad (A2.22)$$

There are several ways of developing the equation $\nabla \times \mathbf{B} = \mu_0 \mathbf{J}$ from eqn (A2.22). One popular method, used, for example, by Pugh and Pugh[7] is to introduce the magnetic scalar potential which is defined, such that *in vacuo*

$$\mathbf{H} = \mathbf{B}/\mu_0 = -\nabla\phi_m$$

Using eqn (A2.22) it can be shown that, for any circuit, the scalar magnetic potential ϕ_m is

$$\phi_m = \frac{I}{4\pi}\Omega$$

where Ω is the solid angle subtended by the complete circuit at the field point.

It can be shown that, if the field point is displaced an amount dl, then

$$\mathbf{H} \cdot d\mathbf{l} = \frac{-I}{4\pi} d\Omega$$

where $d\Omega$ is the change in the solid angle subtended by the circuit at the field point. If $\oint \mathbf{H} \cdot d\mathbf{l}$ is evaluated around a closed path, if the path does not pass through the circuit, the total change in Ω is zero, but, if the path does pass through the circuit, there is a change of 4π in Ω giving

$$\oint \mathbf{H} \cdot d\mathbf{l} = I \tag{A2.35}$$

or, *in vacuo*

$$\oint \mathbf{B} \cdot d\mathbf{l} = \mu_0 I \tag{A2.36}$$

If one has a current distribution, eqn (A2.36) can be generalized to

$$\oint \mathbf{B} \cdot d\mathbf{l} = \mu_0 \int \mathbf{J} \cdot \mathbf{n} dS \tag{A2.37}$$

where $\int \mathbf{J} \cdot \mathbf{n} dS$ must be integrated over a surface bounded by the path of integration for $\oint \mathbf{B} \cdot d\mathbf{l}$. If the path of integration is kept small so that the surface area ΔS is small on the laboratory scale, but kept large on the atomic scale, such that the current density $\mathbf{J}$ can be treated as a continuous function of position, then

$$\oint_{\Delta S} \mathbf{B} \cdot d\mathbf{l} = \mu_0 \mathbf{J} \cdot \mathbf{n} \Delta S$$

From the definition of $\nabla \times \mathbf{B}$ in terms of eqn (A1.12) of Appendix 1, we conclude that

$$\nabla \times \mathbf{B} = \mu_0 \mathbf{J} \tag{A2.38}$$

Another approach, given by Reitz and Milford[8a] is to develop eqn (A2.38) from (A2.22) using vector analysis, assuming that $\nabla \cdot \mathbf{J} = 0$. The condition $\nabla \cdot \mathbf{J} = 0$ implies steady currents in complete circuits. It is implicit in eqn (A2.38), since, taking the divergence of both sides of eqn (A2.38), remembering that the divergence of the curl of any vector is zero, we obtain $\nabla \cdot \mathbf{J} = 0$. Eqn (A2.38) will be extended to cover the case of varying currents in incomplete circuits in Appendix A2(h). It will now be extended to systems containing stationary magnetic materials.

(f) Stationary magnetic materials
When a body is magnetized in the presence of an applied magnetic field, atomic magnetic dipoles are aligned in the direction of the applied magnetic field (for isotropic magnetic materials). The

magnetic dipole moment of an electron arises partly from its orbital motion and partly from its spin. It would be beyond the scope of modern knowledge to give a precise classical model for atomic magnetic dipoles. According to quantum mechanics, what one can observe is that an electron has a magnetic moment. In elementary courses, one of two simple classical models is generally used. According to the Amperian model, the atomic magnetic dipoles can be pictured as little electric current loops, somewhat like the postulated orbit of an electron in Bohr's theory of the hydrogen atom. On the magnetic pole model, the magnetic dipoles are pictured as little bar magnets each consisting of a fictitious north pole and a fictitious south pole. The magnetic moment of a plane current loop is defined as the product of the current and the area of the loop. Using eqn (A2.33), it can be shown (for example, Reitz and Milford[8]), that the vector potential due to a current loop of magnetic moment **m**, at a distance **r** from the dipole large compared with the dimensions of the loop (the dipole approximation), is given by

$$\mathbf{A} = \frac{\mu_0}{4\pi} \frac{\mathbf{m} \times \mathbf{r}}{r^3} \tag{A2.39}$$

The magnetization vector **M** is defined as the magnetic dipole moment per unit volume. If $\mathbf{m}_i$ is the magnetic dipole moment of one of the dipoles in a volume element ΔV, then the magnetization vector **M** is defined as

$$\mathbf{M} = \frac{\sum_{i=1}^{n} \mathbf{m}_i}{\Delta V} \tag{A2.40}$$

where the vector summation is carried out over all the magnetic dipoles in the volume element ΔV. The volume element ΔV can be made small on the laboratory scale, say a 10^{-5} cm cube, and contain many atoms ($\sim 10^9$ for a 10^{-5} cm cube). Thus in a macroscopic theory **M** can be treated as a continuous function of position.

Consider a volume element ΔV at a point inside a magnetized body. The vector potential will be calculated at a point T at a distance **r** from the volume element ΔV. These are similar conditions to those shown in *Figure A2.1(a)*. Proceeding as we did earlier, when we calculated the electric field due to a polarized dielectric, but using eqn (A2.39) instead of eqn (A2.9), we find for the vector potential at T

$$\mathbf{A} = \frac{\mu_0}{4\pi} \int_{V_0} \frac{\mathbf{M} \times \mathbf{r}}{r^3} \, dV$$

APPENDIX 2

This equation (cf. Reitz and Milford[8b]) can be transformed into the expression

$$\mathbf{A} = \frac{\mu_0}{4\pi} \int_{V_0} \frac{\mathbf{\nabla} \times \mathbf{M}}{r} \, dV + \frac{\mu_0}{4\pi} \int_{S_0} \frac{\mathbf{M} \times \mathbf{n}}{r} \, dS \qquad (A2.41)$$

The vector potential of a current system was given by eqn (A2.34). Comparing eqns (A2.34) and (A2.41) it can be seen that for purposes of magnetic field calculation, the magnetized body can be replaced by a fictitious surface current $\mathbf{M} \times \mathbf{n}$ amperes per metre and a fictitious

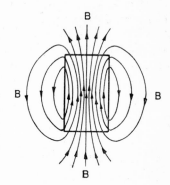

Figure A2.4. The macroscopic magnetic induction $\mathbf{B}$ due to a cylindrical permanent magnet, magnetized uniformly in a direction parallel to its length. The macroscopic field is obtained by averaging the microscopic field over regions of space large on the atomic scale. Irregularities on the atomic scale then average out, giving the smooth continuous magnetic field lines shown. The macroscopic magnetic induction has a finite curl

volume current density $\mathbf{\nabla} \times \mathbf{M}$. The magnetic field of such a fictitious system, calculated using the Biot–Savart law, is the same as the magnetic field of the system of atomic magnetic dipoles, provided the dipole approximation is valid for every single magnetic dipole. The interpretation of the magnetic properties of materials in terms of the properties of electric currents is due to Ampère (Reference: Tricker[2]). Inside the magnetized body, the magnetic field calculated from eqn (A2.41) is equal to the macroscopic magnetic induction $\mathbf{B}$, which is the space average of the microscopic magnetic induction (Reference: Corson and Lorrain[6b]). The microscopic magnetic induction must be averaged over a region of space large on atomic dimensions but small on the laboratory scale to obtain the macroscopic magnetic induction $\mathbf{B}$. There is no need to give a precise model for atomic magnetic dipoles when calculating the macroscopic magnetic induction $\mathbf{B}$. We need only introduce more precise atomic models when calculating the microscopic magnetic field in regions where the dipole approximation, that is eqn (A2.39), breaks down. The *macroscopic* magnetic induction $\mathbf{B}$ due to a cylindrical bar magnet

232

is shown in *Figure A2.4*. The microscopic magnetic induction **b** varies rapidly near and inside atoms, due to the magnetic moments associated with electron orbital motions and the intrinsic magnetic moments associated with electron and nuclear spins. When the microscopic magnetic induction is averaged over a region of space large on atomic dimensions to obtain the macroscopic field, these rapid variations on the atomic scale average out, giving the smooth continuous *macroscopic* magnetic induction field lines shown in *Figure A2.4*. The lines of the macroscopic magnetic induction **B** are closed, so that the macroscopic magnetic induction due to a magnetized medium has a curl.

There is an alternative approach to calculating the macroscopic magnetic fields of magnetic bodies, using the scalar magnetic potential, and mathematical methods similar to those used for dielectrics in Section A2(b). The scalar potential ϕ_m is given by

$$\phi_m = \frac{1}{4\pi}\int_{S_0} \frac{\mathbf{M} \cdot \mathbf{n}\,\mathrm{d}S}{r} + \frac{1}{4\pi}\int_{V_0} \frac{(-\nabla \cdot \mathbf{M})}{r}\,\mathrm{d}V \qquad (A2.42)$$

and **B** is given by

$$\mathbf{B} = -\mu_0\nabla\phi_m + \mu_0\mathbf{M} \qquad (A2.43)$$

The interested reader is referred to Reitz and Milford[8b].

Consider a magnetized body, whose magnetization **M** varies continuously from point to point. For purposes of calculating the *macroscopic* magnetic induction **B**, the magnetized body can be replaced by a fictitious surface current $\mathbf{M} \times \mathbf{n}$, and a fictitious volume current distribution $\mathbf{J}_m = \nabla \times \mathbf{M}$, and the Biot–Savart law, or equations developed from it can then be used to calculate **B**. Consider a surface ΔS, inside the magnetized body. Make ΔS large on the atomic scale, but small on the laboratory scale. Applying Ampère's circuital theorem, eqn (A2.37), to the surface ΔS, we have

$$\oint \mathbf{B} \cdot \mathrm{d}\mathbf{l} = \mu_0\mathbf{J}_m \cdot \mathbf{n}\Delta S$$

Using eqn (A1.12) to define $\nabla \times \mathbf{B}$ we have

$$\nabla \times \mathbf{B} = \mu_0\mathbf{J}_m = \mu_0\nabla \times \mathbf{M} \qquad (A2.44)$$

Hence for a field point inside a magnetized body, in the presence of true electric currents, we have for the curl of the total *macroscopic* magnetic induction,

$$\nabla \times \mathbf{B} = \mu_0(\mathbf{J} + \mathbf{J}_m) \qquad (A2.45)$$

where $\mathbf{J}_m = \nabla \times \mathbf{M}$ is the 'equivalent current density'.

Eqn (A2.45) can be rewritten

$$\mathbf{\nabla} \times \left(\frac{\mathbf{B}}{\mu_0} - \mathbf{M} \right) = \mathbf{J} \tag{A2.46}$$

where $\mathbf{J}$ is the current density due to the actual motion of true electric charge. Now the quantity $(\mathbf{B}/\mu_0 - \mathbf{M})$ appears so frequently in the theory, that it is worthwhile giving it a special name and a symbol. It is called the magnetizing force and denoted by $\mathbf{H}$, so that *by definition*

$$\mathbf{H} = \frac{\mathbf{B}}{\mu_0} - \mathbf{M} \tag{A2.47}$$

Eqn (A2.46) then takes the more concise form

$$\mathbf{\nabla} \times \mathbf{H} = \mathbf{J} \tag{A2.48}$$

One could, of course, always use eqn (A2.46). Eqn (A2.48) holds, for a field point inside a *stationary* magnetic material.

Eqn (A2.47) is often rewritten in the form

$$\mathbf{B} = \mu_r \mu_0 \mathbf{H} \tag{A2.49}$$

where μ_r is the relative magnetic permeability. For paramagnetic and diamagnetic materials μ_r is a constant. For ferromagnetic materials μ_r is a function of $\mathbf{H}$. Ferromagnetic materials also exhibit remanence, in which case μ_r depends on the past history of the material. Eqn (A2.49) is one of the constitutive equations. The value of μ_r depends on the properties of the material at the field point. The case of moving magnets is considered in Chapter 6.

(g) The equation $\mathbf{\nabla} \times \mathbf{E} = -\partial \mathbf{B}/\partial t$

Many introductory texts simply say that the induced e.m.f. in a circuit is equal to $-\mathrm{d}\Phi/\mathrm{d}t$, where $\Phi = \int \mathbf{B} \cdot \mathbf{n}\mathrm{d}S$ is the flux of $\mathbf{B}$ through the circuit. It is important to separate two distinct phenomena which can contribute to induced e.m.f.s, namely motional e.m.f.s and transformer e.m.f.s (Reference: Sears[9]). A motional e.m.f. is generated when a conductor moves in a steady magnetic field (cf. Section 6.4). The e.m.f. arises from the magnetic forces on charges moving with the conductor. In a transformer both the primary and secondary coils are normally stationary, so that in this case the e.m.f. is not associated with the motion of a conductor.

Consider two *stationary* coils *in vacuo* as shown in *Figure A2.5(a)*. There are no motional e.m.f.s in this case. It is found experimentally that when the electric current in the primary is varying with time, e.g. when the key K is opened or closed, a current flows in the secondary. This current flow in the secondary is only present when the current in the primary is varying. The current in the secondary depends on the resistance of the secondary, so that it is an e.m.f. which is induced in the secondary and not a fixed current. It is found experimentally that the e.m.f. in the secondary coil depends on the

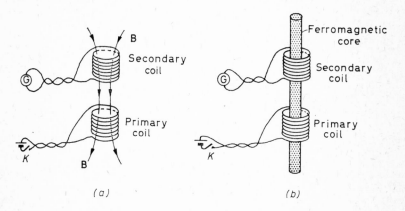

Figure A2.5. (a) A stationary air-cored transformer; (b) a stationary iron-cored transformer

rate of change of magnetic flux through the secondary coil, due to the changes of the current flowing in the primary, that is

$$\varepsilon_{sec} = -\frac{\partial \Phi_{sec}}{\partial t} = -\frac{\partial}{\partial t} \int_{sec} \mathbf{B} \cdot \mathbf{n} dS$$

where $\mathbf{B}$ is the magnetic induction due to the primary. If an alternating current is applied to the primary, an alternating current flows in the secondary. This is the principle of the air-cored transformer. Since there are no moving parts, there are no motional e.m.f.s, so that the current in the secondary must be due to an electric field due to the varying current in the primary, that is, for the secondary coil

$$\varepsilon_{sec} = \oint_{sec} \mathbf{E} \cdot \mathbf{dl} = -\frac{\partial}{\partial t} \int_{sec} \mathbf{B} \cdot \mathbf{n} dS$$

235

Using Stokes' theorem, eqn (A1.25) of Appendix 1,

$$\int_{\text{sec}} (\nabla \times \mathbf{E}) \cdot \mathbf{n} dS = -\frac{\partial}{\partial t} \int_{\text{sec}} \mathbf{B} \cdot \mathbf{n} dS$$

In the limit of a small secondary coil, it follows from the definition of $\nabla \times \mathbf{E}$ in terms of eqn (A1.12) that

$$\nabla \times \mathbf{E} = -\frac{\partial \mathbf{B}}{\partial t} \qquad (A2.50)$$

It is assumed that the electric field given by eqn (A2.50) is present at the position of the secondary coil in *Figure 2.5(a)* and in other parts of space, whether the secondary coil is present or not. The presence of this electric field is confirmed by the action of the Betatron (Reference: Sears[9]). The interpretation of eqn (A2.50) developed in Section 4.4 is that, if the charges in the primary coil in *Figure A2.5(a)* have been moving in the past such that they give rise to a time varying magnetic field at a field point, then the charges in the primary also give an electric field at the same field point, the curl of the electric field intensity being related to the rate of change of the magnetic induction by eqn (A2.50). It is not the changing magnetic flux through the secondary which gives rise to the e.m.f. in the secondary. The two have a common cause, namely the varying current (or moving charges) in the primary.

By considering an example due to Sherwin[10], it is shown in Appendix 5(b) that, in normal air-cored transformers, the induced e.m.f. in the secondary in *Figure A2.5(a)* arises primarily from the velocity dependent term $\mathbf{E}_V$ and not the acceleration dependent term $\mathbf{E}_A$, in the expression for the total electric field of an accelerating charge, namely eqn (3.41). Thus, if eqn (A2.50) is developed from experiments on air-cored transformers carried out at *low* frequency, it is developed, primarily, for the velocity dependent electric field in conditions where the radiation field $\mathbf{E}_A$ is generally negligible. The application of eqn (A2.50) to the acceleration dependent (or radiation) fields of accelerating charges goes beyond the experimental evidence on which eqn (A2.50) is normally developed.

It will now be assumed that there is a ferromagnetic material passing through the primary and secondary coils as shown in *Figure A2.5(b)*. It is found experimentally that this increases the induced e.m.f. in the secondary. The presence of the ferromagnetic material increases the flux of **B** through the secondary. It is found experimentally that the relation

$$\varepsilon = -\partial\Phi/\partial t \qquad (A2.51)$$

236

is still valid. Thus the ferromagnetic material contributes to the electric field at the position of the secondary coil. The electric field due to the current in the primary is still present, but it is augmented by a contribution due to the ferromagnetic material. The increase in the magnetic induction **B** through the secondary coil is due to the alignment of magnetic domains inside the iron core in the direction of the applied magnetic field, and an increase in the size of those magnetic domains parallel to the applied field. The magnetic field of the domains is due partly to the orbital motions of electrons, but with ferromagnetic materials, the main resultant contribution is associated with electron spin (Reference: Bleaney and Bleaney[11]). One cannot give a precise model for electron spin. According to Dirac's wave equation, electrons have an intrinsic angular momentum and an intrinsic magnetic moment, but the theory gives no precise structure for an electron. When the directions of these atomic magnetic moments in space are changed, such that the magnetic field due to them at any field point varies with time, the rotating atomic magnetic moments also give at the field point an electric field which is consistent with

$$\mathbf{\nabla} \times \mathbf{E} = -\partial \mathbf{B}/\partial t \qquad (A2.50)$$

The behaviour of these atomic magnetic moments is consistent with Maxwell's equations. (The fact that a rotating magnetic moment gives an electric field can be illustrated by twisting and rotating a permanent bar magnet in an arbitrary fashion near a stationary closed circuit.) In the case shown in *Figure A2.5(b)*, the e.m.f. and the change of magnetic flux through the secondary are also associated with common causes. The largest contribution is due to rotating atomic magnetic dipoles (mainly electron spins) in the iron core, with a much smaller contribution from the moving charges in the system. If eqn (A2.50) is developed from experiments with iron-cored transformers, it is developed, primarily, as a relation between the electric and magnetic fields of rotating electron spins and not for the linear motions of electric charges. The magnetic moment associated with electron spin behaves in some ways like an ideal permanent magnet.

(h) The equation $\mathbf{\nabla} \times \mathbf{H} = \mathbf{J} + \partial \mathbf{D}/\partial t$

In section A2(e), the equation

$$\mathbf{\nabla} \times \mathbf{B} = \mu_0 \mathbf{J} \qquad (A2.38)$$

was developed from the Biot–Savart law, eqn (A2.22) for steady

237

currents in *complete* circuits. Taking the divergence of both sides of eqn (A2.38), since the divergence of the curl of any vector is zero, we have

$$\mathbf{\nabla} \cdot \mathbf{J} = 0$$

Eqn (A2.38) can only hold for steady currents in complete circuits, and it must be extended in the case of varying currents in incomplete circuits. In elementary texts it is usual to extend eqn (A2.38) by considering a circuit containing a large parallel plate capacitor, as

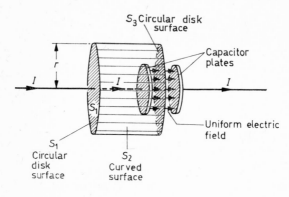

Figure A2.6. The charging up of a large parallel plate capacitor by an alternating current

shown in *Figure A2.6*. It will be assumed that the frequency of the alternating current I, which is charging up the plates, is low enough for the *quasi-stationary approximation* to be valid, so that I can be assumed to have the same value in all parts of the circuit. It will be assumed that the plates of the capacitor are big enough for fringing effects to be neglected. It will be assumed, initially, that the plates are *in vacuo*. Application of Gauss' law shows that, if σ is the charge per unit area on the positive plate, the electric field between the plates is

$$E = \frac{\sigma}{\varepsilon_0} = \frac{Q}{\varepsilon_0 A}$$

where Q is the total charge on the positive plate and A is the area of one of the plates. Let a charge δQ accumulate in a time δt on the positive plate, and let the charge on the negative plate go down by

δQ. This change of charge gives an increment δE in the electric field between the plates, where

$$\delta E = \delta Q / \varepsilon_0 A$$

Hence

$$\frac{dE}{dt} = \frac{1}{\varepsilon_0 A} \frac{dQ}{dt} = \frac{I}{\varepsilon_0 A} \qquad (A2.52)$$

where I is the instantaneous value of the current in the wires. For a field point in empty space, using $\mu_0 = 1/\varepsilon_0 c^2$, eqn (A2.38) can be rewritten

$$\nabla \times H = \varepsilon_0 c^2 \nabla \times B = J \qquad (A2.53)$$

Integrating eqn (A2.53) over the surface of the circular disk shaped surface S_1 shown in *Figure A2.6*, we have

$$\int_{S_1} (\nabla \times H) . \, ndS = \varepsilon_0 c^2 \int_{S_1} (\nabla \times B) . \, ndS = \int_{S_1} J . \, ndS$$

Using Stokes' theorem, eqn (A1.25), we obtain Ampère's circuital law

$$\oint_{S_1} H . \, dl = \varepsilon_0 c^2 \oint_{S_1} B . \, dl = \int_{S_1} J . \, ndS = I \qquad (A2.54)$$

where I is the current in the wire. The integral is evaluated around the circular circumference of S_1 in a clockwise direction as seen from the left-hand side of S_1 in *Figure A2.6*. [Eqn (A2.54) not eqn (A2.53) is sometimes the starting point in elementary treatments.] The line integrals in eqn (A2.54) will now be evaluated around the *same* circular path as before, but the surface integral $\int J . \, ndS$ will be evaluated over the surface S_2, which forms the sides of the cylinder shown in *Figure (A2.6)*, plus the surface S_3 which is a circular disk shaped surface, part of which is between the plates of the capacitor. Now J is zero everywhere on both S_2 and S_3, so that $\oint B . \, dl$ would be zero, if there were no contribution to $\nabla \times B$ and $\nabla \times H$ other than J. However, $\varepsilon_0 c^2 \oint B . \, dl$ should still be equal to I as given by eqn (A2.54). At this stage in the conventional approach, the student is asked to consider $\int \varepsilon_0 (\partial E/\partial t) . \, ndS$ integrated over the surface S_2 and S_3. If it is assumed that there is no fringing field, the electric field is zero outside the space between the plates of the capacitor, and the electric field intensity is uniform over an area A of S_3, where $\partial E/\partial t$ is given by eqn (A2.52). Using eqn (A2.52), we get

$$\int_{(S_2+S_3)} \varepsilon_0 \frac{\partial E}{\partial t} . \, ndS = \int_A \frac{\varepsilon_0 I}{\varepsilon_0 A} dS = I \qquad (A2.55)$$

Thus the correct values for $\oint \mathbf{H} \cdot d\mathbf{l}$ and $\oint \mathbf{B} \cdot d\mathbf{l}$ are obtained, if it is assumed that

$$\oint \mathbf{H} \cdot d\mathbf{l} = \varepsilon_0 c^2 \oint \mathbf{B} \cdot d\mathbf{l} = \int_{(S_2+S_3)} \varepsilon_0 \frac{\partial \mathbf{E}}{\partial t} \cdot \mathbf{n} dS$$

This was an extreme example in which an electric current crossed S_1 but $\mathbf{E} = 0$ everywhere on S_1, whereas no electric current crossed $S_2 + S_3$ but $\partial \mathbf{E}/\partial t$ was finite on part of $S_2 + S_3$. In the conventional approach it is then *suggested* that, in the more general case, when both $\partial \mathbf{E}/\partial t$ and $\mathbf{J}$ are finite on any surface S,

$$\oint_S \mathbf{H} \cdot d\mathbf{l} = \varepsilon_0 c^2 \oint_S \mathbf{B} \cdot d\mathbf{l} = \int_S \mathbf{J} \cdot \mathbf{n} dS + \int_S \left(\varepsilon_0 \frac{\partial \mathbf{E}}{\partial t} \right) \cdot \mathbf{n} dS$$

$$(A2.56)$$

If S is made small enough such that the variations of $\mathbf{J}$ and $\partial \mathbf{E}/\partial t$ over it can be neglected, using eqn (A1.12) of Appendix 1 to define $\nabla \times \mathbf{B}$ and $\nabla \times \mathbf{H}$, we obtain

$$\nabla \times \mathbf{H} = \varepsilon_0 c^2 \nabla \times \mathbf{B} = \mathbf{J} + \varepsilon_0 \frac{\partial \mathbf{E}}{\partial t} \qquad (A2.57)$$

The second term on the right-hand side of eqn (A2.57) is the Maxwell term. If there is a dielectric between the plates of the capacitor in *Figure A2.6*, it is the vector $\mathbf{D}$ between the plates which is related to the true charge on the plates, so that eqn (A2.52) must be rewritten

$$I = A \frac{\partial D}{\partial t} \qquad (A2.58)$$

Thus, when there is a dielectric between the plates of the capacitor in *Figure (A2.6)*, we expect eqn (A2.56) to be replaced by

$$\oint \mathbf{H} \cdot d\mathbf{l} = \varepsilon_0 c^2 \oint \mathbf{B} \cdot d\mathbf{l} = \int \mathbf{J} \cdot \mathbf{n} dS + \int \frac{\partial \mathbf{D}}{\partial t} \cdot \mathbf{n} dS \quad (A2.59)$$

It was shown in Appendix 2(f) that, when stationary magnetic materials are present, the equation $\nabla \times \mathbf{H} = \mathbf{J}$ is still valid. Hence in general, when stationary dielectrics and magnetic materials are present, eqn (A2.59) can be written in the differential form

$$\nabla \times \mathbf{H} = \mathbf{J} + \frac{\partial \mathbf{D}}{\partial t} \qquad (A2.60)$$

The development of eqn (A2.60) in this section from the equation $\mathbf{\nabla} \times \mathbf{H} = \mathbf{J}$, which was developed in Appendix 2(e) for steady currents in complete circuits, was no more than intelligent, inspired guesswork in an example where the quasi-stationary approximation was assumed to be valid. The radiation fields are generally negligible for quasi-stationary conditions. When developed in this way, and in particular when it is applied to the radiation fields of accelerating charges, the validity of eqn (A2.60) really depends on the fact, that predictions based on eqn (A2.60) are found to be in agreement with the experimental results. Eqn (A2.60) is therefore generally checked *a posteriori*. For a discussion of the example illustrated in *Figure A2.6*, when the frequency of the applied e.m.f. is very high and retardation effects and the radiation fields are important, the reader is referred to Feynman, Leighton and Sands[12].

In more advanced texts, the term $\partial \mathbf{D} / \partial t$ in eqn (A2.60) is sometimes developed from the equation of charge continuity. Consider a volume V_0 having a surface area S_0. If electric current is flowing across the surface of V_0, then the total charge inside V_0 must change. If charge is conserved,

$$\int_{S_0} \mathbf{J} \cdot \mathbf{n} \mathrm{d}S \left[= \int_{V_0} \mathbf{\nabla} \cdot \mathbf{J} \mathrm{d}V \right] = -\frac{\partial}{\partial t} \int_{V_0} \rho \mathrm{d}V$$

where Gauss' theorem, eqn (A1.26) of Appendix 1, is used to obtain the expression inside the brackets. If V_0 is made small on the laboratory scale, but large enough on the atomic scale for $\mathbf{J}$ and ρ to be treated as continuous functions of position, we find

$$\mathbf{\nabla} \cdot \mathbf{J} + \frac{\partial \rho}{\partial t} = 0 \qquad (A2.61)$$

This is the equation of charge continuity. Putting $\rho = \mathbf{\nabla} \cdot \mathbf{D}$, we have

$$\mathbf{\nabla} \cdot \left(\mathbf{J} + \frac{\partial \mathbf{D}}{\partial t} \right) = 0$$

Now the divergence of the curl of any vector is zero so that we can put $\mathbf{J} + \partial \mathbf{D} / \partial t$ equal to the curl of a vector, which *suggests* extending the equation

$$\mathbf{\nabla} \times \mathbf{H} = \mathbf{J} \qquad (A2.62)$$

into

$$\mathbf{\nabla} \times \mathbf{H} = \mathbf{J} + \frac{\partial \mathbf{D}}{\partial t} \qquad (A2.60)$$

Taking the divergence of both sides of eqn (A2.60) shows that the equation of charge continuity has been incorporated into eqn (A2.60). Hence, eqn (A2.60) is a *plausible* extension of eqn (A2.62), when there is a varying charge distribution at the field point. However, it is *assumed* in classical electromagnetism that when $\mathbf{J} = 0$, the equation

$$\nabla \times \mathbf{H} = \frac{\partial \mathbf{D}}{\partial t}$$

is valid. In such conditions, for example for a field point in empty space, the question of charge conservation at the field point does not arise. Eqn (A2.60) was again developed by intelligent guesswork, and must be checked *a posteriori* when developed in this way.

Writing eqn (A2.60) in full we have

$$\varepsilon_0 c^2 \nabla \times \mathbf{B} = \varepsilon_0 \frac{\partial \mathbf{E}}{\partial t} + \frac{\partial \mathbf{P}}{\partial t} + \nabla \times \mathbf{M} + \mathbf{J} \qquad \text{(A2.63)}$$

where $\mathbf{P}$ and $\mathbf{M}$ are the polarization and magnetization at the field point. Both $\mathbf{E}$ and $\mathbf{B}$ are the macroscopic fields at the field point. The relativistic interpretation of eqns (A2.60) and (A2.63) is developed in Section 4.9.

When developing Maxwell's equations, many texts go further than developing eqn (A2.60) by intelligent guesswork. Many texts try and interpret the Maxwell term $\varepsilon_0(\partial \mathbf{E}/\partial t)$ in terms of 'the continuity of total electric current flow'. In *Figure A2.6* the current flowing into the capacitor is I. If the plates are *in vacuo*, there is no movement of electric charge between the plates, but, according to eqn (A2.52) $\varepsilon_0 A(\partial E/\partial t)$ is numerically equal to I. Many texts say that the electric current is 'completed' by a 'displacement current' in the *empty space* between the plates. This gives some students the erroneous impression that the term $\varepsilon_0(\partial E/\partial t)$ in empty space behaves in every way like an electric current and gives rise to a magnetic field according to the Biot–Savart law. It is shown in Chapter 4 that the displacement 'current' or Maxwell term does not generate the magnetic field. The electric and magnetic fields are both due to a common cause, namely the moving charges in the system. It is only in eqn (A2.63) that the $\varepsilon_0(\partial \mathbf{E}/\partial t)$ term appears to behave like an electric current. It is shown in Section 4.9 that this is largely fortuitous, since the roles of the Maxwell term, $\varepsilon_0(\partial E/\partial t)$, and the $\mathbf{J}$ term in eqn (A2.63) are completely different.

One of the reasons that the Maxwell term is sometimes 'interpreted' in terms of 'current flow' is a historical one. In Maxwell's time, the mechanical ether theories were in vogue. According to these theories, it was assumed that electric and magnetic forces as well as light waves were transmitted by a hypothetical ether which was assumed to have mechanical properties. It was assumed that one could have forces on the ether present in empty space. Maxwell used precise mechanical models of the ether at various times (Whittaker[1]). The ether was treated as a dielectric. To quote Heaviside, 'Ether is dielectric'. In the nineteenth century, it was assumed that the role of the term $\varepsilon_0(\partial E/\partial t)$ in the ether was similar to the role of the term $\partial P/\partial t$ in a solid dielectric. Maxwell assumed that the current in a circuit containing a capacitor was completed by an actual displacement of charges in the hypothetical ether. The continuity of 'total current' in this way was one of the clues which led Maxwell to introduce the term $\partial D/\partial t$ (Reference: Bork[13]). For this reason the term came to be known as the displacement 'current'.

The mechanical ether theories were, to a large extent, abandoned at the end of the nineteenth century, when people found that it was impossible to measure any of the hypothetical mechanical properties attributed to the ether, or even to show the existence of a preferred reference frame in which the hypothetical ether was assumed to be at rest. According to contemporary classical electromagnetism there is no magnetic force $\varepsilon_0(\partial E/\partial t) \times B$ on empty space, and the displacement 'current' produces no magnetic field according to the Biot–Savart law.

Integrating the equation of charge continuity, $\nabla . J = -\partial\rho/\partial t$ over a finite volume, and applying eqn (A1.26) and the equation $\nabla . D = \rho$, we obtain

$$\int J . n dS = -\frac{\partial}{\partial t}\int \rho dV = -\frac{\partial}{\partial t}\int \nabla . D dV = -\frac{\partial}{\partial t}\int D . n dS$$

$$(A2.64)$$

According to the above set of equations, if a net electric current is leaving a surface, the total electric charge inside the surface is changing. This change in the magnitude of the total charge inside the surface leads to a change in the flux of D (and of E) from the surface. For example, every charge q leaving the surface changes the electric flux leaving the surface by $-q/\varepsilon_0$. That is, the electric current flow gives a $\partial\rho/\partial t$, which in turn gives a $\partial D/\partial t$, such that eqn (A2.64) is satisfied. There is no need to think of $\varepsilon_0(\partial E/\partial t)$ as an electric 'current' to interpret eqns (A2.64) and (A2.63).

243

APPENDIX 2

(*i*) *Discussion*

A survey has been given in this Appendix of the experimental evidence on which Maxwell's equations are normally developed in the form

$$\nabla \cdot \mathbf{D} = \nabla \cdot (\varepsilon_0 \mathbf{E} + \mathbf{P}) = \rho \qquad (A2.19)$$

$$\nabla \cdot \mathbf{B} = 0 \qquad (A2.29)$$

$$\nabla \times \mathbf{E} = -\frac{\partial \mathbf{B}}{\partial t} \qquad (A2.50)$$

$$\nabla \times \mathbf{H} = \mathbf{J} + \frac{\partial \mathbf{D}}{\partial t} \qquad (A2.60)$$

or

$$\varepsilon_0 c^2 \nabla \times \mathbf{B} = \mathbf{J} + \nabla \times \mathbf{M} + \varepsilon_0 \frac{\partial \mathbf{E}}{\partial t} + \frac{\partial \mathbf{P}}{\partial t} \qquad (A2.63)$$

These equations hold for the macroscopic (or local space and time average) electric intensity **E** and macroscopic magnetic induction **B** inside *stationary* materials. The other quantities ρ, **J**, **P**, **M**, **D** and **H** are also macroscopic quantities. In practice Maxwell's equations are applied in a far wider experimental context, than the experimental evidence on which they are normally developed. For example, they are applied to the radiation fields of accelerating charges. It is found that Maxwell's equations are valid in this wider context. For a fuller discussion of this point the reader is referred to Section 4.13.

A discussion of how various macroscopic electromagnetic phenomena are related to the electric and magnetic fields of accelerating charges is given in Appendix 3(d) [cf. eqns (A3.28) and (A3.29)].

REFERENCES

[1] WHITTAKER, E. T. *A History of the Theories of Aether and Electricity.* Thomas Nelson, Edinburgh, 1951

[2] TRICKER, R. A. R. *Early Electrodynamics, the First Law of Circulation.* Pergamon Press, Oxford, 1965

[3] PLIMPTON, S. J. and LAWTON, W. E. *Phys. Rev.* **50** (1936) 1066

[4] MAGIE, W. F. *Source Book in Physics.* p. 408. McGraw-Hill, New York, 1935

[5] ROSSER, W. G. V. *Introduction to the Theory of Relativity.* p. 471. Butterworths, London, 1964

6 CORSON, D. R. and LORRAIN, P. *Introduction to Electromagnetic Fields and Waves.* (a) Ch. 3, (b) Ch. 7. W. H. Freeman and Company, San Francisco, 1962
7 PUGH, E. M. and PUGH, E. W. *Principles of Electricity and Magnetism.* p. 247. Addison-Wesley, Reading, Mass., 1960
8 REITZ, J. R. and MILFORD, F. J. *Foundations of Electromagnetic Theory.* (a) Appendix 3, (b) Ch. 10. Addison-Wesley, Reading, Mass., 1960
9 SEARS, F. W. *Principles of Physics II, Electricity and Magnetism.* Ch. 12. Addison-Wesley, Reading, Mass., 1946
10 SHERWIN, C. W. *Basic Concepts in Physics.* Ch. 5. Holt, Rinehart and Winston, New York, 1961
11 BLEANEY, B. I. and BLEANEY, B. *Electricity and Magnetism.* 2nd. Ed., p. 634. Oxford University Press, 1965
12 FEYNMAN, R. P., LEIGHTON, R. B. and SANDS, M. *The Feynman Lectures on Physics.* Vol. 2, pp. 23–2. Addison-Wesley, Reading, Mass., 1964
13 BORK, A. M. *Amer. J. Phys.* 31 (1963) 854

APPENDIX 3

THE FIELDS OF ACCELERATING CHARGES

In Chapter 3, the expressions for the electric and magnetic fields of a 'point' charge moving with *uniform* velocity were developed from Coulomb's law using the transformations of the theory of special relativity and the principle of constant electric charge. In Chapter 4, Maxwell's equations were developed from these expressions for the special case of a system of 'point' charges moving with uniform velocities. In this Appendix, the more general case of accelerating charges will be considered. We shall start by using *conventional* electromagnetism to indicate how eqns (3.24) and (3.27) must be extended when a charge is accelerating. In the light of the viewpoint developed in Chapter 4, it will then be illustrated how Maxwell's equations should be interpreted, when they are applied to a system of accelerating charges.

(a) Electric field of an accelerating charge

Some of the effects of the acceleration of a 'point' charge on its electric field will be illustrated by a simple example. Consider a 'point' charge q which is at rest at the point A in *Figure A3.1(a)*, until the time $t = 0$. Let the charge q then have a uniform acceleration **a** for a short time Δt, until it reaches the point B with a velocity $u = a\Delta t$, which is very much less than c. The distance AB is equal to $\frac{1}{2}a(\Delta t)^2$. Let the charge q continue to move with uniform velocity u for a time t very much greater than Δt, such that the charge is at the point C in *Figure A3.1(a)* at the time $t+\Delta t$.

Consider the shape of a typical electric field line at the time $t+\Delta t$, when the charge q is at C. The electric field at a radial distance greater than $c(t+\Delta t)$ from A 'left' the charge when the charge was stationary at A. This portion of the electric field diverges radially from A and the field strength is given by Coulomb's law. This portion of the electric field is represented by RS in *Figure A3.1(a)*. The electric field at a radial distance less than ct from B, which is represented by CP in *Figure A3.1(a)*, diverges radially from C the 'present' position of the charge, that is, the position of the charge at the time $t+\Delta t$. For $u \ll c$, according to eqn (3.24) the electric intensity for the portion CP of the field line is also given to a good approximation by Coulomb's law, so that CP is approximately

246

parallel to *RS*. The electric field associated with the period when the charge was accelerating is confined to the region bounded by the arcs *PQ* and *RT*. It is reasonable to join *P* and *R* to represent that part of the electric field line associated with the period when *q* was

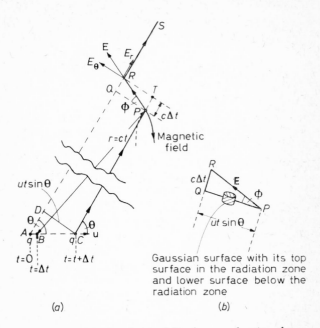

Figure A3.1. (a) The electric field of an accelerating charge. The charge accelerates from rest for a short time Δt, and then moves with uniform velocity **u** for a very much longer time t from B to C. A typical electric field line at the time $t + \Delta t$, when the charge is at C, is CPRS. The portion PR is associated with the period when the charge was accelerating, and leads to the radiation field. (b) The calculation of the radial component of the electric field in the radiation zone

accelerating. It will be assumed that *CPRS* is a typical electric field line at the time $t + \Delta t$ when *q* is at *C*. The portion *PR* will now be considered in more detail.

If $u \ll c$, then $CP \gg AC$, so that the arcs *PQ* and *TR* can be treated as straight lines. In the triangle *ACD* in *Figure A3.1(a)*

$$AC = \tfrac{1}{2}a(\Delta t)^2 + ut \simeq ut$$

$$CD = AC \sin \theta = ut \sin \theta = PQ$$

247

APPENDIX 3

In the triangle PQR, shown enlarged in *Figure A3.1(b)*

$$\tan \phi = \frac{QR}{QP} = \frac{c\Delta t}{ut \sin \theta} \tag{A3.1}$$

Resolving the electric field in the region PR into a component E_r parallel to QR and a component E_θ parallel to PQ, we have

$$\tan \phi = E_r/E_\theta \tag{A3.2}$$

From eqns (A3.1) and (A3.2)

$$E_\theta = \frac{ut \sin \theta}{c\Delta t} E_r$$

If $\Delta t \ll t$ and $u \ll c$, then $AP \simeq CP \simeq BP = r$. Now $u = a\Delta t$, and $t = r/c$ is the time the fields take to propagate the distance r from B, the retarded position of the charge, to the field point P. Hence,

$$E_\theta = \frac{ar \sin \theta}{c^2} E_r \tag{A3.3}$$

In classical electromagnetism it is assumed that the equation $\mathbf{V}.\mathbf{E} = \rho/\varepsilon_0$ is valid for the fields of accelerating charges, so that in empty space $\mathbf{V}.\mathbf{E} = 0$. Draw a cylindrical Gaussian surface as shown in *Figure A3.1(b)*. The flux of $\mathbf{E}$ into the Gaussian surface should be equal to the flux of $\mathbf{E}$ out of the surface. If its height is very small compared with its radius, the flux of $\mathbf{E}$ through the curved surface is negligible. Hence, the radial component of the electric field intensity in the region PQ must be continuous, and given by its value just below PQ, which is given to a good approximation by Coulomb's law. Hence,

$$E_r = q/4\pi\varepsilon_0 r^2$$

Substituting for E_r in eqn (A3.3)

$$E_\theta = \frac{qa \sin \theta}{4\pi\varepsilon_0 rc^2} \tag{A3.4}$$

Notice E_θ is proportional to $1/r$ whereas E_r is proportional to $1/r^2$. At large distances, E_θ predominates giving rise to the radiation field. Eqn (A3.4) can be rewritten in the vector form

$$\mathbf{E}_A \simeq \frac{q}{4\pi\varepsilon_0 c^2 r^3} \mathbf{r} \times (\mathbf{r} \times \mathbf{a}) \tag{A3.5}$$

where **a** is the acceleration of the charge, $u \ll c$ and **r** is a vector from the retarded position of the charge to the field point P, where the electric field is determined at the time $t + \Delta t$. The fields in the region between the arcs PQ and TR, associated with the period when the charge was accelerating, will be referred to as the radiation zone. For a discussion of the case when **u** the velocity of the charge is comparable with c, the reader is referred to Tessman and Finnell[4].

(b) Magnetic field of an accelerating charge

Associated with the electric radiation field, there is also a magnetic radiation field. If it is assumed that $\mathbf{V} \cdot \mathbf{B}$ is zero, then the magnetic field lines are expected to be continuous. For the special case shown in *Figure A3.1(a)*, when **a** is parallel to **u** and $u \ll c$, by symmetry, the magnetic field lines should be closed circles concentric with the direction of **a**. Maxwell's equations will now be used to develop an approximate expression for the magnetic radiation field $\mathbf{B}_A$.

Consider the case illustrated previously in *Figure A3.1*, and shown again on a smaller scale in *Figure A3.2*. The charge starts from rest at the point A at $t = 0$, accelerates for a very short time Δt with uniform acceleration **a** reaching a speed

$$u = a\Delta t \tag{A3.6}$$

The charge then carries on with speed u and reaches the point C at a time $t = (r/c) + \Delta t$. The fields **E** and **B** are propagated from the retarded position of the charge with a speed c in empty space, and take a time r/c to reach P from A, where in *Figure A3.2*, r is the distance AP. Up until the time $t = r/c$, the magnetic field at the field point P in *Figure A3.2* is zero, and the electric field lines diverge radially from A, since they are associated with the period when the charge was stationary. Consider the surface of the spherical cap having P on its periphery, as shown in *Figure A3.2*. The solid angle subtended by this surface at A is $2\pi(1 - \cos\alpha)$, where in *Figure A3.2* α is the angle PAD. Up until the time $t = r/c$ the flux of **E** through the spherical cap is $q(1 - \cos\alpha)/2\varepsilon_0$.

After the time $t = (r/c) + \Delta t$, that is, after the radiation zone passes P, the electric and magnetic fields at P are those appropriate to a charge moving with a uniform velocity $u = a\Delta t$. The lines of **E** then diverge from the present position of the charge, which at the time $t = (r/c) + \Delta t$ is the point C. According to eqn (3.24), if $u \ll c$, the electric field intensity at P is still given to a good approximation by Coulomb's law. Hence at $t = (r/c) + \Delta t$, the flux of **E** through the surface of the spherical cap is $q(1 - \cos\alpha')/2\varepsilon_0$, where α' is equal to the angle PCD in *Figure A3.2*. The change in the flux of **E** through the surface of the spherical cap is

$$\Delta\Psi = \frac{q}{2\varepsilon_0}(1-\cos\alpha') - \frac{q}{2\varepsilon_0}(1-\cos\alpha)$$

$$= -\frac{q}{2\varepsilon_0}(\cos\alpha' - \cos\alpha) = -\frac{q}{2\varepsilon_0}\Delta(\cos\alpha)$$

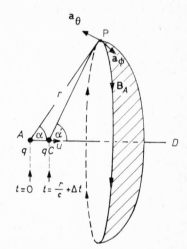

Figure A3.2. Calculation of the magnetic field of an accelerating charge. The charge accelerates for a very short time Δt from A with acceleration a. It then moves with uniform velocity $u = a\Delta t$ for a very much longer time reaching C at a time $t = (r/c) + \Delta t$

This change in the flux of **E** takes place in the time Δt the radiation zone takes to pass P. Its origin is associated with the acceleration of the charge at its retarded position between $t = 0$ and $t = \Delta t$. During the time Δt the radiation zone takes to pass P

$$\frac{\Delta\Psi}{\Delta t} = -\frac{q}{2\varepsilon_0}\frac{d(\cos\alpha)}{dt} = \frac{q}{2\varepsilon_0}\sin\alpha\left(\frac{d\alpha}{dt}\right) \tag{A3.7}$$

Now in *Figure A3.2*, the charge moves from A to C in the time $t = r/c + \Delta t \simeq r/c$. Hence

$$AC \simeq ut \simeq ur/c$$

Now in the triangle APC, the angle APC is equal to $\alpha' - \alpha = \Delta\alpha$. If $\Delta\alpha$ is small, $\sin(\Delta\alpha) \simeq \Delta\alpha$ and $PC \simeq r$. Hence

$$\Delta\alpha \simeq \frac{(AC)\sin\alpha}{PC} \simeq \frac{ur\sin\alpha}{rc} = \frac{u\sin\alpha}{c}$$

But from eqn (A3.6), $u = a\Delta t$, so that

$$\frac{\Delta\alpha}{\Delta t} = \frac{a\sin\alpha}{c}$$

Substituting in eqn (A3.7)

$$\frac{d\Psi}{dt} = \frac{qa}{2\varepsilon_0 c}\sin^2\alpha$$

Using Maxwell's equations, eqn (4.134), integrating over the surface of the spherical cap in *Figure A3.2*, and applying Stokes' theorem, eqn (A1.25), we obtain

$$\frac{1}{c^2}\int\frac{\partial\mathbf{E}}{\partial t}\cdot\mathbf{n}dS = \frac{1}{c^2}\frac{\partial\Psi}{\partial t} = \int(\nabla\times\mathbf{B})\cdot\mathbf{n}dS = \oint\mathbf{B}\cdot d\mathbf{l}$$

By symmetry **B** should have the same value along the periphery of the spherical cap. Hence,

$$2\pi r\sin\alpha B_A = \frac{1}{c^2}\frac{\partial\Psi}{\partial t} = \frac{qa}{2\varepsilon_0 c^3}\sin^2\alpha$$

or,

$$B_A = \frac{qa}{4\pi\varepsilon_0 c^3 r}\sin\alpha \tag{A3.8}$$

Using spherical polars, measuring the polar angle $\theta(=\alpha)$ from the direction of **a** and **u**, we can write eqns (A3.4) and (A3.8) in the form

$$\mathbf{E}_A \simeq \mathbf{a}_\theta E_\theta = \mathbf{a}_\theta\frac{qa\sin\theta}{4\pi\varepsilon_0 c^2 r} \tag{A3.9}$$

$$\mathbf{B}_A \simeq \mathbf{a}_\phi B_\phi = \mathbf{a}_\phi\frac{qa\sin\theta}{4\pi\varepsilon_0 c^3 r} \tag{A3.10}$$

where $\mathbf{a}_\theta$ and $\mathbf{a}_\phi$ are unit vectors in the directions of increasing θ and ϕ respectively. Notice

$$E_A = cB_A$$

When $u \ll c$, the fields $\mathbf{E}_A$ and $\mathbf{B}_A$ are perpendicular to each other and to **r**. The rate of emission of electromagnetic radiation can be calculated using the Poynting vector, $\mathbf{S} = \mathbf{E}\times\mathbf{H}$. For $u \ll c$,

$$S = EH = \varepsilon_0 c^2 EB = \frac{q^2 a^2\sin^2\theta}{16\pi^2\varepsilon_0 c^3 r^2}$$

The intensity of radiation in different directions is proportional to $\sin^2\theta$, where θ is the angle between **a** and **r**. To obtain the rate of loss of energy of the accelerated charge, we must integrate over the surface of a sphere, giving

$$-\frac{dU}{dt} = \frac{q^2 a^2}{6\pi\varepsilon_0 c^3}$$

For **a** parallel to **u** and $u \ll c$, eqn (A3.8) can be rewritten

$$\mathbf{B}_A = \frac{\mathbf{r} \times \mathbf{E}_A}{rc} = \frac{q\mathbf{a} \times \mathbf{r}}{4\pi\varepsilon_0 c^3 r^2} \tag{A3.11}$$

During the period it was accelerating, the charge had a finite velocity, so that in addition to $\mathbf{B}_A$ one expects a contribution to the total magnetic field at P in *Figure A3.1(a)* associated with the velocity of the charge at its retarded position. Hence, for $u \ll c$, one expects four fields, namely $\mathbf{E}_V$ and $\mathbf{B}_V$ depending on the velocity of the charge at its retarded position, and $\mathbf{E}_A$ and $\mathbf{B}_A$ depending on the acceleration of the charge at its retarded position.

The full expressions for the electric and magnetic fields of an accelerating charge moving with a velocity comparable with the velocity of light were developed from Gauss' law by Frisch and Wilets[1] using the transformations of the theory of special relativity and the principle of constant electric charge. Effectively, Frisch and Wilets assumed that the total flux of **E** from an accelerating charge was always equal to q/ε_0. Frisch and Wilets[1] showed that, in general,

$$\mathbf{E} = \mathbf{E}_V + \mathbf{E}_A \tag{A3.12}$$

where,

$$\mathbf{E}_V = \frac{q}{4\pi\varepsilon_0 s^3}\left[\mathbf{r} - \frac{r\mathbf{u}}{c}\right]\left[1 - \frac{u^2}{c^2}\right] \tag{A3.13}$$

$$\mathbf{E}_A = \frac{q}{4\pi\varepsilon_0 s^3 c^2}\left\{[\mathbf{r}] \times \left(\left[\mathbf{r} - \frac{r\mathbf{u}}{c}\right] \times [\mathbf{a}]\right)\right\} \tag{A3.14}$$

and

$$\mathbf{B} = [\mathbf{r}] \times [\mathbf{E}]/[rc] = \mathbf{B}_V + \mathbf{B}_A \tag{A3.15}$$

where,

$$\mathbf{B}_V = \frac{q[\mathbf{u}] \times [\mathbf{r}]}{4\pi\varepsilon_0 c^2 s^3}\left[1 - \frac{u^2}{c^2}\right] \tag{A3.16}$$

$$\mathbf{B}_A = \frac{q[\mathbf{r}]}{4\pi\varepsilon_0 c^3 s^3 [r]} \times \left\{[\mathbf{r}] \times \left(\left[\mathbf{r} - \frac{r\mathbf{u}}{c}\right] \times [\mathbf{a}]\right)\right\} \tag{A3.17}$$

252

and
$$\dot{s} = [r - \mathbf{r} \cdot \mathbf{u}/c] \qquad (A3.18)$$

These equations can also be developed from conventional electromagnetism in the way outlined in Section 5.8. All the terms in eqns (A3.12) to (A3.18) namely [r], [u] and [a] refer to the retarded position of the charge as illustrated in *Figure 3.5(b)* of Chapter 3. If [a] is zero, then $\mathbf{E}_A$ and $\mathbf{B}_A$ are both zero, and eqns (A3.12) and (A3.15) reduce to eqns (3.39) and (3.40) respectively, which are the same as eqns (3.24) and (3.27) respectively for the electric and magnetic fields of a 'point' charge moving with uniform velocity. Thus the first terms on the right-hand side of eqns (A3.12) and (A3.15) namely $\mathbf{E}_V$ and $\mathbf{B}_V$, are similar to the fields of a 'point' charge moving with uniform velocity. For example, the contribution to the total electric intensity due to $\mathbf{E}_V$ diverges radially from the position the accelerating charge would have had, if it had carried on with uniform velocity u. That is, the contribution to the electric intensity associated with $\mathbf{E}_V$ diverges radially from the 'projected' position of the charge (cf. *Figure 3.5(b)* of Chapter 3). However, $\mathbf{E}_V$ and $\mathbf{B}_V$ are not carried along convectively with the charge, since, if the charge is accelerating, its velocity is changing and the values of $\mathbf{E}_V$ and $\mathbf{B}_V$ are varying with time at a fixed distance from the moving charge. If they are present, $\mathbf{E}_A$ and $\mathbf{B}_A$ predominate at large distances from the charge. If $u \ll c$, eqns (A3.14) and (A3.17) reduce to eqns (A3.5) and (A3.11), the approximate expressions for $\mathbf{E}_A$ and $\mathbf{B}_A$ developed earlier in this Appendix. When [a] is parallel to [u] the radiation fields become

$$\mathbf{E}_A = \frac{q}{4\pi\varepsilon_0 c^2 s^3} [\mathbf{r}] \times ([\mathbf{r}] \times [\mathbf{a}])$$

$$\mathbf{B}_A = \frac{q[r]}{4\pi\varepsilon_0 c^3 s^3} [\mathbf{a}] \times [\mathbf{r}]$$

These equations differ from the low velocity approximations in the terms in s^3. In the limit $u \to 0$ the electric radiation field is symmetrical in the forward and backward direction, but due to the term in s^3 in the denominator, the electric and magnetic fields and the intensity of radiation peaks more and more in the direction of u, as u increases. For a discussion of the applications of eqns (A3.12) and (A3.15) the reader is referred to Panofsky and Phillips[2].

(c) Maxwell's equations applied to accelerating charges
In classical electromagnetism it is assumed that the *total* fields $\mathbf{E} = \mathbf{E}_V + \mathbf{E}_A$ and $\mathbf{B} = \mathbf{B}_V + \mathbf{B}_A$ for an accelerating charge satisfy

Maxwell's equations, which for a field point in empty space take the form

$$\mathbf{V} \cdot \mathbf{E} = 0 \qquad\qquad (A3.19)$$

$$\mathbf{V} \cdot \mathbf{B} = 0 \qquad\qquad (A3.20)$$

$$\mathbf{V} \times \mathbf{E} = -\partial \mathbf{B}/\partial t \qquad\qquad (A3.21)$$

$$\varepsilon_0 c^2 \mathbf{V} \times \mathbf{B} = \varepsilon_0 (\partial \mathbf{E}/\partial t) \qquad\qquad (A3.22)$$

These expressions should be valid for the *total* fields given by eqns (A3.12) and (A3.15). According to eqns (A3.19) and (A3.20), away from the accelerating charge, the lines of $\mathbf{E}$ and the lines of $\mathbf{B}$ are continuous. The interpretation of eqn (A3.21) is similar to that developed in Section 4.4, for the case of a charge moving with uniform velocity. If the accelerating charge has been moving in the past, such that its total magnetic field at any field point is varying with time, then that charge also gives rise to an electric field at that field point, the curl of the total electric field intensity $\mathbf{E}$ being equal to minus the time rate of change of the total magnetic induction $\mathbf{B}$ at the fixed field point. Similarly, if at any field point, the total electric field due to the accelerating charge is varying with time, then the accelerating charge also gives rise to a magnetic field at that field point, the curl of $\mathbf{B}$ being equal to $(1/c^2)\partial \mathbf{E}/\partial t$. This interpretation of Maxwell's equations is similar to that developed in Chapter 4 for a charge moving with uniform velocity, though due to the acceleration of the charge, the magnitudes of $\mathbf{E}$ and $\mathbf{B}$ and the shapes of the field lines are different in the two cases. Maxwell's equations are again relations between the field quantities $\mathbf{E}$ and $\mathbf{B}$ at the field point.

Now consider a field point inside a system of accelerating point charges, which build up a macroscopic charge and current distribution. The macroscopic electric intensity $\mathbf{E}$ and the macroscopic magnetic induction $\mathbf{B}$ are again the local space and time averages of the microscopic fields. The vectors $\mathbf{V} \times \mathbf{E}$ and $\mathbf{V} \times \mathbf{B}$ can again be defined in terms of the microscopic fields using eqn (A1.12) keeping ΔS small on the laboratory scale, but large on the atomic scale. Similarly, $\mathbf{V} \cdot \mathbf{B}$ and $\mathbf{V} \cdot \mathbf{E}$ can be defined in terms of the microscopic fields using eqn (A1.10) keeping ΔV large on the atomic scale but small on the laboratory scale. If a 'point' charge is accelerating, according to eqn (A3.12) the total flux of $\mathbf{E}$ from the charge is equal to q/ε_0. For example, consider a spherical Gaussian surface surrounding an accelerating 'point' charge. The flux of the total electric intensity from this surface will be evaluated. Let the centre

254

of the spherical Gaussian surface be the *retarded* position of the charge, appropriate to the time when $\int \mathbf{E} \cdot \mathbf{n} dS$ is evaluated over the surface of the sphere at a fixed time. For the conditions illustrated in *Figure 3.5(b)* of Chapter 3, the centre of the Gaussian spherical surface would be R, the retarded position of the charge, whereas the field point P in *Figure 3.5(b)* would be on the surface of the Gaussian spherical surface, which would have a radius $[\mathbf{r}]$. Since the velocity of the charge is $< c$, both the present and projected positions of the charge are inside the spherical Gaussian surface. Since $\mathbf{E}_V$ diverges from the projected position of the charge, following the method of Section 4.2, $\int \mathbf{E}_V \cdot \mathbf{n} dS = q/\varepsilon_0$. According to eqn (A3.14), $\mathbf{E}_A$ is perpendicular to $[\mathbf{r}]$, so that for this Gaussian surface $\int \mathbf{E}_A \cdot \mathbf{n} dS$ is zero. Hence $\int \mathbf{E} \cdot \mathbf{n} dS = q/\varepsilon_0$, where $\mathbf{E} = \mathbf{E}_V + \mathbf{E}_A$ is the total electric intensity. Proceeding precisely as in Section 4.2, one can then show that

$$\mathbf{V} \cdot \mathbf{E} = \rho/\varepsilon_0 \qquad (A3.23)$$

Since $\mathbf{B} = [\mathbf{r}] \times \mathbf{E}/[rc]$, $\mathbf{B}$ is perpendicular to $[\mathbf{r}]$ so that $\int \mathbf{B} \cdot \mathbf{n} dS$ evaluated over the spherical Gaussian surface is zero. In the absence of magnetic monopoles, this leads to the equation

$$\mathbf{V} \cdot \mathbf{B} = 0 \qquad (A3.24)$$

If eqn (A3.21) applies to the fields of an accelerating 'point' charge, application of the principle of superposition in the way described in Section 4.4 gives the relation

$$\mathbf{V} \times \mathbf{E} = -\partial \mathbf{B}/\partial t \qquad (A3.25)$$

between the total macroscopic fields $\mathbf{E}$ and $\mathbf{B}$. Similarly, applying the principle of superposition, one might expect from eqn (A3.22),

$$\varepsilon_0 c^2 \mathbf{V} \times \mathbf{B} = \varepsilon_0 (\partial \mathbf{E}/\partial t) \qquad (A3.26)$$

relating the total macroscopic fields $\mathbf{E}$ and $\mathbf{B}$. However, if an accelerating charge crosses the surface ΔS used to determine $\oint \mathbf{B} \cdot \mathbf{dl}$ in the definition of $\mathbf{V} \times \mathbf{B}$ in terms of eqn (A1.12), then there is a discontinuity in the flux of $\mathbf{E}$ through ΔS. Even though the acceleration of the charge affects the electric field due to the charge in different directions of space, it was shown earlier that the total flux of the total electric intensity $\mathbf{E}$ from an accelerating charge is still q/ε_0, so that the discontinuity in the flux of $\mathbf{E}$ when the accelerating charge crosses ΔS is still $-q/\varepsilon_0$, just as in the case of a charge moving with uniform velocity. It follows from the argument given in Section 4.9 that, for a system of accelerating point charges also, the $\mathbf{J}$ term must be added to eqn (A3.26) to allow for the fact that, though the dis-

255

continuities in electric flux through the surface contribute to the rate of change of electric flux through that surface, the discontinuities in the electric flux do not contribute to $c^2 \oint \mathbf{B} \cdot d\mathbf{l}$ when eqn (A3.26) is applied. Hence, eqn (A3.26) becomes

$$\varepsilon_0 c^2 \mathbf{V} \times \mathbf{B} = \varepsilon_0 \partial \mathbf{E}/\partial t + \mathbf{J} \qquad (A3.27)$$

Eqns (A3.23), (A3.24), (A3.25) and (A3.27) are Maxwell's equations relating the macroscopic fields $\mathbf{E}$ and $\mathbf{B}$ due to a system of accelerating point charges. The effects of magnetization and polarization at field points inside *stationary* dielectrics can be treated precisely as in Appendix 2, giving eqns (4.139), (4.140), (4.141) and (4.142). It is postulated in relativistic electromagnetism that these equations are also valid at field points inside materials moving with uniform velocity (cf. Chapter 6).

(d) Origins of macroscopic electromagnetic phenomena

It should be stressed that Maxwell's equations, eqns (A3.23), (A3.24), (A3.25) and (A3.27) are relations between the *total* fields $\mathbf{E}$ and $\mathbf{B}$ at the field point. In general, one cannot separate $\mathbf{E}_V$ and $\mathbf{B}_V$ on the one hand and $\mathbf{E}_A$ and $\mathbf{B}_A$ on the other and treat them as separate pairs for the application of Maxwell's equations. For example, the magnetic radiation field $\mathbf{B}_A$ was calculated earlier in Appendix 3(b) from the change in the flux of $\mathbf{E}_V$ through the circular disk-shaped surface in *Figure A3.2*. The electric radiation field $\mathbf{E}_A$ was also calculated from the changes in $\mathbf{E}_V$. Actually, Frisch and Wilets[1] developed all the fields from Gauss' law. However, in a large range of phenomena, either the velocity dependent terms (quasi-stationary conditions) or the acceleration dependent (or radiation) fields (high frequencies at large distances from the source) predominate.

Following Sherwin[3], in the limit $u \ll c$, eqns (A3.12) and (A3.15) can be rewritten in the approximate form

$$\mathbf{E} = \frac{q}{4\pi\varepsilon_0} \left[\underbrace{\frac{\mathbf{r}}{r^3}}_{\substack{E_S \\ \text{electro-} \\ \text{statics}}} + \underbrace{\underbrace{\frac{(3\mathbf{u} \cdot \mathbf{r})\mathbf{r}}{cr^4} - \frac{\mathbf{u}}{cr^2}}_{E^0{}_V} + \underbrace{\frac{\mathbf{r} \times (\mathbf{r} \times \mathbf{a})}{c^2 r^3}}_{\substack{E_A \\ \text{electromagnetic} \\ \text{radiation}}}}_{E_V, \text{ transformers, inductance}} \right] \qquad (A3.28)$$

$$\mathbf{B} = \frac{q}{4\pi\varepsilon_0} \left[\underbrace{+ \frac{\mathbf{u} \times \mathbf{r}}{c^2 r^3}}_{\substack{B_V, \text{ Biot–Savart} \\ \text{law, motors,} \\ \text{generators}}} + \underbrace{\frac{\mathbf{a} \times \mathbf{r}}{c^3 r^2}}_{\substack{B_A \text{ electromagnetic} \\ \text{radiation}}} \right] \qquad (A3.29)$$

In eqns (A3.28) and (A3.29), [r], [u] and [a] refer to the retarded position of the charge.

Eqns (A3.12) and (A3.15) or eqns (A3.28) and (A3.29) summarize classical electromagnetism from the atomistic viewpoint. The macroscopic fields of macroscopic charge and current distributions can be interpreted as the superposition of these fields for moving atomic charges (plus contributions from intrinsic magnetic moments such as electron spin). These resultant fields satisfy Maxwell's equations. Eqns (A3.13), (A3.14), (A3.16) and (A3.17) for E_V, E_A, B_V and B_A respectively are sometimes referred to as the four fields of dynamic electricity.

Sometimes E_V is split in two parts, giving to first order u/c,

$$E_V = E_S + E_V^0 = \frac{q[r]}{4\pi\varepsilon_0[r]^3} + \frac{q}{4\pi\varepsilon_0}\left[\frac{(3u \cdot r)r}{cr^4} - \frac{u}{cr^2}\right] \quad (A3.30)$$

where E_S is the electrostatic field, whereas E_V^0 is that part of E_V which depends on the velocity of the charge. If the electrostatic field is separated from the velocity dependent terms in eqn (A3.13), we have five fields in all, namely E_s, E_V^0, B_V, E_A and B_A (Reference: Sherwin[3]). The origins of various macroscopic electromagnetic phenomena are shown in eqns (A3.28) and (A3.29).

It is shown in Appendix 4 that the magnetic field B_V gives rise to the Biot–Savart law, which is used to calculate the magnetic fields for quasi-stationary conditions. Thus B_V gives the main contribution to that part of the magnetic field in normal a.c. circuits which arises from electric currents, and thus with the Lorentz force accounts for the action of dynamos (motional e.m.f.) and electric motors. (In practice, ferromagnetic materials are generally used to increase the magnetic fields in dynamos and motors. The extra magnetic fields in these cases arise primarily from the alignment of electron spins in the direction of the applied magnetic field.)

If $u = 0$, the term E_V given by eqn (A3.13) gives Coulomb's law and accounts for electrostatics. It is illustrated in Appendix 5, that the term E_V gives rise to the induced electric field in air-cored transformers, mutual inductances etc. Thus the terms E_V and B_V developed from Coulomb's law, using relativity theory in Chapter 3, that is eqns (3.24) and (3.27), account for most of the electric and magnetic effects associated with d.c. and low frequency a.c. circuits.

The terms B_A and E_A give rise to the radiation fields. The radiation fields are proportional to r^{-1}, whereas both E_V and B_V are proportional to r^{-2}. If they are present, the radiation fields predominate at large distances from the charge. They are important at high frequencies and at large distances from the source.

APPENDIX 3

REFERENCES

[1] FRISCH, D. H. and WILETS, L. *Amer. J. Phys.* **24** (1956) 574
[2] PANOFSKY, W. K. H. and PHILLIPS, M. *Classical Electricity and Magnetism.* 2nd Ed., Ch. 20. Addison-Wesley, Reading, Mass., 1964
[3] SHERWIN, C. W. *Basic Concepts in Physics.* Holt, Rinehart and Winston, New York, 1961
[4] TESSMAN, J. R. and FINNELL, J. T. *Amer. J. Phys.* **35** (1967) 523

APPENDIX 4

THE BIOT–SAVART LAW

The Biot–Savart law is used to calculate the magnetic fields due to *steady* conduction currents in *complete* circuits. The frequency of the commercial electricity supply is 50 cycles per second, so that for a.c. experiments on the laboratory scale, the changes in current are generally slow enough for retardation effects to be neglected when calculating **B**. At such frequencies, it is generally assumed that the current has the same value in all parts of the circuit, and the Biot–Savart law is applied to calculate the associated magnetic fields. These conditions are generally called quasi-stationary conditions.

Now the conduction electrons in a metal generally have kinetic energies up to ~ 4 eV so that their velocities are up to $\sim c/250$. Their mean drift velocity under the influence of an electric field is very much less than this, about 10^{-4} metres/second in a typical conductor (see Section 2.2). If it is assumed that $\beta^2 = u^2/c^2 \ll 1$, eqn (3.27) becomes

$$\mathbf{B} \simeq \frac{q\mathbf{u} \times \mathbf{r}}{4\pi\varepsilon_0 c^2 r^3} \qquad (A4.1)$$

Consider an element $d\mathbf{l}_1$ of a thin conductor which makes up circuit number 1, as shown in *Figure A4.1*. Let the steady conduction current in circuit 1 be I_1. The element $d\mathbf{l}_1$ is small on the laboratory scale, but is large enough on the atomic scale to contain many charge carriers. Let the resultant current flow in the element dl_1 be parallel to the y axis as shown in *Figure A4.1*. Let the total number of charge carriers, of charge q coulombs each, inside the element dl_1 be n_1. Let the velocity of the charge carrier i be $\mathbf{u}_i$. Since there is no resultant current flow in the x and z directions,

$$\sum_{i=1}^{n_1} (u_i)_x = \sum_{i=1}^{n_1} (u_i)_z = 0 \qquad (A4.2)$$

In the y direction

$$\sum_{i=1}^{n_1} (u_i)_y = n_1 \bar{u}_1; \quad \text{or} \quad \bar{u}_1 = \frac{1}{n_1} \sum_{i=1}^{n_1} (u_i)_y \qquad (A4.3)$$

where $\bar{u}_1$ is the mean drift velocity of the charge carriers inside

259

$I_1 dl_1$. In eqns (A4.2) and (A4.3) the summation is over the n_1 charges inside dl_1. Now the current I_1 which is in the $+y$ direction is the charge passing any point in the circuit per second, so that

$$(I_1)_y = qN_1\bar{u}_1 \tag{A4.4}$$

where N_1 is the total number of charge carriers per metre length of

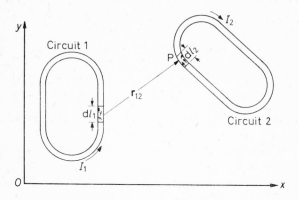

Figure A4.1. The forces between two complete circuits carrying steady currents. The thicknesses of the conductors are negligible compared with r_{12}. The current element $I_1 dl_1$ is parallel to the y axis

conductor 1. (For conduction electrons, both q and $\bar{u}_1$ are negative.) Substituting for $\bar{u}_1$ from eqn (A4.3) into eqn (A4.4), we have

$$(I_1)_y = q\frac{N_1}{n_1} \sum_{i=1}^{n_1} (u_i)_y \tag{A4.5}$$

But $n_1 = N_1 dl_1$. Hence eqn (A4.5) can be rewritten

$$(I_1)_y = \frac{q}{dl_1} \sum_{i=1}^{n_1} (u_i)_y \tag{A4.6}$$

Eqn (A4.6) is a special case, for which $(I_1)_x = (I_1)_z = 0$, of the equation

$$I_1 dl_1 = q \sum_{i=1}^{n_1} \mathbf{u}_i \tag{A4.7}$$

where the summation is over all the n_1 charge carriers inside the length dl_1 of conductor number 1. If the charges do not all have the same charge, eqn (A4.7) can be generalized to

$$I_1 dl_1 = \sum_{i=1}^{n_1} q_i u_i \qquad (A4.8)$$

According to eqn (A4.1) the contribution of one of the charges q_i inside the length dl_1 of conductor 1 to the magnetic induction at the field point P at a distance $\mathbf{r}_{12}$ away, as shown in *Figure A4.1*, is

$$\mathbf{B}_i \simeq \frac{q_i \mathbf{u}_i \times \mathbf{r}_{12}}{4\pi\varepsilon_0 c^2 r_{12}^3} \qquad (A4.1)$$

Summing for all the n_1 charge carriers inside the length dl_1, since for thin conductors they are all at approximately the same distance r_{12} from the field point P, we have

$$\sum_{i=1}^{n_1} \mathbf{B}_i \simeq \sum_{i=1}^{n_1} \frac{q_i \mathbf{u}_i \times \mathbf{r}_{12}}{4\pi\varepsilon_0 c^2 r_{12}^3} \simeq \left(\sum_{i=1}^{n_1} q_i \mathbf{u}_i \right) \times \frac{\mathbf{r}_{12}}{4\pi\varepsilon_0 c^2 r_{12}^3}$$

Using eqn (A4.8) we obtain for the magnetic field due to $I_1 dl_1$

$$d\mathbf{B}_1 \simeq \frac{I_1 dl_1 \times \mathbf{r}_{12}}{4\pi\varepsilon_0 c^2 r_{12}^3} \qquad (A4.9)$$

The total magnetic induction due to circuit 1 is

$$\mathbf{B}_1 \simeq \frac{1}{4\pi\varepsilon_0 c^2} \oint_1 \frac{I_1 dl_1 \times \mathbf{r}_{12}}{r_{12}^3} \qquad (A4.10)$$

Eqns (A4.9) and (A4.10) are generally known as the Biot–Savart law. Readers should bear in mind that eqn (A4.10) is not exact. To be more precise one should use eqn (3.44) for the magnetic fields of individual charges allowing for terms in β^2 and also for the accelerations of the charge carriers. (Refer to Problem A4.1.)

Consider the current element $I_2 dl_2$, which forms part of circuit number 2 as shown in *Figure A4.1*. By analogy with eqn (A4.8)

$$I_2 dl_2 = \sum_{j=1}^{n_2} q_j \mathbf{u}_j$$

where the summation is over all the n_2 charges inside $I_2 dl_2$. If $d\mathbf{B}_1$ is the magnetic field due to $I_1 dl_1$ at the position of $I_2 dl_2$, the force on any one of the charges q_j inside $I_2 d\mathbf{l}_2$ due to $d\mathbf{B}_1$ is $q_j \mathbf{u}_j \times d\mathbf{B}_1$. The total magnetic force on $I_2 d\mathbf{l}_2$ due to $I_1 d\mathbf{l}_1$ is

$$d\mathbf{f}_2 = \sum_{j=1}^{n_2} q_j \mathbf{u}_j \times d\mathbf{B}_1 = \left(\sum_{j=1}^{n_2} q_j \mathbf{u}_j \right) \times d\mathbf{B}_1$$

$$= I_2 d\mathbf{l}_2 \times d\mathbf{B}_1 \qquad (A4.11)$$

Strictly, this force acts only on the moving charges. It is transferred to the conductor as a whole by collisions between the charge carriers and the lattice of the solid conductor. Using eqn (A4.9),

$$d\mathbf{f}_2 \simeq \frac{I_2 d\mathbf{l}_2 \times (I_1 d\mathbf{l}_1 \times \mathbf{r}_{12})}{4\pi\varepsilon_0 c^2 r_{12}^3} \qquad (A4.12)$$

Thus the total magnetic force on circuit number 2 is

$$\mathbf{f}_2 \simeq \frac{I_1 I_2}{4\pi\varepsilon_0 c^2} \oint_2 \oint_1 \frac{d\mathbf{l}_2 \times (d\mathbf{l}_1 \times \mathbf{r}_{12})}{r_{12}^3} \qquad (A4.13)$$

Eqn (A4.13) is generally known as Grassmann's equation[1]. It is a combination of the Biot–Savart law, eqn (A4.9) and the Lorentz force, eqn (A4.11). Some textbooks, such as Panofsky and Phillips[2], start electromagnetism by quoting eqn (A4.13) in the form

$$\mathbf{f}_1 = \frac{\mu_0 I_1 I_2}{4\pi} \oint_2 \oint_1 \frac{d\mathbf{l}_2 \times (d\mathbf{l}_1 \times \mathbf{r}_{12})}{r_{12}^3} \qquad (A4.14)$$

Eqns (A4.13) and (A4.14) are consistent if

$$\mu_0 = 1/\varepsilon_0 c^2, \quad \text{or} \quad c = 1/(\mu_0 \varepsilon_0)^{1/2} \qquad (A4.15)$$

The constant μ_0 is generally introduced into electromagnetism since, in the early stages of a course on electricity, electromagnetism and electrostatics are treated as separate subjects. Eqn (A4.15) is implicit in the theory of special relativity. When electromagnetism is approached via relativity μ_0 is a superfluous constant, and in Chapter 4, $1/\varepsilon_0 c^2$ was generally used in its place. The use of $1/\varepsilon_0 c^2$ in eqn (A4.14) would illustrate from the outset the second order magnitude of the magnetic forces between electric currents. Experiments with current balances of different geometrical configurations have shown that, for steady conduction currents in complete circuits, eqn (A4.14) is correct, in integral form, to better than 1 part in 10^5.

If one applies eqn (A4.12), to 'isolated' current elements, the forces predicted do not obey Newton's third law. For example, in *Figure (A4.2)*, according to the Biot–Savart law, the current element $I_1 dl_1$ produces no magnetic field at the position of the current element $I_2 dl_2$, whereas the latter current element produces a magnetic induction equal to $\mu_0 I_2 dl_2 / 4\pi x^2$ at the position of $I_1 dl_1$. Hence, the magnetic force on $I_1 dl_1$ is equal to $\mu_0 I_1 I_2 dl_1 dl_2 / 4\pi x^2$ newtons, whereas there is no magnetic force on $I_2 dl_2$, so that action and reaction are not equal and opposite in this case.

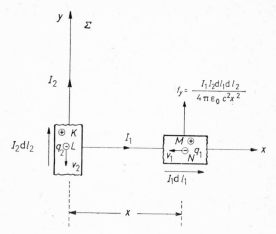

Figure A4.2. Two current elements: $I_2 dl_2$ is at the origin and consists of a stationary positive charge q_2 and denoted K and a charge L equal to $-q_2$ moving with uniform velocity v_2; $I_1 dl_1$ consists of a charge q_1 at rest and denoted by M and a charge $-q_1$ moving with uniform velocity v_1 and denoted by N

It will be assumed that each of the current elements in *Figure A4.2* consists of a positive charge at rest and a negative charge moving with uniform velocity. It is necessary to consider both the electric and the magnetic forces between the charges. By selecting suitable axes in each case, the forces between any two charges can be calculated using eqns (3.13), (3.14) and (3.15) or by using eqns (3.24) and (3.27) for **E** and **B** in the expression for the Lorentz force, which is equivalent to using eqns (3.13), (3.14) and (3.15). The results are given in Table A4.1. It can be seen from Table A4.1 that the *total* force (magnetic plus electric) between any pair of charges deviates from action and reaction by terms of the order of v^2/c^2 only. Deviations of the order of v^2/c^2 from Newtonian mechanics are common in the theory of special relativity and one is prepared to accept them.

The combined force on K plus L, that is on $I_2 dl_2$ is zero, whilst the resultant force on M plus N that is $I_1 dl_1$ is given by

$$f_x = f_z = 0$$

$$f_y = \frac{q_1 q_2}{4\pi\varepsilon_0 x^2} \frac{1}{\sqrt{1 - v_2^2/c^2}} \frac{v_1 v_2}{c^2}$$

Table A4.1*

Charges	f_x	f_y
M on K	$\dfrac{-q_1 q_2}{4\pi\varepsilon_0 x^2}$	—
M on L	$\dfrac{+q_1 q_2}{4\pi\varepsilon_0 x^2}$	—
N on K	$\dfrac{+q_1 q_2}{4\pi\varepsilon_0 x^2}(1 - v_1^2/c^2)$	—
N on L	$\dfrac{-q_1 q_2}{4\pi\varepsilon_0 x^2}(1 - v_1^2/c^2)$	—
K on M	$\dfrac{+q_1 q_2}{4\pi\varepsilon_0 x^2}$	—
K on N	$\dfrac{-q_1 q_2}{4\pi\varepsilon_0 x^2}$	—
L on M	$\dfrac{-q_1 q_2}{4\pi\varepsilon_0 x^2} \dfrac{1}{\sqrt{1 - v_2^2/c^2}}$	—
L on N	$\dfrac{+q_1 q_2}{4\pi\varepsilon_0 x^2} \dfrac{1}{\sqrt{1 - v_2^2/c^2}}$	$\dfrac{+q_1 q_2}{4\pi\varepsilon_0 x^2} \dfrac{1}{\sqrt{1 - v_2^2/c^2}} \dfrac{v_1 v_2}{c^2}$

* The signs on the components refer to the directions of the forces in *Figure A4.2.* e.g.—f_x means a force in the negative O_x direction.

If $v_2 \ll c$

$$f_y \simeq \frac{I_1 dl_1 I_2 dl_2}{4\pi\varepsilon_0 c^2 x^2} \tag{A4.16}$$

Eqns (A4.12) and (A4.16) are not exact, but they are accurate enough for normal conduction currents to illustrate how deviations from Newton's third law may be interpreted in terms of the theory of special relativity, provided both the electric and magnetic forces between moving charges are included. If the current elements are electrically neutral, the electric forces cancel leaving only the second order magnetic forces, which arise from deviations from Newtonian mechanics.

In this Appendix, only the magnetic field associated with the velocity dependent term, given by eqns (3.45) and (A4.1) was considered. A discussion of the radiation magnetic field due to a magnetic dipole is given in Appendix 5, cf. eqn (A5.41). The radiation magnetic field is important at high frequencies, at large distances from the source. The resultant radiation magnetic fields are generally negligible for steady currents in complete circuits, which are the conditions in which the Biot–Savart law is applicable. (Reference: Rosser[3].)

REFERENCES

[1] GRASSMANN, H. *Ann. Phys. Chem.* **64** (1845) 1
[2] PANOFSKY, W. K. H. and PHILLIPS, M. *Classical Electricity and Magnetism.* 2nd Ed., Ch. 7. Addison-Wesley, Reading, Mass., 1962
[3] ROSSER, W. G. V. *Contemp. Phys.* **3** (1961) 28

PROBLEMS

Problem A4.1—Using eqn (3.27), show that eqn (A4.9) can be rewritten in the more accurate form

$$d\mathbf{B}_1 \simeq \sum_{i=1}^{n_1} \frac{q_i \mathbf{u}_i \times \mathbf{r}_{12}}{4\pi\varepsilon_0 c^2 r_{12}^3} \{1 + \beta_i^2(\tfrac{3}{2}\sin^2\theta_i - 1)\}$$

where the summation is over all the n_1 charges in $I_1 d\mathbf{l}_1$ and θ_i is the angle between $\mathbf{u}_i$ and $\mathbf{r}_{12}$. Assume that the velocities of individual charge carriers is $\sim c/250$, but that their mean drift velocity is only $\sim 10^{-4}$ metres/second. Show that the resultant contribution of the term proportional to β^2 inside the brackets is negligible compared with unity.

Using eqn (3.46) show that for normal steady conduction currents, the resultant magnetic field associated with the accelerations of the charge carriers is generally negligible (Reference: Rosser[3]).

Problem A4.2—Show that, if the magnetic forces between two circuits are given by eqn (A4.14) namely

$$\mathbf{f}_2 = \frac{\mu_0 I_1 I_2}{4\pi} \oint_2 \oint_1 \frac{d\mathbf{l}_2 \times (d\mathbf{l}_1 \times \mathbf{r}_{12})}{r_{12}^3}$$

then the forces between the two *complete* circuits satisfy Newton's third law (Reference: Panofsky and Phillips[2]).

APPENDIX 5

ELECTRIC FIELDS ASSOCIATED WITH COMPLETE ELECTRICAL CIRCUITS

There are two main types of electric fields associated with conduction current flow in electrical circuits. Their properties will be considered separately.

(a) Electric current flow and Ohm's law
For a wide range of conductors, Ohm's law is valid, that is

$$\frac{V}{I} = R \qquad (A5.1)$$

where V is the potential difference across the conductor, I is the current in the conductor and R, the ratio of V and I, is called the resistance. If Ohm's law is obeyed R is a constant. In this section, we shall consider how such a potential difference can arise. Consider a conductor of length l, area of cross section A, carrying a current I in a direction parallel to its length. If there is no seat of e.m.f. in the conductor, V is numerically equal to El, where E is the electric field intensity in the conductor. The resistance of the conductor is $R = l/\sigma A$, where σ is the conductivity. Substituting in eqn (A5.1) we find, for the current density $J(=I/A)$,

$$J = \sigma E$$

or in general

$$\mathbf{J} = \sigma \mathbf{E} \qquad (A5.2)$$

If Ohm's law is valid, σ is a constant. Ohm's law is not a universal law of nature, but an experimental law valid for some conductors only. In general, σ may be a function of $\mathbf{E}$. For single crystals σ may be a tensor. Eqn (A5.2) is a constitutive equation, σ depending on the properties of the conducting medium. In eqn (A5.2), $\mathbf{E}$ is the total electric field at the point and is the resultant of the electric field associated directly with the seat of e.m.f. and the electric field due to the surface charge distributions, which help to sustain the electric current flow in a conductor.

As a first introduction to a seat of e.m.f. it has become fashionable to use an idealized Van de Graaff generator. In this example, the

266

moving belt of the Van de Graaff generator literally lifts charges mechanically from a low to a higher electric potential and in this way can maintain the current flow in an external circuit. When a Van de Graaff is operating on open circuit, there are electric charges of opposite signs on the positive and negative terminals. [The leakage current will be ignored at present.] The charges on the terminals give rise to an electric field, which will be denoted by E_0, and which

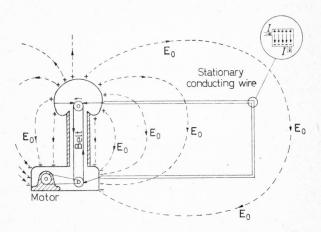

Figure A5.1. *The Van de Graaff generator gives rise to an external electric field E_0, which in turn gives rise to a conduction current in the stationary conducting wire. When the conduction current is steady, surface charge distributions help to guide the current flow in a direction parallel to the wire. When the current is steady, the resultant electric field inside the wire is parallel to the wire. In the inset, it is assumed that there is a cube of copper at the bend in the conducting wire*

extends into the space outside the terminals as shown by the dotted electric field lines in *Figure A5.1*. For purposes of discussion, assume that a conducting wire is brought up 'instantaneously' so as to join the terminals of the Van de Graaff generator. The electric field E_0 due to the charges on the terminals extends into the connecting wire and will act on the free electrons in the connecting wire. There will be a conduction current given by $J = \sigma E$ where J is the conduction current density, σ the conductivity and E is the instantaneous value of the electric field inside the conductor. In the present idealized case, the initial direction of current flow will be parallel to E_0, the electrostatic field due to the charges on the Van de Graaff generator

267

terminals. In general E_0 will not be parallel to the connecting wire. (In practice there will be transient currents as the wire is brought up to join the terminals). The moving charges constituting the conduction current in the connecting wire will build up charge distributions on the surface of the wire, which will give rise to electric fields which, in the steady state, will prevent current flow in a direction perpendicular to the connecting wire. Inside the wire, when the steady state is reached the resultant electric field E due to the source of e.m.f. and the surface charge distributions is parallel to the wire. Thus during the transient state, charge distributions are built up on the surface of the connecting wire which serve to 'guide' the steady electric conduction current along the length of the connecting wire. Under the influence of the electric field, due to the charges on the terminals of the generator and the surfaces of the connecting wire, electrons will flow from the connecting wire into the positive terminal of the generator and from the negative terminal of the generator into the connecting wire, thereby reducing the charges on the terminals. This loss of charge is compensated, to some extent, by the electric charges carried mechanically by the belt of the Van de Graaff generator from one terminal to the other. A state of *dynamic* equilibrium is reached when the charge carried by the belt per second is equal to the conduction current flowing in the connecting wire. At 'dynamic' equilibrium, the conduction current flow is parallel to the wire, and the magnitude of the conduction current is given by $J = \sigma E$, where E is now the local electric field due to both the charges on the terminals of the Van de Graaff and on the surface of the connecting wire. The transient currents flow to the surface of the connecting wire until such a time that the potential drop along the wire is consistent with Ohm's law.

The existence of electric fields associated with these surface charge distributions is demonstrated by the photographs due to Jefimenko[1], an example of which is shown in *Figure A5.2*. Jefimenko used two dimensional printed circuits to demonstrate the electric fields associated with current carrying conductors of various geometrical configurations connected to a Van de Graaff generator. The case shown in *Figure A5.2* is similar to the experimental situation shown in *Figure A5.1*. The photograph in *Figure A5.2* confirms that the electric field inside the conductor is parallel to the conductor. The photograph shows that the electric fields extend into the space outside the conductor. This is to be expected, since according to the boundary conditions (Problem 4.13) the tangential component of E is continuous. At the surface, there is a discontinuity in the normal component of D equal to the charge per unit area on the surface of the conductor, so that outside the conductor the electric field has both a normal and a tangential component. It is left as an exercise for the reader to sketch the surface charges expected in *Figure A5.1* and the electric fields associated with them. In practice, the electric fields in the space *outside* the conductors are normally neglected, since the current flow due to them is negligible, since the ratio of the conductivities of a typical metal and a non conductor such as air is $> 10^{20}$.

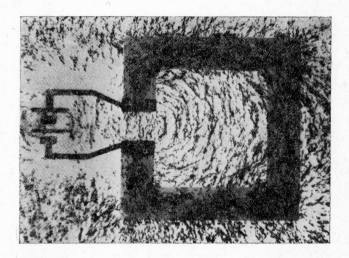

Figure A5.2. An example of the electric fields associated with steady electric currents due to Jefimenko[1]. A Van de Graaff was used as the source of e.m.f., and grass seeds used to show the electric field. Notice that 'inside' the conductor the electric field is parallel to the conductor. The electric field extends into the space outside the conductor (Reproduced by permission of the American Journal of Physics)

The surface charges play a vital role in making an electric current flow. Consider a wire a few miles long connected to the Van de Graaff shown in *Figure A5.1*. On this scale $\mathbf{E}_0$, the electric field due to the charges on the terminals of the Van de Graaff, goes down inversely as the cube of the distance from the generator. Yet when the steady state is reached, the current flow in a uniform wire is the same in all parts of the wire, indicating that the total electric field inside the wire is then uniform and parallel to the wire, even at large distances, where the electric field due directly to the source of e.m.f. is negligible. In this case the electric field at a point at large distances from the source of e.m.f. is due almost entirely to surface charges in the near vicinity of the point. Initially, during the transient state, the electric field $\mathbf{E_0}$ due to the source of e.m.f. gives rise to surface charges in the vicinity of the source of e.m.f. These surface charges then give rise to electric fields which are propagated from these charges with a velocity c in empty space. However, this electric field decreases as $1/r^2$ (or $1/r$ for the radiation field). Thus the main effects due to the surface charges produced near the source of e.m.f. are fairly localized and give rise to surface charge distributions just along the wire, which then give rise to more surface charge distributions further along the wire, so that the surface charge distributions are built up progressively along the wires. The rate of build up of these surface charges depends on the inductance and capacitance per unit length of the wire. Thus, though electric fields themselves can be propagated with a velocity c, the build up of the surface charges along the wires which give the main contribution to the local electric field generally takes place at a slower speed, depending on the capacitance and inductance of the wire.

Problem A5.1—There is a charge density of ρ_0 inside a *homogenous*, stationary conductor of conductivity σ and dielectric constant ε_r. How long does it take the volume charge distribution to disappear? Where does it go?

(*Answer:* $\rho = \rho_0 \exp[-\sigma t/\varepsilon_r \varepsilon_0]$. *Hint:* Take the divergence of $\mathbf{J} = \sigma \mathbf{E}$, substitute in the equation of charge continuity, and use $\nabla \cdot \mathbf{E} = \rho/\varepsilon_r \varepsilon_0$. For copper, $\varepsilon_r \sim 1$, $\sigma = 5 \cdot 7 \times 10^7$ mhos/metre. Hence, the relaxation time $\varepsilon_r \varepsilon_0 / \sigma$ is $\sim 10^{-19}$ seconds. The charge goes to the surface. In a moving conductor, $\mathbf{J} = \sigma(\mathbf{E} + \mathbf{u} \times \mathbf{B})$, and one can have volume charge distributions of the type shown in *Figure 6.9* of Chapter 6.)

The magnitudes of the surface charge distributions are *extremely* small. To obtain a rough estimate of their magnitudes, consider a copper cube inserted in the corner of the conducting wire in *Figure A5.1*, as shown in the inset. Consider the steady state when the conduction current enters the cube at one corner and leaves at the opposite corner at right angles to the direction of entry, as shown in the inset. If the conduction current is to turn through 90° the electric field inside the conducting rail must also turn through 90°. Most of the necessary positive surface charge distributions will be on the top of the cube and most of the negative charge distributions will be at the bottom of the cube. In order to obtain a *rough* estimate of the magnitudes of the charge distributions, it will be assumed that the conduction current is uniform across the cube. The electric field in this case would also be uniform. If the current density is $\mathbf{J}$, and the area of cross section A, following Hammond[6], we have, using eqn (A5.2)

$$J = I/A = \sigma E, \quad \text{or} \quad E = I/\sigma A$$

Applying Gauss' law (cf. a parallel plate capacitor) if Q is the charge on the top surface of the cube, assuming E is uniform, and neglecting fringing effects

$$E = Q/\varepsilon_r \varepsilon_0 A$$

After substituting for E, for copper, the macroscopic charge is

$$Q = \left(\frac{\varepsilon_r \varepsilon_0}{\sigma}\right) I = \text{relaxation time} \times \text{current} = 1\cdot5 \times 10^{-19} I.$$

For a current $I = 100$ amperes, $Q = 1\cdot5 \times 10^{-17}$ coulomb. Since the electronic charge is $1\cdot6 \times 10^{-19}$ coulomb, Q is about 100 positive ions or missing electrons on the top surface of the cube in the inset of *Figure A5.1*, with about 100 electrons on the bottom surface. For $I = 1$ ampere, $Q = 1\cdot5 \times 10^{-19}$ coulomb or an average of about 1 missing electron on the top surface. For $I = \text{lmA}$, $Q = 1\cdot5 \times 10^{-22}$ coulomb or an *average* of 1/1,000 of a missing electron on the top surface. Thus one needs only an excess of a few tens of electrons on the surface of the connecting wire to divert all the electric current used in a normal house, from the main supply cables from the generating station. One needs macroscopic charge distributions equal to only a fraction of the charge of an electron to divert the current flow in various parts of a transistor radio.

The requirement of an average of less than one electron on the surface of the copper block in the inset of *Figure A5.1* certainly requires some imagination, since we cannot, as yet, have charges less than the electronic charge of $1\cdot6 \times 10^{-19}$ coulomb. The charge on the surface of the conductor is the resultant of the positive charges on the lattice ions and the negative charges on the conduction electrons inside the conductor. The conduction electrons undergo random thermal motions with speeds $\sim c/250$ superimposed on their mean drift velocity $\sim 10^{-4}$ metres/second under the influence of the electric field. Consequently, on the microscopic time scale, due to the motions of conduction electrons there are enormous fluctuations in the resultant charge on the surface of the conductor. These fluctuations on the microscopic time scale average out in the times of $> 10^{-10}$ seconds that can be measured experimentally, corresponding to changes in conduction current $< 10^{10}$ Hz. It is the resultant macroscopic charge distribution (which is a time average of the microscopic charge distribution) which is equal to a resultant charge less than the charge on one electron. Due to the motions of the conduction electrons, even when the conduction current is steady, there are fluctuations in the surface charge distributions on the microscopic time scale. It is the resultant macroscopic (or time average) charge distribution which is constant when the conduction current is steady. (For a somewhat similar discussion of atomic time scales, compare the discussion of Ampère's circuital theorem in Section 4.9.)

(b) *Origin of induced electric fields* (*electromagnetic induction*)

In this section, the electric fields actually due to conduction currents in complete electrical circuits will be calculated from the eqn (3.41) for the electric field of a moving point charge, namely,

$$E = E_V + E_A \qquad (A5.3)$$

where,

$$E_V = \frac{q}{4\pi\varepsilon_0 s^3}\left[r - \frac{ru}{c}\right]\left[1 - \frac{u^2}{c^2}\right] \qquad (A5.4)$$

$$E_A = \frac{q}{4\pi\varepsilon_0 s^3 c^2}\left\{[r] \times \left(\left[r - \frac{ru}{c}\right] \times [a]\right)\right\} \qquad (A5.5)$$

and

$$s = \left[r - \frac{r \cdot u}{c}\right] \qquad (A5.6)$$

Eqns (A5.4) and (A5.5) relate the electric field to the retarded position of the charge [cf. Sections 3.3, 5.8 and Appendix 3(b)].

Consider a steady conduction current flowing in an electrically neutral *complete* circuit. [In this section, the electric fields due directly to the source of e.m.f. and to the surface charge distributions associated with conduction current flow, discussed in the previous section, will be ignored.] If it is assumed that the charge density is zero in all parts of the circuit, the scalar potential ϕ due to the actual conduction current is zero, and, if the conduction current in the circuit is steady, the vector potential at any fixed field point due to the conduction current is constant. Hence

$$E = -\nabla\phi - \frac{\partial A}{\partial t} = 0$$

That is, the electric field actually due to a *steady* conduction current in an electrically neutral *complete* circuit is zero. In order to get zero electric field for a complete circuit one must, in general, include both E_V and E_A (Reference: Baker[2]). It will be shown that the conduction current in various parts of the circuit can give rise to

272

electric fields. It is the resultant electric field due to the whole circuit which is zero, when the conduction current is steady. When the conduction current in the circuit is varying, the electric fields due to the various parts of the circuit do not compensate each other, and there is a resultant induced electric field. Since the fields of each moving charge satisfy the equation $\mathbf{V} \times \mathbf{E} = -\partial \mathbf{B}/\partial t$, the resultant electric and magnetic fields of the complete circuit must also satisfy this relation. In practice, the use of this equation is the most convenient way of calculating induced e.m.f.s. In order to illustrate in one simple case how induced e.m.f.s arise from an atomistic viewpoint we shall consider an example given by Sherwin[3]. Consider the circuit

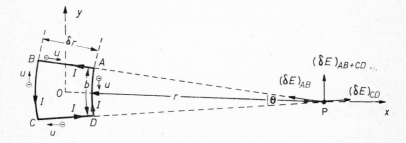

Figure A5.3. The calculation of the electric fields at the field point P due to a conduction current flowing in the coil ABCD. The current flows from A to B to C to D to A. It is assumed that the positive ions in the coil are at rest and that the electrons move with constant speed u from A to D to C to B to A. At large distances, the coil can be treated as a magnetic dipole, of dipole moment m = IA = Ibδr. The direction of the vector **r** is from the coil ABCD to the field point P (after Sherwin[3])

ABCD shown in Figure A5.3. The resultant electric field at the field point P, will be calculated, when a steady conduction current I flows from A to B to C to D. Let the conduction current be due to electrons moving with a velocity **u** which is opposite in direction to I. At present it will be assumed that the speed of the electrons remains constant. It will be assumed that there are an equal number of singly charged positive ions at rest in the circuit, so that the circuit is electrically neutral at all points.

Let AD and BC be arcs of circles, with centres at the field point P and radii r and r+δr respectively, where r ≫ δr. Let B be on PA produced and let C be on PD produced so that

$$AB = CD = \delta r \qquad (A5.7)$$

273

Let the length of $AD = b$ so that the length of CB is

$$CB = b(1 + \delta r/r) \qquad (A5.8)$$

The acceleration dependent fields of the moving electrons will be neglected at present. Eqn (A5.4) applies to the retarded positions of the charges. Let N be the number of stationary positive ions, of charge $+q$ each, per unit length and let N be the number of moving electrons, of charge $-q$ each, per unit length, measured at a fixed time.

Consider the section BA of the circuit shown in *Figure A5.3*. Consider the imaginary 'information collecting sphere' which collapses with velocity c to reach the field point P when the fields are required at P. The number of stationary positive ions 'counted' by the 'information collecting sphere' in the section BA is $N\delta r$. The total number of electrons 'counted' in the section BA is less than this, since the electrons are moving towards P. According to eqn (5.31), the number of electrons passed by the 'information collecting sphere' between B and A is $N\delta r(1 - u/c) = N\delta r(1 - \beta)$. [Those electrons which were at a distance $u\delta r/c$ from A when the 'information collecting sphere' was at B, reach A when the 'information collecting sphere' reaches A, so that the number counted is the number of electrons that were in the length $\delta r(1 - u/c)$, when the 'information collecting sphere' was at B.] For the section BA, the velocity $\mathbf{u}$ of the electrons is parallel to $\mathbf{r}$, as shown in *Figure A5.3*, so that $\mathbf{u} \cdot \mathbf{r} = ur$. We have, putting $\beta = u/c$

$$s = [r - \mathbf{r} \cdot \mathbf{u}/c] = r[1 - \beta]$$

$$\left[\mathbf{r} - \frac{r\mathbf{u}}{c} \right] = \mathbf{r}[1 - \beta]$$

Substituting in eqn (A5.4), we have for the electric intensity due to one of the electrons in BA

$$\mathbf{E} = -\frac{q\mathbf{r}(1 - \beta)(1 - \beta^2)}{4\pi\varepsilon_0 r^3 (1 - \beta)^3} = -\frac{q\mathbf{r}(1 + \beta)}{4\pi\varepsilon_0 r^3 (1 - \beta)} \qquad (A5.9)$$

where $\mathbf{r}$ is a vector from the charge to the field point P, and $-q$ is the charge of the electron. Since the number of electrons passed by the 'information collecting sphere' is $N\delta r(1 - \beta)$, the total electric field due to all the electrons in AB is

$$(\delta \mathbf{E}_-)_{AB} \simeq -\frac{qN\mathbf{r}_0(1 + \beta)\delta r}{4\pi\varepsilon_0 r_0^3} \qquad (A5.10)$$

ELECTRIC FIELDS OF ELECTRICAL CIRCUITS

where $r_0 = (r + \delta r/2)$, is the distance of the middle point of AB from P. The electric intensity due to the stationary positive charges in AB is

$$(\delta E_+)_{AB} \simeq +\frac{qN r_0 \delta r}{4\pi\varepsilon_0 r_0^3} \qquad (A5.11)$$

Adding eqns (A5.10) and (A5.11), we obtain since $r_0 \simeq r$

$$(\delta E)_{AB} = -\frac{qN\delta r}{4\pi\varepsilon_0 r_0^2}\beta \simeq -\frac{qN\delta r}{4\pi\varepsilon_0 r^2}\beta$$

This is in the direction from P to A, as shown in *Figure A5.3*.

Applying a similar argument to the section CD, allowing for the fact that in this case **u** is anti-parallel to **r**, and the number of electrons counted by the 'information collecting sphere' is $N\delta r(1 + \beta)$, we have

$$(\delta E)_{CD} = +\frac{qN\delta r}{4\pi\varepsilon_0 r^2}\beta$$

in the direction from CD to P as shown in *Figure A5.3*. The resultant of $(\delta E)_{AB}$ and $(\delta E)_{CD}$ is

$$(\delta E)_{AB+CD} = \frac{qN\delta r}{4\pi\varepsilon_0 r^2}\beta\theta = \frac{qNb\delta r}{4\pi\varepsilon_0 r^3}\beta \qquad (A5.12)$$

where $\theta = AD/r = b/r$. The resultant electric intensity is in the $+y$ direction in *Figure A5.3*.

Now consider the section AD. All the positive and negative charges are at the same distance r from the field point, so that the numbers of both positive and negative charges counted by the 'information collecting sphere' are equal to Nb. In this case **u** the velocity of the electrons, which is from A to D, is perpendicular to **r**, so that **u . r** is zero, and eqn (A5.4) can be rewritten, for any one of the electrons in AD, in the form

$$E_V = -\frac{q}{4\pi\varepsilon_0 r^3}\left(\mathbf{r} - \frac{r\mathbf{u}}{c}\right)(1 - \beta^2)$$

Adding for all the Nb negative charges in AD, we have

$$(\delta E_-)_{AD} = -\frac{qNb}{4\pi\varepsilon_0 r^3}\left(\mathbf{r} - \frac{r\mathbf{u}}{c}\right)(1 - \beta^2) \qquad (A5.13)$$

For the stationary positive charges in AD

$$(\delta E_+)_{AD} = +\frac{qNb\mathbf{r}}{4\pi\varepsilon_0 r^3} \qquad (A5.14)$$

275

APPENDIX 5

Adding eqns (A5.13) and (A5.14), we obtain

$$(\delta \mathbf{E})_{AD} = -\frac{qNb}{4\pi\varepsilon_0 r^3}\left[-\frac{r\mathbf{u}}{c}(1-\beta^2)-\mathbf{r}\beta^2\right]$$

Neglecting terms of order β^2 and β^3 compared with β, we have, since $\mathbf{u}$ is in the direction from A to D,

$$(\delta E)_{AD} = -\frac{qNb}{4\pi\varepsilon_0 r^2}\beta \qquad (A5.15)$$

This is in the negative y direction in *Figure A5.3*.

To calculate $(\delta E)_{BC}$ from eqn (A5.15) it must be remembered that b must be changed into $b(1+\delta r/r)$ and r into $r(1+\delta r/r)$, whilst the direction of $\mathbf{u}$ the velocity of the electrons is reversed. Hence,

$$(\delta E)_{BC} = \frac{qNb\beta(1+\delta r/r)}{4\pi\varepsilon_0 r^2(1+\delta r/r)^2} \simeq \frac{qNb\beta}{4\pi\varepsilon_0 r^2}\left(1-\frac{\delta r}{r}\right) \qquad (A5.16)$$

This is in the $+y$ direction in *Figure A5.3*. Adding eqns (A5.15) and (A5.16) gives

$$(\delta E)_{AD+BC} = -\frac{qNb\delta r}{4\pi\varepsilon_0 r^3}\beta \qquad (A5.17)$$

This field is in the $-y$ direction in *Figure A5.3*. Comparing eqns (A5.12) and (A5.17), it can be seen that $(\delta E)_{AB+CD}$ is equal and opposite to $(\delta E)_{AD+BC}$, so that, to first order of β, the electric intensity at P is zero. It can be seen that the currents in the various portions AB, BC, CD and DA of the circuit give rise to electric fields at P. It is the *resultant* electric intensity due to the *complete* circuit which is zero, when the current is *steady*.

In the above derivation, the radiation fields were neglected. If $u \ll c$, from eqn (A.5.5),

$$\mathbf{E}_A \simeq \frac{q}{4\pi\varepsilon_0 c^2}\frac{[\mathbf{r}]\times\{[\mathbf{r}]\times[\mathbf{a}]\}}{[r^3]} \qquad (A5.18)$$

Expanding the triple vector product, using eqn (A1.6),

$$\mathbf{E}_A \simeq \frac{q}{4\pi\varepsilon_0[r^3]c^2}[\mathbf{r}(\mathbf{r}\cdot\mathbf{a})-\mathbf{a}(\mathbf{r}\cdot\mathbf{r})] = \frac{q}{4\pi\varepsilon_0[r^3]c^2}[\mathbf{r}\,ar\cos\theta-\mathbf{a}r^2]$$

where θ is the angle between $\mathbf{r}$ and $\mathbf{a}$. Hence

$$\mathbf{E}_A \simeq \frac{q}{4\pi\varepsilon_0[r^3]c^2}[r^2\mathbf{a}_{\parallel}-r^2\mathbf{a}] \simeq -\frac{q}{4\pi\varepsilon_0 c^2}\frac{[\mathbf{a}_\perp]}{[r]} \qquad (A5.19)$$

276

ELECTRIC FIELDS OF ELECTRICAL CIRCUITS

where [a] is the acceleration of the charge at its retarded position, and [a$_\perp$] is the component of [a] perpendicular to [r]. It was assumed in *Figure A5.3*, that the electrons moved with constant speed u. The electrons do have a centripetal acceleration u^2/r in the portions AD and BC, but since this acceleration is parallel to [r], according to eqn (A5.19) their acceleration dependent electric fields at P are zero. It is left as an exercise for the reader to show that the fields associated with accelerations of the electrons around the corners in *Figure A5.3*, are negligible, to the order of accuracy we are working to.

It will now be assumed that the conduction current is varying with time. This varying current gives a varying magnetic field at the field point P, and according to Maxwell's equations there should be a resultant electric field at P associated with this varying magnetic field. The origin of this electric field will now be traced. Let the electrons in the circuit $ABCD$ have a *uniform* acceleration a parallel to their velocities. Let the electrons have a speed u, when the 'information collecting sphere' passes BC. When the 'information collecting sphere' passes AD at a time $\delta r/c$ later, the speed of the electrons in AD is $u+a\delta r/c$. Eqns (A5.16) and (A5.15) now become

$$(\delta E)_{BC} = \frac{qNbu}{4\pi\varepsilon_0 r^2 c}\left(1-\delta r/r\right) \text{ in the } +y \text{ direction} \qquad (A5.20)$$

and

$$(\delta E)_{AD} = -\frac{qNb(u+a\delta r/c)}{4\pi\varepsilon_0 r^2 c} \text{ in the } -y \text{ direction} \qquad (A5.21)$$

Adding eqns (A5.20) and (A5.21), we have

$$(\delta E)_{AD+BC} = -\frac{qNb}{4\pi\varepsilon_0 r^2 c}\left(\frac{u\delta r}{r}+\frac{a\delta r}{c}\right) \qquad (A5.22)$$

This field is in the $-y$ direction.

Now the velocity of the electrons is varying as the 'information collecting sphere' moves along BA and CD. If $\bar{u} = (u+a\delta r/2c)$ is the mean velocity of the electrons in AB and CD measured by the 'information collecting sphere', then eqn (A5.12) can be rewritten

$$(\delta E)_{AB+CD} = \frac{qNb\delta r}{4\pi\varepsilon_0 r^3 c}\left(u+\frac{a\delta r}{2c}\right) \qquad (A5.23)$$

This is in the $+y$ direction. Adding eqns (A5.22) and (A5.23), we obtain for the resultant electric field at P,

$$E_y = -\frac{qNb}{4\pi\varepsilon_0 r^2 c^2}\left(a\delta r - \frac{a(\delta r)^2}{2r}\right)$$

or,

$$E_y \simeq -\frac{qNba\delta r}{4\pi\varepsilon_0 r^2 c^2} \qquad (A5.24)$$

Now the conduction current I is given by

$$I = qNu$$

Hence, $\partial I/\partial t^* = qN\partial u/\partial t^* = qNa$, where $t^* = t - r/c$ is the retarded time appropriate to the measurement of the electric field at P at a time t. The area of the circuit $ABCD$ is given by

$$S = b\delta r$$

Hence,

$$E_y = -\frac{S\partial I/\partial t^*}{4\pi\varepsilon_0 c^2 r^2} \qquad (A5.25)$$

Now the magnetic moment of the circuit is defined as

$$m = IS \qquad (A5.26)$$

The direction of the magnetic moment $\mathbf{m}$ is in the $+z$ direction in *Figure A5.3*.

Using $\mu_0 = 1/\varepsilon_0 c^2$, eqn (A5.25) becomes

$$E_y = -\frac{\mu_0 S\partial I/\partial t^*}{4\pi r^2} = -\frac{\mu_0 \dot{m}}{4\pi r^2} \qquad (A5.27)$$

The origin of the electric field given by eqns (A5.24), (A5.25) and (A5.27) is due to the fact that, when the 'information collecting sphere' passes AD, the electrons are moving at a higher speed than when it passes BC. This electric field is associated with the *velocity* dependent electric field given by eqn (A5.4). The electric field given by eqn (A5.27) arises from retardation effects.

It will now be shown that the electric field given by eqn (A5.27) is consistent with Maxwell's equations, namely with eqn (4.137). At large distances from the coil, the lines of $\mathbf{E}$ should be circles. Choose cylindrical co-ordinates r, ϕ, z with the coil $ABCD$ as origin. Let the z axis be perpendicular to the plane of the coil in *Figure A5.3*. Let ϕ be measured from the x axis. In cylindrical co-ordinates the

z component of $\mathbf{\nabla} \times \mathbf{E}$ at the field point P is

$$(\mathbf{\nabla} \times \mathbf{E})_z = \frac{1}{r}\left[\frac{\partial}{\partial r}(rE_\phi) - \frac{\partial E_r}{\partial \phi}\right] \qquad \text{(A5.28)}$$

(Reference: Pugh and Pugh[4].) According to eqn (A5.27), the electric field at P has the components

$$E_\phi = -\frac{S(\partial I/\partial t^*)}{4\pi\varepsilon_0 c^2 r^2}; \quad E_r = 0$$

Hence,

$$(\mathbf{\nabla} \times \mathbf{E})_z = -\frac{S}{4\pi\varepsilon_0 c^2 r}\frac{\partial}{\partial r}\left(\frac{1}{r}\frac{\partial I}{\partial t^*}\right)$$

It must be remembered that, if r is varied at a fixed time when calculating $\mathbf{\nabla} \times \mathbf{E}$, a change in r corresponds to a change in the appropriate retarded time t^*. Since t^* equals $t - r/c$, $(\partial t^*/\partial r)$ is equal to $-1/c$. Hence,

$$(\mathbf{\nabla} \times \mathbf{E})_z = -\frac{S}{4\pi\varepsilon_0 c^2 r}\left[-\frac{1}{r^2}\frac{\partial I}{\partial t^*} + \frac{1}{r}\frac{\partial^2 I}{\partial t^{*2}}\frac{\partial t^*}{\partial r}\right]$$

$$(\mathbf{\nabla} \times \mathbf{E})_z = \frac{S(\partial I/\partial t^*)}{4\pi\varepsilon_0 c^2 r^3} + \frac{S(\partial^2 I/\partial t^{*2})}{4\pi\varepsilon_0 c^3 r^2} \qquad \text{(A5.29)}$$

The magnetic induction at the field point P in *Figure A5.3* will now be calculated using the equation

$$\mathbf{B} \simeq \frac{q\mathbf{u} \times \mathbf{r}}{4\pi\varepsilon_0 c^2 r^3} \qquad \text{(3.31)}$$

assuming that $u \ll c$. The magnetic induction at P, due to the current in AB and CD is zero. The magnetic induction at P due to the moving electrons (of charge $-q$ each) in BC is in the $+z$ direction and given by

$$(\delta B)_{BC} = \frac{qNb(1 + \delta r/r)u}{4\pi\varepsilon_0 c^2 r^2(1 + \delta r/r)^2} \simeq \frac{qNbu}{4\pi\varepsilon_0 c^2 r^2}\left(1 - \frac{\delta r}{r}\right) \qquad \text{(A5.30)}$$

The magnetic induction at P due to the current in AD is

$$(\delta B)_{AD} = -\frac{qNb(u + a\delta r/c)}{4\pi\varepsilon_0 c^2 r^2} \qquad \text{(A5.31)}$$

Adding eqns (A5.30) and (A5.31), remembering that $S = b\delta r$, $qNu = I$ and $qNa = \partial I/\partial t^*$, we have

$$B_z = -\frac{SI}{4\pi\varepsilon_0 c^2 r^3} - \frac{S(\partial I/\partial t^*)}{4\pi\varepsilon_0 c^3 r^2} \qquad \text{(A5.32)}$$

279

For the magnetic field of a *stationary* coil, differentiating the current in the coil partially with respect to time t at a fixed field point is the same as differentiating with respect to retarded time t^* at the retarded time t^*. Hence,

$$-\frac{\partial B_z}{\partial t} = +\frac{S(\partial I/\partial t^*)}{4\pi\varepsilon_0 c^2 r^3} + \frac{S(\partial^2 I/\partial t^{*2})}{4\pi\varepsilon_0 c^3 r^2} \qquad (A5.33)$$

Comparing eqns (A5.29) and (A5.33) shows that

$$(\boldsymbol{\nabla} \times \mathbf{E})_z = -\partial B_z/\partial t$$

This is in agreement with Maxwell's equations. Only the velocity dependent electric and magnetic fields $\mathbf{E}_V$ and $\mathbf{B}_V$ given by eqns (3.42) and (3.45), were used to determine the expressions for the electric and magnetic fields of the magnetic dipole $ABCD$ in *Figure A5.3*.

In practice, for quasi-stationary conditions it is almost invariably simpler to follow the conventional macroscopic approach and use Maxwell's macroscopic equations, rather than start with the fields of moving 'point' charges. A brief outline of the conventional macroscopic approach will now be given. Firstly, the magnetic field at the point P in *Figure A5.3* will be calculated using the Biot–Savart law, eqn (A4.10) of Appendix 4 or eqn (A2.22) of Appendix 2. For quasi-stationary conditions, that is ignoring retardation effects, and assuming that the current I has the same value in all parts of the circuit, the reader can use eqn (A4.10) to show that

$$B_z = -\frac{SI}{4\pi\varepsilon_0 c^2 r^3} = -\frac{\mu_0 m}{4\pi r^3} \qquad (A5.34)$$

(Alternatively, the reader can put $\partial I/\partial t^*$ equal to zero in eqn (A5.32) for low frequency quasi-stationary conditions.) The reader will recognize eqn (A5.34) as the Gauss B position for a magnetic dipole. The direction of the magnetic field is given by the corkscrew rule. It follows from the discussion of Section 4.4 of Chapter 4 that, whenever a system of moving charges gives rise to a varying magnetic field at a point, they also give rise to an electric field. There is no need to enquire how this resultant electric field arises. From eqn (4.59), we know that $\boldsymbol{\nabla} \times \mathbf{E}$ is equal to $-\partial \mathbf{B}/\partial t$. Hence,

$$(\boldsymbol{\nabla} \times \mathbf{E})_z = -\frac{\partial B_z}{\partial t} = +\frac{S(\partial I/\partial t)}{4\pi\varepsilon_0 c^2 r^3} = \frac{\mu_0 \dot{m}}{4\pi r^3} \qquad (A5.35)$$

This is the same as the first term in eqn (A5.29). If it is assumed that the current in the coil $ABCD$ in *Figure A5.3* is given by

$I = I_0 \sin 2\pi f t^*$, the ratio of the second to the first term on the right-hand side of eqn (A5.29) is of order $2\pi f r/c$. For $f = 50$ Hz and $r = 1$ metre this ratio is $\sim 10^{-6}$, so that for quasi-stationary conditions the second term can generally be neglected. Similarly, the second terms on the right-hand sides of eqns (A5.32) and (A5.33) can be neglected for quasi-stationary conditions. Eqns (A5.34) and (A5.35) derived in the conventional macroscopic way are generally sufficiently accurate for quasi-stationary conditions. Thus for quasi-stationary conditions, when calculating the magnetic induction, retardation effects can be neglected, that is the current can be assumed to have the same value in all parts of the circuit, when calculating the magnetic induction using the Biot–Savart law. However, it was shown earlier that the induced electric field given by eqn (A5.27) arises from retardation effects, even though it can be calculated from an expression for the magnetic field, in which retardation effects are neglected, The resultant induced electric field at P in *Figure A5.3* is due to the fact that the electric current in the circuit $ABCD$ is greater when the 'information collecting sphere' passes AD than BC.

In practice, in quasi-stationary problems, one is generally interested in the e.m.f. in a secondary coil, and one generally uses

$$\oint \mathbf{E} \cdot d\mathbf{l} = -\frac{\partial \Phi}{\partial t} \tag{A5.36}$$

where Φ is the total magnetic flux through the secondary, calculated using the Biot–Savart law, assuming that, at any instant, the current has same value in all parts of the primary. Consider a circle of radius r, centre the coil $ABCD$ in *Figure A5.3*. The net magnetic flux through this circle is upwards, towards the reader. Since the lines of $\mathbf{B}$ are continuous, this flux is numerically equal to the magnetic flux which crosses the xy plane downwards in *Figure A5.3* at a radial distance greater than r. Hence, using eqn (A5.34)

$$\Phi = \int_r^\infty \frac{\mu_0 m}{4\pi r^3} 2\pi r dr = \frac{\mu_0 m}{2r}$$

If the induced electric field lines are circles, using eqn (A5.36) we have

$$\oint \mathbf{E} \cdot d\mathbf{l} = 2\pi r E = \mu_0 \dot{m}/2r$$

or

$$E = \frac{\mu_0 \dot{m}}{4\pi r^2}$$

281

This is consistent with eqn (A5.27). The direction of the induced electric field can be determined using Lenz's law. The calculation of induced e.m.f. and induced electric fields in this Appendix, from an atomistic point of view, will probably strike the reader as being elaborate and complicated compared with the conventional macroscopic approach based on Maxwell's equations, or even more so compared with using eqn (A5.36). This illustrates the usefulness of eqn (A5.36). Having seen in one *simple* case how the induced electric field arises from the atomistic point of view, the reader is advised to stick to the conventional approach for quasi-stationary conditions.

The radiation fields at P in *Figure A5.3* will now be considered. Assume that the electrons have a uniform acceleration $\mathbf{a}$. In eqn (A5.19), $\mathbf{a}_\perp$ is the component of $\mathbf{a}$ perpendicular to $\mathbf{r}$. The accelerations of the electrons in BA and CD are parallel and anti-parallel to $\mathbf{r}$ respectively so that the electric radiation field associated with them is zero at the field point P. In AD and BC, the accelerations of the electrons are perpendicular to $\mathbf{r}$, so that $\mathbf{a}_\perp$ equals $\mathbf{a}$. Since the 'information collecting sphere' passes all the charges in AD at the same time, using eqn (A5.19) we find that the acceleration dependent electric field at P in *Figure A5.3* is

$$(\delta E)_{AD} = -\frac{Nbqa}{4\pi\varepsilon_0 c^2 r} \tag{A5.37}$$

For negative charges accelerating in the direction from A to D, this field is in the $-y$ direction at P. Similarly,

$$(\delta E)_{BC} = +\frac{Nb(1+\delta r/r)qa}{4\pi\varepsilon_0 c^2 r(1+\delta r/r)} = \frac{Nbqa}{4\pi\varepsilon_0 c^2 r} \tag{A5.38}$$

This is opposite in direction to the field given by eqn (A5.37). Adding eqns (A5.37) and (A5.38), it can be seen that the resultant electric radiation field at P is zero, if a is constant. To obtain a resultant electric radiation field at P, we need a finite $\partial^2 I/\partial t^{*2}$, in which case a in eqn (A5.37) must be changed to $(a+\dot{a}\delta r/c)$. The resultant electric radiation field at P is then

$$(E_y)_{rad} = -\frac{Nbq\dot{a}\delta r}{4\pi\varepsilon_0 c^2 rc}$$

$$= -\frac{S\partial^2 I/\partial t^{*2}}{4\pi\varepsilon_0 c^3 r} = -\frac{\mu_0\ddot{m}}{4\pi cr} \tag{A5.39}$$

This result is in agreement with the well-known result that the radiation fields at large distances from a magnetic dipole are proportional to $\ddot{m}$. Actually, in the general case

$$\mathbf{E}_{\text{rad}} = -\frac{\mu_0 \ddot{\mathbf{m}} \times \mathbf{r}}{4\pi cr^2} \tag{A5.40}$$

This reduces to eqn (A5.39), when $\mathbf{r}$ is perpendicular to $\mathbf{m}$. The ratio of the radiation electric field given by eqn (A5.39) to the induced field given by eqn (A5.27) is $r\ddot{m}/(\dot{m}c)$. Hence, the radiation field predominates at high frequencies at large distances from the source.

According to eqn (3.46), the radiation magnetic field of an accelerating 'point' charge is

$$\mathbf{B}_A = \frac{[\mathbf{r}] \times \mathbf{E}_A}{[rc]} \simeq \frac{q[\mathbf{a}] \times [\mathbf{r}]}{4\pi\varepsilon_0 c^3 [r^2]}$$

The radiation magnetic field at P due to the accelerating electrons in AB and CD is zero. The radiation magnetic fields due to the accelerating negative electrons in BC and AD are $+qNba/4\pi\varepsilon_0 c^3 r$ and $-qNb(a+\dot{a}\delta r/c)/4\pi\varepsilon_0 c^3 r$ respectively, giving

$$(B_z)_{\text{rad}} = -\frac{S(\partial^2 I/\partial t^{*2})}{4\pi\varepsilon_0 c^4 r} = -\frac{\mu_0 \ddot{m}}{4\pi c^2 r} \tag{A5.41}$$

This is consistent with the following general expression for the magnetic radiation field of a magnetic dipole,

$$\mathbf{B}_{\text{rad}} = -\frac{\mu_0 \mathbf{r} \times (\ddot{\mathbf{m}} \times \mathbf{r})}{4\pi c^2 r^3} \tag{A5.42}$$

Notice, $E_{\text{rad}} = cB_{\text{rad}}$. When $r \gg b$, the Poynting vector at a point in the xy plane, given by EH or EB/μ_0, is equal to $\mu_0(\ddot{m})^2/16\pi^2 c^3 r^2$. It follows from eqns (A5.40) and (A5.42) that, in general, the Poynting vector is proportional to $\sin^2\theta$, where θ is the angle between $\mathbf{m}$ and $\mathbf{r}$. The radiation fields of a magnetic dipole are similar to the radiation fields of an electric dipole, but with the roles of $\mathbf{E}$ and $\mathbf{B}$ interchanged. In general, the most convenient way of calculating the radiation from an antenna is to use the retarded potentials, eqns (5.44) and (5.45), developed in Section 5.7 of Chapter 5.

Using eqns (A5.28) and (A5.39), in cylindrical co-ordinates, we have

$$(\nabla \times \mathbf{E})_z = \frac{1}{r}\frac{\partial}{\partial r}(rE_\phi) = -\frac{\mu_0}{4\pi cr}\frac{\partial}{\partial r}(\ddot{m})$$

$$= -\frac{\mu_0}{4\pi cr}\dddot{m}\frac{\partial t^*}{\partial r} = \frac{\mu_0\dddot{m}}{4\pi c^2 r}$$

This is equal to $-\partial B_z/\partial t$, where B_z is given by eqn (A5.41). This is in agreement with Maxwell's equations.

The total electric field at P is obtained by adding eqns (A5.27) and (A5.39), and the total magnetic field by adding eqns (A5.32) and (A5.41). We have made many approximations in this Appendix. Our aim was to illustrate the atomistic origins of some macroscopic electromagnetic phenomena. In order to confirm that the fields we obtained are applicable to a stationary magnetic dipole, the reader can assume that the current in $ABCD$ varies harmonically with time, and show that, the fields developed in the text are then consistent with the standard expressions for the fields of a magnetic dipole given, for example, by Stratton[5]. As a further exercise, the reader can check that the fields given by Stratton, satisfy Maxwell's equations and check that the approximations and formulae used to calculate $\nabla \times E$ in the text are valid.

REFERENCES

[1] JEFIMENKO, O. *Amer. J. Phys.* **30** (1962) 19
[2] BAKER, D. A. *Amer. J. Phys.* **32** (1964) 153
[3] SHERWIN, C. W. *Basic Concepts of Physics*. Holt, Rinehart and Winston, New York, 1961
[4] PUGH, E. M. and PUGH, E. W. *Principles of Electricity and Magnetism*. p. 75. Addison-Wesley, Reading, Mass., 1960
[5] STRATTON, J. A. *Electromagnetic Theory*. p. 437. McGraw-Hill, New York and London, 1941
[6] HAMMOND, P. *Electromagnetism for Engineers*. p. 155. Pergamon Press, London, 1964

APPENDIX 6

THE MAGNETIC FIELDS DUE TO DISPLACEMENT 'CURRENTS'

Consider a classical 'point' charge q moving in empty space with uniform velocity u, where $u \ll c$. It will be assumed initially that the classical 'point' charge is a small conducting sphere. A similar problem was treated in 1881 by FitzGerald[1] who showed that

> From these it appears at once that the magnetic effect of the displacement currents is nil . . . .
>
> It may be worth while remarking that no effect except light, has ever yet been traced to the displacement currents assumed by Maxwell, in order to assume all currents to flow in closed circuits. It has not, as far as I am aware, been ever actually demonstrated that open circuits such as Leyden jar discharges, produce exactly the same effects as closed circuits, and until some such effect of displacement currents is observed, the whole theory of them will be open to question.

Let the charge be at the origin of the inertial frame Σ at the time the fields are measured. If $u \ll c$, the electric field at a point x', y', z' at a distance $\mathbf{r}'$ from the origin is given to a good approximation by Coulomb's law (cf. eqn 3.24),

$$\mathbf{E}(\mathbf{r}') = \frac{q\mathbf{r}'}{4\pi\varepsilon_0 r'^3} = -\mathbf{V}'\phi \qquad (A6.1)$$

where ϕ is the scalar potential, and $\mathbf{V}'$ stands for $\mathbf{i}(\partial/\partial x') + \mathbf{j}(\partial/\partial y') + \mathbf{k}(\partial/\partial z')$.

If the charge is moving, the displacement 'current' at a point x', y', z' in empty space is

$$\mathbf{J}'(\mathbf{r}') = \varepsilon_0 \frac{\partial \mathbf{E}(\mathbf{r}')}{\partial t} \qquad (A6.2)$$

For purposes of discussion it will now be assumed that this displacement current *'produces'* a magnetic field in other parts of space which can be calculated using the Biot–Savart law. If this were true, the magnetic field at a field point P having co-ordinates x, y, z at a distance $\mathbf{r}$ from the origin would be

$$\mathbf{B}(\mathbf{r}) = \int \frac{\mathbf{J}'(\mathbf{r}') \times (\mathbf{r} - \mathbf{r}')\mathrm{d}V'}{4\pi\varepsilon_0 c^2 |\mathbf{r} - \mathbf{r}'|^3} \qquad (A6.3)$$

285

The integration must be over the whole of space. Since,

$$\left| \mathbf{r} - \mathbf{r}' \right| = \sqrt{(x-x')^2 + (y-y')^2 + (z-z')^2}$$

$$\mathbf{V}' \left(\frac{1}{|\mathbf{r} - \mathbf{r}'|} \right) = \frac{(\mathbf{r} - \mathbf{r}')}{|\mathbf{r} - \mathbf{r}'|^3}$$

Hence, eqn (A6.3) can be rewritten

$$\mathbf{B}(\mathbf{r}) = \frac{1}{4\pi\varepsilon_0 c^2} \int \mathbf{J}'(\mathbf{r}') \times \mathbf{V}' \left(\frac{1}{|\mathbf{r} - \mathbf{r}'|} \right) dV' \tag{A6.4}$$

From eqn (A1.21) of Appendix 1 for any scalar ψ and vector $\mathbf{A}$,

$$\mathbf{V} \times \psi \mathbf{A} = \psi \mathbf{V} \times \mathbf{A} - \mathbf{A} \times \mathbf{V} \psi \tag{A1.21}$$

Putting $\mathbf{A} = \mathbf{J}'$ and $\psi = 1/|\mathbf{r} - \mathbf{r}'|$, eqn (A1.21) becomes

$$\mathbf{V}' \times \left(\frac{\mathbf{J}'}{|\mathbf{r} - \mathbf{r}'|} \right) = \left(\frac{1}{|\mathbf{r} - \mathbf{r}'|} \right) \mathbf{V}' \times \mathbf{J}' - \mathbf{J}' \times \mathbf{V}' \left(\frac{1}{|\mathbf{r} - \mathbf{r}'|} \right)$$

Now, using eqn (A6.2)

$$\mathbf{V}' \times \mathbf{J}' = \mathbf{V}' \times \varepsilon_0 \frac{\partial \mathbf{E}}{\partial t} = \varepsilon_0 \frac{\partial}{\partial t} (\mathbf{V}' \times \mathbf{E})$$

From eqn (A6.1), if $u \ll c$, $\mathbf{E} \simeq -\mathbf{V}'\phi$. Hence, since the curl of a gradient is zero,

$$\mathbf{V}' \times \mathbf{J}' = -\varepsilon_0 \frac{\partial}{\partial t} \mathbf{V}' \times (\mathbf{V}'\phi) = 0$$

Hence, eqn (A6.4) can be rewritten

$$\mathbf{B}(\mathbf{r}) = \frac{-1}{4\pi\varepsilon_0 c^2} \int \mathbf{V}' \times \frac{\mathbf{J}'(\mathbf{r}')}{|\mathbf{r} - \mathbf{r}'|} dV' \tag{A6.5}$$

From eqn (A1.22), for two vectors $\mathbf{A}$ and $\mathbf{C}$, we have

$$\mathbf{V} \cdot (\mathbf{A} \times \mathbf{C}) = \mathbf{C} \cdot (\mathbf{V} \times \mathbf{A}) - \mathbf{A} \cdot (\mathbf{V} \times \mathbf{C}) \tag{A1.22}$$

Let $\mathbf{C}$ be a constant vector; $\mathbf{V} \times \mathbf{C}$ is then zero. Integrating eqn (A1.22) over a finite volume, since $\mathbf{V} \times \mathbf{C}$ is zero,

$$\int \mathbf{V} \cdot (\mathbf{A} \times \mathbf{C}) dV = \int \mathbf{C} \cdot (\mathbf{V} \times \mathbf{A}) dV$$

Applying Gauss' mathematical theorem, eqn (A1.26), to the left-hand side, we have

$$\int (\mathbf{A} \times \mathbf{C}) \cdot \mathbf{n} dS = \int \mathbf{C} \cdot (\nabla \times \mathbf{A}) dV \qquad (A6.6)$$

Now from eqn (A1.5)

$$(\mathbf{A} \times \mathbf{C}) \cdot \mathbf{n} = (\mathbf{n} \times \mathbf{A}) \cdot \mathbf{C} = -\mathbf{C} \cdot (\mathbf{A} \times \mathbf{n})$$

Hence, eqn (A6.6) can be rewritten

$$-\int \mathbf{C} \cdot (\mathbf{A} \times \mathbf{n}) dS = \int \mathbf{C} \cdot (\nabla \times \mathbf{A}) dV$$

Since, the scalar product obeys the distributive law of addition, and $\mathbf{C}$ is a constant vector

$$-\mathbf{C} \cdot \int \mathbf{A} \times \mathbf{n} dS = \mathbf{C} \cdot \int (\nabla \times \mathbf{A}) dV$$

Hence

$$\int \mathbf{A} \times \mathbf{n} dS = -\int (\nabla \times \mathbf{A}) dV \qquad (A6.7)$$

Put $\mathbf{A} = \mathbf{J}'/|\mathbf{r}-\mathbf{r}'|$ in eqn (A6.7) and substitute in eqn (A6.5) to obtain

$$\mathbf{B}(\mathbf{r}) = +\frac{1}{4\pi\varepsilon_0 c^2} \int \frac{\mathbf{J}'(\mathbf{r}')}{|\mathbf{r}-\mathbf{r}'|} \times \mathbf{n} dS'$$

Using eqn (A6.2),

$$\mathbf{B}(\mathbf{r}) = +\frac{\varepsilon_0}{4\pi\varepsilon_0 c^2} \frac{\partial}{\partial t} \int \frac{\mathbf{E}(\mathbf{r}')}{|\mathbf{r}-\mathbf{r}'|} \times \mathbf{n} dS' \qquad (A6.8)$$

Near infinity $\mathbf{E}(\mathbf{r}')$ is proportional to $1/r'^2$ so that $\mathbf{E}(\mathbf{r}')/|\mathbf{r}-\mathbf{r}'|$ is approximately proportional to $1/r'^3$ at infinity and the integral over the surface at infinity is zero. If the 'point' charge is treated as a little conducting sphere, there is a discontinuity in the electric field at the surface. The electric field is normal to the surface of the conducting sphere so that $\mathbf{E}(\mathbf{r}') \times \mathbf{n}$ is zero, just outside the sphere. Hence, the integral on the right-hand side of eqn (A6.8) is zero over a surface just outside the conductor. Inside the conducting sphere $\mathbf{E} = 0$. Hence, from eqn (A6.8),

$$\mathbf{B}(\mathbf{r}) = 0$$

Thus, if it were assumed that the displacement 'current' in empty space, due to the varying electric field of the moving 'classical point charge', did give a magnetic field according to the Biot–Savart law, the resultant magnetic field at any field point due to all the displacement 'currents' in the whole of space would be zero. Thus, if one did include the displacement 'current' term $\varepsilon_0(\partial E/\partial t)$ in eqn (4.120) its net effect would be zero, so that its inclusion in eqn (4.120) does not matter in practice. However, it was shown in Chapter 4, that it is wrong in principle to say that the displacement 'current' (or Maxwell term) $\varepsilon_0(\partial E/\partial t)$ *produces* a magnetic field. It is associated with a magnetic field, since a moving charge gives both an electric and a magnetic field. If we are given the curl of **B**, we can calculate $\partial E/\partial t$ and vice versa.

It is left as an exercise for the reader to consider more realistic models for 'point' charges. For example, experiments on the scattering of high energy electrons by protons have shown that the electric charge of a proton is measured to be distributed in space (Hill[2]). In this case there would be no discontinuity in **E** at the 'surface' of the 'point' charge and one would only have to consider the integral in eqn (A6.8) over a surface at infinity. For a discussion of the case of electric circuits in the presence of dielectric materials the reader is referred to Whitmer[3].

REFERENCES

[1] FITZGERALD, G. F. *Proc. Roy. Dublin Soc.* **3** (1881) 250
[2] HILL, R. D. *Tracking Down Particles.* p. 122. W. A. Benjamin, New York, 1963
[3] WHITMER, R. M. *Amer. J. Phys.* **33** (1965) 481

INDEX

Accelerating charge
  electric intensity, 41, 42, 246–249, 251
  energy emitted, 251, 252
  general case, 252, 253
  magnetic induction, 41, 42, 249–253
  Maxwell's equations, 253–256
Action and reaction, 262–265
Adams, N. I., 148
Alväger, T., 4
Ampère, A. M., 125, 232
Ampere, definition of, 227
Ampère's circuital theorem, 22, 99–102, 229
  microscopic interpretation of, 100–102
Ampère's formula for the forces between current elements, 226
Amperian model of magnetic materials, 126, 231, 232
Atomistic origin of electromagnetic phenomena, 256, 257

Baker, D. A., 272
Bar magnets, forces between, 125
Becker, R., 173
Betatron, 236
Biot–Savart law, 38, 71, 102, 106–108, 242, 243, 256, 259, 265, 280, 281, 285–288
  capacitor charging, 108–110
  capacitor discharging, 127
  conventional approach, 225–227
  curl of magnetic induction, 229–234
  divergence of magnetic induction, 226–228
  Heaviside's rational current element, 111–114
  line of charge, 84–95
  magnetic analogue, 125
  Maxwell term, 106, 107
  point charge, 78–84
  quasi-stationary conditions, 225, 259
  via relativity, 259–265

Bleaney, B., 237
Bleaney, B. I., 237
Bork, A. M., 243
Born, M., 121
Boundary conditions, 126
Bucherer, A. H., 17

Capacitor
  charging, 108–110, 237–240
  discharging, 127
Charge, constancy of, 16, 17, 26, 29–34, 46, 166
Charge density
  definition, 47, 165
  transformation of, 165–173
Charge, equation of continuity of, 241, 243
Charge, fields of accelerating, 41, 42, 246–257
Charge moving with uniform velocity
  electric flux from, 52
  electric intensity, 33–38
  fields of using field transformations, 161
  forces between moving charges, 29–34
  magnetic induction, 33–35
  magnetic induction from Maxwell's equations, 78–84
  magnetic induction from the Biot–Savart law, 78–84
Conduction currents
  forces between, 23, 24
  mean drift velocity of conduction electrons, 23, 259, 271
Conductivity, electrical, 117, 183, 266
Conductor, moving, dipole moment of, 169–173
Constitutive equations, 117, 152, 175, 182, 183, 224, 234
Continuity of charge, 241, 243
Convection currents, forces between
  Maxwell's equations, 20–24
  special relativity, 25–28

289

INDEX